Values, Payments and Institutions for Ecosystem Management

Values, Payments and Institutions for Ecosystem Management

A Developing Country Perspective

Edited by

Pushpam Kumar

Chief, Ecosystem Services Unit, United Nations Environment Programme (UNEP), Kenya

Ibrahim Thiaw

United Nations Assistant Secretary-General and Deputy Executive Director, United Nations Environment Programme (UNEP), Kenya

With

Tom Barker (*Technical Editor*)

University of Liverpool, UK

United Nations Environment Programme (UNEP)

Edward Elgar

Cheltenham, UK • Northampton, MA, USA

© United Nations Environment Programme 2013

All rights reserved. No part of this publication may be reproduced, stored in a retrieval system or transmitted in any form or by any means, electronic, mechanical or photocopying, recording, or otherwise without the prior permission of the publisher.

Published by
Edward Elgar Publishing Limited
The Lypiatts
15 Lansdown Road
Cheltenham
Glos GL50 2JA
UK

Edward Elgar Publishing, Inc.
William Pratt House
9 Dewey Court
Northampton
Massachusetts 01060
USA

A catalogue record for this book
is available from the British Library

Library of Congress Control Number: 2013943227

This book is available electronically in the ElgarOnline.com
Economics Subject Collection, E-ISBN 978 1 78195 369 3

ISBN 978 1 78195 368 6

Typeset by Servis Filmsetting Ltd, Stockport, Cheshire
Printed and bound in Great Britain by T.J. International Ltd, Padstow

Contents

List of contributors	vii
Foreword by Achim Steiner	ix
Preface	xi

1. Values, payments and institutions for ecosystem management: a developing country perspective — 1
   Pushpam Kumar and Ibrahim Thiaw

2. Making payments for ecosystem services work — 16
   Rodrigo Arriagada and Charles Perrings

3. Valuing ecosystem services: benefits, values, space and time — 58
   Brendan Fisher, Ian Bateman and R. Kerry Turner

4. Managing trade-offs in ecosystem services — 70
   Thomas Elmqvist, Magnus Tuvendal, Jagdish Krishnaswamy and Kristoffer Hylander

5. Revisiting the relationship between equity and efficiency in payments for ecosystem services — 90
   Unai Pascual, Roldan Muradian, Luis C. Rodriguez and Anantha K. Duraiappah

6. Are the amounts of payments for environmental services enough to contribute to poverty alleviation efforts in developing countries? — 109
   Luis C. Rodriguez, Unai Pascual and Roldan Muradian

7. Unifying environmental and social protection: learning from PES and CCT in developing countries — 120
   Luis C. Rodriguez, Unai Pascual, Roldan Muradian, Nathalie Pazmino and Stuart Whitten

8. Exploring the potential of payments for ecosystem services for *in-situ* agrobiodiversity conservation — 150
   Ulf Narloch, Adam G. Drucker and Unai Pascual

9	Paying for international environmental public goods *Rodrigo Arriagada and Charles Perrings*	172
10	Institution and ecosystem functions: the case of Keti Bunder, Pakistan *John M. Gowdy and Aneel Salman*	192
11	How ecosystem-based restoration can yield a double dividend of adaptation to climate change and enhancement of ecosystem services *James Blignaut*	215
12	The ethical foundations of cultural diversity in ecosystems and their role in economic valuation *Ian Parker*	259
13	Lessons learned and conclusions *Pushpam Kumar*	279

Index   291

Contributors

Rodrigo Arriagada
Department of Agricultural Economics, Pontificia Universidad Catolica de Chile and ecoSERVICES Group, School of Life Sciences, Arizona State University, USA

Ian Bateman
Centre for Social and Economic Research on the Global Environment, School of Environmental Sciences, University of East Anglia, UK

James Blignaut
ASSET Research, Beatus and Department of Economics, University of Pretoria, South Africa

Adam G. Drucker
Bioversity International, Italy

Anantha K. Duraiappah
International Human Dimension Programme (IHDP), Bonn, Germany

Thomas Elmqvist
Department of Systems Ecology, Stockholm University, Sweden and Stockholm Resilience Centre, Stockholm University, Sweden

Brendan Fisher
Woodrow Wilson School of Public and International Affairs, Princeton University, USA

John M. Gowdy
Rittenhouse Professor of Humanities and Social Science, Department of Economics, Rensselaer, Polytechnic Institute, USA

Kristoffer Hylander
Department of Botany, Stockholm University, Sweden

Jagdish Krishnaswamy
Ashoka Trust for Research in Environment and Ecology (ATREE), Bangalore, India

Pushpam Kumar
United Nations Environment Programme (UNEP), Nairobi, Kenya

Roldan Muradian
Centre for International Development Issues (CIDIN), Radboud University of Nijmegen, Netherlands

Ulf Narloch
World Conservation and Monitoring Centre (WCMC), Cambridge, UK

Ian Parker
Discourse Unit, Manchester, UK

Unai Pascual
Department of Land Economy. University of Cambridge, Cambridge, UK and Basque Centre for Climate Change, Spain

Nathalie Pazmino
The World Bank, Sydney, Australia

Charles Perrings
ecoSERVICES Group, School of Life Sciences, Arizona State University, USA

Luis C. Rodriguez
Commonwealth Scientific and Industrial Research Organisation (CSIRO) Sustainable Ecosystems, Canberra, Australia

Aneel Salman
Centre for Public Policy and Governance, Forman Christian University, Pakistan

Ibrahim Thiaw
United Nations Environment Programme (UNEP), Nairobi, Kenya

R. Kerry Turner
Centre for Social and Economic Research on the Global Environment, School of Environmental Sciences, University of East Anglia, UK

Magnus Tuvendal
Department of Systems Ecology, Stockholm University, Sweden

Stuart Whitten
CSIRO Ecosystem Sciences, Canberra, Australia

Foreword

One of the serious downsides of conventional economics is the way it is leading to the over-consumption of the planet's productive, life-support systems. Sound and solid economic evaluation of ecosystem services, such as those generated by forests or freshwaters, holds real promise in reversing these trends.

Capturing costs and benefits can also assist towards objectivity and credibility in the overall management strategy of ecosystems and will also assist policy-makers in estimating the impacts of policy reforms in different sectors and the implications for different societal groups.

The importance of economic valuation was highlighted in the discourse at the United Nations Conference on Sustainable Development (Rio+20), held in Rio de Janeiro, Brazil, in June 2012 – 20 years after the 1992 Earth Summit in the same city. Its outcome document, *The Future We Want*, emphasizes the importance of evaluating environmental and economic factors in countries' efforts in promoting an inclusive green economy, and its integration into decision-making processes. It also touches upon the need to undertake scientific analysis on the costs and benefits of green economy policies in the context of sustainable development and poverty eradication.

In addition to the critical role of economic valuation of ecosystem services in decision-making processes, innovative policies such as payments for ecosystem services (PES) have been shown to have strong potential in catalysing the successful maintenance of healthy, functioning ecosystems – in part by providing incentives for local people to be involved while contributing to human well-being and poverty eradication.

The task of making PES schemes effective, however, requires careful design and institutional arrangements to ensure compliance, efficiency and equity. This book, *Values, Payments and Institutions for Ecosystem Management: A Developing Country Perspective*, edited by Kumar and Thiaw, brings together useful insights from leading thinkers and exponents of the theory and practice of economic valuation of ecosystem services and PES implementation.

By providing the real-world examples and lessons, the book will guide policy-makers and experts in their efforts in exploring and applying these pathways and tools in the larger context of development policies of nations and the pursuit of a sustainable century.

Achim Steiner
United Nations Under-Secretary-General
and
Executive Director United Nations Environment Programme

Preface

The outcome of the United Nations Conference on Sustainable Development (UNCSD) in Rio de Janeiro in 2012 (Rio+20) 'The Future We Want' describes the importance of an Inclusive Green Economy in the context of sustainable development and poverty eradication. The consensus emerging from world leaders is clear and loud for broader measures of progress to complement conventional indices such as gross domestic product (GDP) and sustainable management of ecosystems and economy. There have been persistent efforts over the last two decades to reform our national accounting system through 'green accounting' or 'inclusive wealth' measures. It is expected that one of these new measures will finally provide a yardstick for sustainable development. These improved measures of national performance will also reflect our institutional and political commitments through the Agenda 21 document 'Our Common Future' of the 1992 Earth Summit at Rio and 'The Future We Want' document accepted at Rio+20 in 2012. The United Nations, encouraged by the general success of *quantitative* targets for the Millennium Development Goals in focusing policy-making and civil society discourse on specific, tangible and measurable results towards solving poverty, is now formulating 'Sustainable Development Goals', starting in 2015. Developing the accounts of society so that they measure what matters becomes even more crucial in that context.

Economic valuation of ecosystem services plays a pivotal role not only in resolving the conflicts and trade-offs amongst services but greatly helps in integrating physical accounts with the existing accounts of the economy. The integration facilitated through valuation provides a direction towards sustainability of economic development and ecosystems. Capturing and demonstrating the value of contributions of ecosystems to the economy, which in turn strengthens the constituents and determinants of human well-being, go a long way in operationalization of discourse on sustainability of society and economy. Valuation of ecosystem services and biodiversity provides an effective tool to assist decision-makers in designing cost-effective response policies for integrating changing ecosystem services into a decision-making framework.

The emerging importance and use of innovative tools like payments for ecosystem services (PES) hinges upon the robustness of economic valuation, as only a credible value would facilitate a transaction between the beneficiaries and providers of ecosystem services. In fact, transactions between the beneficiaries and providers of ecosystem services although widely used require objective criteria, scientific understanding and a transdisciplinary approach to make it a credible response tool for ecosystem management. Some of the quintessential aspects of PES are how the value of the service is measured. Although a promising tool for ecosystem management, PES suffers from several limitations. Most of the limitations arise from the necessary preconditions required for a transaction to take place between the parties involved. For example, inadequate understanding of what are being bought and sold, and long-term implications for local livelihoods and resource rights. This happens when the clarity of the service is less evident. Further, limited access to information about payments for ecosystem services, lack of financing for PES assessment, limited bargaining power to influence, shape or enforce rules and contracts, a limited asset base to absorb risks, invest time and resources in management, limited organization or outreach to attract buyers, and a lack of efficient intermediary institutions to reduce transaction costs are all well documented in the literature.

This volume covers some of these critical issues in valuation and payments, which are central to ecosystem management. The role of institutions especially in developing countries has also been adequately covered to make the volume useful and practice centred.

We hope this volume comprising some seminal work by the acknowledged experts of the themes will be a welcome addition to the ongoing debate on ecosystem management in developing parts of the world.

Pushpam Kumar
Ibrahim Thiaw

1. Values, payments and institutions for ecosystem management: a developing country perspective

Pushpam Kumar and Ibrahim Thiaw

Ecosystem services are the benefits provided for human well-being from functioning ecosystems (Millennium Ecosystem Assessment 2005; Nelson et al. 2009). Ecosystems themselves are highly valuable natural capital, and ecosystem services provide direct and indirect benefits essential to sustain and fulfil life, as well as supporting important parts of the economy (Millennium Ecosystem Assessment 2005; Huberman 2008; Nelson et al. 2009). Indeed, the short-term economic benefits of many of the anthropogenic activities currently degrading ecosystems are eclipsed by the long-term economic, ecological and social values of the services that they provide (Sharma et al. 2008; Wunder et al. 2008).

There are four main categories of ecosystem services: provisioning, regulating, supporting and cultural. Provisioning services such as food, water, fuel and fibre have a clear market value and are, therefore, easier to quantify (Tallis and Kareiva 2005; Muradian and Kumar 2009). Regulating (e.g. flood control and preventing soil erosion), supporting (e.g. nutrient cycling and soil formation) and cultural services (e.g. recreation and spiritual appreciation) have greater uncertainty with regards to biophysical production functions and are, therefore, more difficult to capture in terms of economics (Kumar and Kumar 2008).

Following the United Nations' Millennium Ecosystem Assessment (2005), ecosystem services are being used as a conceptual framework to understand nature's value and to provide guidance for environmental management (Redford and Adams 2009). The application of this approach aims to reconcile disparities between ecosystem service provision and individual livelihoods. It attempts to promote greater equity in natural capital distribution and to recognize the trade-offs between natural resource availability and resource use (Wunder 2005; McShane et al. 2011). Indeed, the disproportionate dependence of the rural poor on ecosystem services means that the decline in the provision and availability of these services will have a significant and direct threat to their welfare and livelihoods (Ferraro 2002; Grieg-Gran et al. 2005; Daily et al. 2011; Tallis et al. 2011).

Compensation, or payments for ecosystem services (PES), is an economics-based policy approach that uses compensatory means to resolve conflicts between the integrity of natural resources and their commercial exploitation. External beneficiaries of services, such as downstream water users, are contracted to pay 'service providers' (e.g. local landowners) on the condition that they adopt land practices and resource uses that restore or maintain ecosystem conservation (Wunder 2005).

Payments for ecosystem services are based on the economic argument that the degradation of ecosystems and the loss of services they provide is a result of market failure. While ecosystem services provide essential benefits to individual well-being, many of these are external to the market system (Millennium Ecosystem Assessment 2005; Kemkes et al. 2010; Goulder and Kennedy 2011). Although natural capital, such as timber, can be a market good with a commercial value, non-market goods, such as nutrient cycling, carbon sequestration and watershed retention, become positive 'externalities', or uncompensated benefits (Landell-Mills and Porras 2002).

In addition, non-market goods are often public goods (non-excludable and non-rival). Since beneficiaries lack the incentive to pay suppliers, they can become 'free-riders' which undermines the formation of markets. This market failure leads to the under-valuing of, and inadequate investment in, the protection of ecosystems, and has serious impacts on both human and environmental well-being (Landell-Mills and Porras 2002; Kemkes et al. 2010).

To complicate matters, the physical structure of the environment, such as forests and water resources, which provide many of these ecosystem services as public goods, is often privately owned (Kemkes et al. 2010). Policy is, therefore, needed to enable the provision of ecosystem services by private landowners to be paid.

PES act as a mechanism to enable non-market values of ecosystem services (the 'externalities') to be included in economic incentives for local actors to provide such services (Engel et al. 2008). Most PES schemes utilize Wunder's (2005) basic principles of PES as a voluntary transaction: 'An ecosystem service is bought by one or more buyers from one or more providers, if and only if a provider secures the provision of the service (conditionality)' (Wunder 2005; Kemkes et al. 2010). Once a payment is made, landowners can be compensated for the voluntary provision of ecosystem services. These 'conditional' payments are a central feature of PES. They establish an incentive structure for both providers and beneficiaries

(Wunder 2005). However, most PES schemes are designed for a specific and local context, so very few PES programmes actually fulfil these criteria (Wunder et al. 2008).

There are many types of PES that focus on specific ecosystem services, but due to their commercial scale, only four main schemes have emerged: carbon sequestration, watershed protection, biodiversity protection and landscape beauty (Wunder 2005).

1. Carbon sequestration: Paying landowners to maintain or plant additional trees to store and sequester carbon may halt deforestation, maintain slope stability and indirectly protect high altitude watersheds.
2. Watershed protection: Direct watershed protection allows participants to pay upstream landowners for best practice land use that limits deforestation and land degradation, in order to reduce risks such as floods and soil erosion while maintaining the quality of aquifers.
3. Biodiversity protection: Protection of biodiversity can be conducted via payments to set aside or restore biological corridors, or to protect species directly.
4. Landscape beauty: PES schemes to maintain landscape beauty may, for example, involve tourism operators who pay local communities not to hunt wildlife in areas used for wildlife viewing (Wunder 2005).

PES schemes are financed in a number of ways (as discussed in the case study section, below). Schemes such as the Vittel PES scheme in France (Perrot-Maitre 2006) are user-financed, while in China's Sloping Land Conversion Programme (Bennett 2008), third parties initiate conditional payments for landowners to provide ecosystem services. Government-financed programmes are often large scale and focus on multiple services with 'bundled' payments by users. This enables users to access a greater diversity of funding sources, thus making conservation over conversion a more competitive land-use option (Asquith et al. 2008).

There are, however, numerous trade-offs in PES design. Focusing on the provision of one ecosystem service may result in trade-offs in the provision of others. For example, maximizing carbon sequestration may not improve hydrological services or promote high biodiversity, but it can minimize transaction costs with a broad selection of providers over a much wider region (Wunder 2005). In addition, larger scale programmes can benefit from reduced transaction costs with a broad selection of providers. That permits multiple objectives to be pursued, such as poverty reduction and conservation (Wunder et al. 2008).

ECONOMIC VALUATION OF ECOSYSTEM SERVICES AND PES

Ecosystems are highly interdependent. Therefore, their valuation requires an understanding of ecology and their functioning, and how these relate to their ecology and function. The creation of a payments system to incentivize communities to conserve ecosystems, rather than to selling or exploiting the resources, has been developed alongside attempts to value ecosystem services. Indeed, it is essential to highlight the economic benefits of ecosystems functioning for both policy and decision-making. It identifies trade-offs between current market systems that favour conversion, and PES that promote conservation (Fisher et al. 2009; Kemkes et al. 2010).

Economic valuation of environment is an inherently anthropogenic concept. There are multiple meanings of what denotes 'value', such as market value, intrinsic use and non-use value, option value, existence value and vicarious values (Kumar and Kumar 2008). Monetizing these values is predominantly based on either ecological valuation (excluding human needs and wants) or economic valuation (based on consumer preference) (Spangenberg and Settele 2010). Capturing *all* the economic value of an ecosystem is also difficult, as many values are almost impossible to monetize (Kemkes et al. 2010). Subjective interpretations of value are also highly significant in management decisions, i.e. what is valuable to whom, and what is prioritized?

While potentially subjective, based on a number of assumptions, PES valuing mechanisms are preferable to a lack of valuation and accountability, which is currently driving ecosystem degradation and incorrect priorities in policy (Spangenberg and Settele 2010). Valuation of ecosystem services highlights considerations on economic efficiency, since it estimates the costs incurred by providers, and thus the amount of compensation needed through PES. This estimated value can identify economic trade-offs between and across different users and providers of ecosystem services. For example, many individuals depend on forest resources for income, subsistence and ecosystem services.

In rationalizing resource-use decisions, individuals will always seek to maximize their utility. Some have proposed that economic incentives such as PES can act as redistributive mechanisms between different social groups, and that the inequality between rural and urban communities can frame the issue of benefits originating from different ecosystems (Muradian and Kumar 2009).

Appraisal of PES: Some Success Stories

Payments are not always an appropriate or effective policy tool. There is no uniform application of PES, and its implementation is highly context specific (Kemkes et al. 2010). Adequate institutional arrangements, political support, and the individual characteristics of ecosystems will determine how effective payments can be. To determine how payments will interact within a market system, policy-makers must distinguish the scale at which they are made and whether the ecosystems themselves are a common pool resource, public goods or market goods.

Furthermore, incentives to provide ecosystem services do not necessarily equate to permanence. They do not guarantee long-term service provision over generations of services (Wunder et al. 2008). Consideration of these limitations is vital to ensure that PES schemes are designed to maximize efficiency, and do not create perverse incentives, which distort markets or encourage environmental degradation to occur within PES schemes (Pagiola and Platais 2007). Still, there are numerous examples of successful design and implementation of PES schemes around the world. The PES approach has been implemented in both developed and developing countries. Its application has been diverse, ranging from small-scale watersheds to entire nations (Kemkes et al. 2010).

The Los Negros user-financed PES scheme was established in 2003 and is facilitated by a local non-governmental organization (NGO): Fundación Natura Bolivia. By 2007, 46 farmer-landowners in the community of Santa Rosa were enrolled and paid (by the local Pampagrande Municipality using external donor support from the US Fish and Wildlife Service) to conserve the forest and biodiversity. Non-cash payments are made at an agreed annual payment of one artificial beehive per 10 hectares of forest protected per year, in addition to the value of technical apicultural training (Asquith et al. 2008). This scheme aims to protect the Upper Los Negros watershed (2,774 hectares) and local biodiversity, which provides benefits to local water users for consumption and irrigation (Asquith et al. 2008). Payments are conditional on yearly inspection of land use (in which tree cutting, hunting and forest clearance are prohibited) as stipulated in landowner contracts.

PROFAFOR is a long-established, user-financed carbon sequestration programme in Ecuador, which started in 1993. The scheme was implemented by PROFAFOR, an Ecuadorian company created by service buyers, and pays private landowners and communities in long-term contracts (15, 20 and 99 years) to fix carbon through reforestation and afforestation in an area covering 22,287 hectares as of year 2005 (Wunder and Alban 2008). The buyer is FACE Foundation, a Dutch electricity consortium. Other ecosystem benefits are gained alongside global climate change mitigation.

The Sloping Land Conversion Programme is one of the first PES schemes in China (Bennett 2008). This public scheme was initiated and financed by the Central Government as a pilot project in 1999. It had developed into a full-scale scheme by 2002. The programme focuses on the urgent problem of soil erosion and watershed protection, but also includes poverty alleviation as part of its goals. Payments directly engage rural households to increase forest cover through reforestation and afforestation, as well as the retirement of croplands on steeply sloping and marginal lands (Asquith et al. 2008; Bennett 2008). By 2005, nearly 7.2 million hectares of cropland were retired from agricultural production and 4.92 million hectares had been reforested. This has also benefited downstream water consumers and timber harvesters.

The Pago por Servicios Ambientales programme in Costa Rica is a country-wide project established in 1996 that covers 270,000 hectares. In the mid-1990s, Costa Rica had one of the world's highest deforestation rates. In response, the government designed payments within the framework of environmental, forestry, and biodiversity legislation, to be run by a semi-autonomous agency, FONAFIFO.

Some 'PES-like' schemes, such as CAMPFIRE in Zimbabwe, can offer insights into how early wildlife service initiatives have evolved. Originally designed as a community-based natural resource management programme, CAMPFIRE attempts to realign economic distortions that drive the 'misuse' of natural resources and to provide economic incentives to conserve wildlife (Child 1996). Initiated by the Zimbabwe Park Authority and various NGOs in 1989, the project utilized the high and tangible value of wildlife to sell access to safari operators to communal lands covering 14.4 million hectares (Child 1996; Frost and Bond 2008; Wunder 2008). In terms of PES attributes, beneficiaries (private safari companies and international donors) pay local communities, via district councils, for (auctioned) access to, and conservation of, natural landscapes. While there are success stories on execution of PES schemes in many parts of the world, it is clear that successful application of PES for management of ecosystem services requires specific institutions, especially in developing countries.

In this volume, Arriagada and Perrings (Chapter 2) analyse the different PES schemes currently being operated in order to identify the most efficient approaches, considering both 'process' (how the scheme works) and 'outcome' (what the scheme produces). The authors then go about answering three questions: 'What are PES schemes?', 'What can they be expected to achieve?' and 'What lessons can be learned from experiences to date on the efficient design of future PES schemes?'

The authors highlight the fact that many methods, including taxes, subsidies, user charges, access fees, penalties for non-compliance and, more recently, PES have been developed to address the market failures behind the collapse of ecosystem services as outlined in the Millennium Ecosystem Assessment (2005). It is stated that the majority of current PES schemes are modest in scale, and use local level arrangements involving spontaneous, private market-type arrangements, such as those of nature-based tourism. Additionally, PES schemes have been developed that offer financial incentives for local actors to provide a wide range of ecosystem services that lie outside normal market transactions, such as the four main ecosystem services covered by PES schemes (watershed services, carbon sequestration, landscape beauty, and biodiversity conservation; see above).

Expectations of PES schemes depend upon whether they are able to deliver outcomes, measured in terms of the flow of services, which are better than the outcomes in the absence of such schemes. Many PES schemes have effects on land-use decisions by those who are not directly involved in the scheme. These spillovers or 'leakages' can be either negative or positive. If negative, the aim should be to provide benefits sufficient to offset the spillovers. Calculations of the net benefits (additionality) of PES schemes should consider both positive and negative spillovers.

Lessons to be learned from existing PES schemes, as detailed by Arriagada and Perrings (Chapter 2), include the understanding that both national governments and international representative bodies need to be involved in globally relevant PES schemes such as Reducing Emissions from Deforestation and Forest Degradation (REDD); that PES schemes will increasingly be asked to assure provision of services that provide benefits at much larger scales; that the design of PES schemes should be complemented by the measurement of ecosystem services produced through the schemes; and ultimately, that it is necessary to ensure that PES schemes deliver additional benefits relative to the status quo.

Fisher et al. (Chapter 3) outline the theoretical discourse and provide an analysis of some of the important issues that practitioners must consider prior to attempting to link ecological and economic models via valuation approaches. The analysis is centred on three major themes: 'services versus benefits', 'price versus value' and 'here and now versus there and then'. Through consideration of the distinction between services and benefits, the chapter presents a mechanism by which the concept of ecosystem services can be made operational in the context of economic valuation. The authors go on to demonstrate the difference between price and value, the latter being the focus for decision-making. Finally, they discuss the temporal and spatial considerations pertaining to the development and implementation of valuation exercises. An understanding of the messages in this chapter is critical for any valuation work regarding ecosystem services.

Elmqvist et al. (Chapter 4) present an ecological perspective on regulating services, demonstrating the role of economics in developing ways to manage trade-offs between provisioning and regulating services. Building on the analysis of a hypothetical scenario looking at coffee production, the chapter proposes a general framework for managing these trade-offs using a landscape-based approach.

In Chapter 5, Pascual et al. offer a framework to understand the relationship between efficiency and equity in payment mechanisms for ecosystem services, devoting particular attention to the role of the institutional setting, social perceptions about economic fairness, and uncertainty and interactions between agents, including power relations. It is pointed out that the mainstream interpretation of PES as a mere market-based instrument to internalize environmental externalities overlooks the complex interactions that take place between agents, and determines both equity and efficiency outcomes. Some key questions regarding PES are raised, including: What are the key factors conditioning the relationship between equity and efficiency in PES schemes? What is the role of social perception about fairness in the performance of PES? What is the contribution of a political economy approach for understanding the relationship between efficiency and equity in PES?

The authors argue that trade-offs between efficiency and equity should be considered as a structural feature of PES schemes. It is stated that practitioners, particularly in developing countries, are usually confronted with this kind of trade-off, and that they are not able to detach efficiency from equity considerations since, in reality, PES schemes tend to be part of broader rural development interventions. This chapter also argues that the pro-poor approach, though pertinent, is limited in scope, particularly for addressing the relationship between equity and efficiency, primarily because it does not tackle the equity and economic fairness dimension of PES in a comprehensive way. Pascual et al. (Chapter 5) argue that a comprehensive theoretical framework is needed for understanding the interactions that shape the relationship between equity and efficiency in PES schemes. The authors state that such frameworks should rest on institutional and political economy insights, and should aim to analyse, among other things, the role of the adopted notions of economic fairness criteria, possible efficiency-equity trade-offs, and power relations between stakeholders.

Rodriguez et al. (Chapter 6) discuss the links between ecosystem services and poverty, and investigate whether tools such as PES and Conditional

Cash Transfers (CCT) can help to alleviate poverty by making targeted payments. It is stated that while PES schemes are conservation instruments, they are able to affect people's livelihoods indirectly, and this has been the starting point for the exploration of their potential as a poverty reduction tool. Because of this potential, attention has been drawn from donors for the use of PES as potential pro-poor payment mechanisms.

The authors outline the fact that, within CCT programmes, the amount of payment needed to have a positive impact on poverty is estimated at the design stage. This, therefore, represents the level of external support required to have a significant effect on poverty reduction. As such, it is possible to compare this figure with the amount likely to be required by a PES scheme within the same country and region. Consequently, it is possible to estimate whether PES can make a significant contribution to poverty reduction initiatives. This hypothesis was tested by selecting ten countries from Africa, Latin America and Asia, where both PES schemes and CCT have been implemented. It was shown from these country examples that, in most cases, the PES amounts were within range of the CCT payments and that the differences between PES and CCT payments were not significant. Therefore, it is suggested that PES could play a significant contribution to poverty alleviation if properly designed and targeted.

In Chapter 7, Rodriguez et al. evaluate the potential of combining PES and CCT to form a unified scheme for environmental and social protection. It is stated that although these instruments are designed to achieve distinct objectives, they have sufficient similarities and challenges in their design and implementation phases to allow the development of a unified framework and payment scheme.

In the light of identifying the potential to use PES and CCT jointly, the authors build on this concept and propose a joint payment scheme entitled Payments for Environmental and Poverty Alleviation Services (PEPAS). The authors believe it is possible to identify target areas and households where a joint programme might be implemented, based on the relationships between the levels of poverty and provision of ecosystem services, respectively, defined by poverty lines and environmental metrics. Overlapping areas indicate where both programmes may operate jointly, and it is in these areas where unified payment schemes could be implemented. The authors outline the fact that PEPAS schemes are not expected to follow a 'one-size-fits-all' design, but will be more diverse, reflecting the differences in setting up both the conservation and poverty alleviation goals, the relative importance of the two goals, and the particular environmental, socio-economic and political context in which payments are to be implemented.

The design of PEPAS schemes should include conditional payments for both environmental and social objectives, whereby payments are designed to adjust individual behaviour towards the delivery of conservation and poverty alleviation targets. It is pointed out, however, that the inclusion of both environmental and socio-economic conditions can be expensive. Payments transferred in PEPAS schemes should ensure that costs incurred by beneficiaries, in terms of environmental and social actions (e.g. land management practices or sending children to school), are compensated in full. Finally, the authors highlight the fact that PEPAS schemes have the potential to be cheaper when compared with the aggregated costs of separate PES and CCT programmes.

In Chapter 8, Narloch et al. discuss the application of PES-type programmes in the context of *in-situ* (on-farm) conservation of plant and animal genetic resources (PAGR). In the light of this, the authors develop the concept of Payments for Agro-biodiversity Conservation Services (PACS), which seek to tackle market failures associated with the public goods characteristics of agro-biodiversity, by increasing the private benefits from utilizing threatened, but valuable, crop varieties or livestock breeds on the farm, through individual-based or community-based reward mechanisms. It is suggested that instruments such as PACS might provide policy-makers with an effective tool for conserving PAGR.

It is stated that one of the main sources of PAGR loss occurs when farmers replace traditional crop and livestock varieties with more financially profitable, 'improved' varieties. To counter this, the authors propose that a scheme such as PACS could pay farmers for conserving traditional varieties and provide compensation based on yield gaps between traditional and improved varieties. Narloch et al. (Chapter 8) predict that a scheme such as PACS would deliver multiple benefits for the conservation of PAGR. This is not only because on-farm conservation would result in the sustained utilization of threatened crop and livestock varieties within traditional farming systems (thus conserving specific agricultural practices), but it would also maintain local traditions, culture and agricultural knowledge.

The authors point out that while PACS schemes seem to provide a theoretically straightforward solution to the problem of agro-biodiversity losses, on-the-ground implementation of such schemes is much more complex. For example, some generic constraints still to be overcome in implementing such programmes might include the setting up of new institutions to administer negotiations, transactions, monitoring and enforcement; land tenure issues; the involvement of intermediaries to act as transfer agents, or brokers to establish and maintain contacts between different actors; the creation of institutional arrangements to deal with baselines, verification of service delivery and sanctions for non-compliance (with regard to monitoring and enforcement of PACS contracts); and challenges in establishing strong scientific foundations. It is concluded that a tool such as PACS, while offering useful potential for policy-makers, needs further research and development owing to the many potential constraints and general lack of field experience and research in this area.

In Chapter 9, Arriagada and Perrings highlight the issues of paying for international environmental public goods that are currently undersupplied. They define international public goods as those goods with benefits that extend to people in multiple countries (and often across multiple generations); where the goods generated in one county generate spillover effects beyond that country's boundaries; and where goods cover more than one group of countries: where they benefit not only a broad spectrum of countries but also a broad spectrum of the global population, as well as meeting the needs of the both present and future generations. It is stated that the beneficiaries of these international public goods can include national populations and their representatives, nation states, transnational corporations and NGOs. Providing some examples of international environmental public goods, the authors state that these can include strictly global goods, 'such as conservation of the genetic diversity on which all future evolution depends, the mitigation of climate change, the control of emerging infectious diseases, and the management of sea areas beyond national jurisdictions'. One such international environmental public goods described by Arriagada and Perrings (Chapter 9) is REDD, as an example of a situation whereby local providers are compensated through a payment scheme for any additional costs they might incur in supplying international environmental public goods, which contribute to meeting global demands.

Regulating and supporting ecosystem services are, according to the authors, those most often responsible for supplying international public goods. The main issues that Arriagada and Perrings focus on in Chapter 9, with regard to international public goods supply, are how to deliver environmental public goods that are provided or generated at particular locations but offer benefits over wider areas, and which goods generate local benefits that are below the local cost of supply. International environmental public goods of this nature are often accounted for by individual countries that are globally significant for the provision of specific services (e.g. countries in tropical regions with high levels of biodiversity). Chapter 9, therefore, proposes that national contributions to international environmental public goods of this sort should be targeted. Such targeting may come from direct investment for the supply of goods (e.g. from the

Global Environment Facility (GEF)), or in the form of payments for the benefits of the provision of goods (e.g. PES schemes). It is pointed out that GEF is under-resourced and offers only weak targeting, and, therefore, that PES 'may become the dominant mechanism for assuring local provision of international environmental public goods'.

Chapter 10 (Gowdy and Salman) discusses ways to value the services of ecosystems and biodiversity from several perspectives, including market economics, human well-being and ecosystem functioning, and explores the hierarchies of services provided and the complexities involved in economic evaluation of ecological systems. Particular focus is placed upon the value of regulating services in the villages of Keti Bunder in, for example, maintaining water quality and quantity, erosion control, water purification, regulation of human disease vectors, provision of biological species, pollination and storm protection.

Blignaut (Chapter 11) discusses how ecosystem-based restoration activities can contribute to adaptation to climate change. The author introduces the concepts of climate change adaptation in conjunction with the restoration of natural capital, discussing the value to be gained in the restoration of degraded land and the potential it has to contribute to human well-being and adaptation to the impacts of climate change. The author describes the two strategies in the human response to climate change: mitigation and adaptation. While climate change mitigation can contribute to getting society onto a low carbon trajectory, the author argues that it is climate change adaptation that has the greater potential for dealing with extreme weather events brought about by climate change in terms of the vulnerability and resilience of people and communities. Examples are provided of local adaptation to climate change, including educational programmes, health interventions (for example, the distribution of mosquito nets), improved early warning systems and disaster management, and investments in the built environment (for example, raised roads and bridges).

The author proposes that the restoration of natural capital will begin to provide beneficial returns immediately, such as the conservation of ecosystems and their goods and services, regardless of the occurrence of extreme weather events. Natural capital restoration is defined as 'any activity that integrates investment in and replenishment of natural stocks to improve the flows of ecosystem goods and services, while enhancing all aspects of human well-being' (Aronson et al. 2007).

The author poses the question of 'whether climate change policy can be designed in such a way that it also addresses the ills of poverty and environmental degradation?' In response to this, it is stated that restoration of degraded land can and should be integrated into international climate change policies. The author expands on this by highlighting the potential benefits that can be had from adaptation strategies; namely that restoration reduces people's vulnerabilities by acting as a buffer against losses of ecosystem goods and services; that it is a highly valuable insurance policy against future disasters; that it safeguards genetic diversity of biological material; and that it sequesters carbon and, in so doing, contributes to climate change mitigation. The benefits of restored, intact, natural capital are said by Blignaut to be far-reaching: providing protection to infrastructure and communities; delivering ecosystem goods and services; conserving genetic resources; buffering the effects of climate change; and, at the same time, benefiting human well-being by protecting against poverty and death. It can, therefore, be seen that restoration of degraded land can be an effective climate change adaptation strategy that can and should be integrated into policies and interventions on the international scene.

Finally, in Chapter 12, Parker sets out to explore the feasibility of constructing policy-relevant foundations for the comprehension of the social dimensions involved in the economic valuation of ecosystems and biodiversity. In particular, the author attempts to answer how people living in ecosystems relate to the domain of the 'economic' as the concern of structurally oriented social policy. By reviewing the ethical paradigms of diversity and separation, and the ethics of evaluation and morality in policy initiatives, Parker attempts to comprehend the nature of economic valuation in different cultural contexts. By analysing these ethical paradigms in parallel with a number of different studies of cultural diversity, Parker is able to consider the contradictions between and within communities, and the conceptual problems of these Western ethical paradigms. It is then proposed that principles for a debate-mediated conception of human wants and needs, which respects and extends cultural diversity, would be the preferable cultural stand-point for policy-makers when considering the economic valuation of resources. Concluding, Parker reasons that taking a step back from the partial, limited and reductive notion of economic valuation would not mean that one single alternative should be implemented in its place. Rather, it means that social actors would have the capacity to contest policy and construct a new democratic foundation with inclusive cultural diversity.

We hope that the concepts and collective experiences given in these chapters will provide a fresh perspective on the nature of economic valuation and institutional arrangements in the context of PES in developing countries, and will provide critical insights to decision-makers, researchers and conservation managers.

REFERENCES

Aronson, J., S. Milton and J.N. Blignaut (eds) (2007). *Restoring Natural Capital: Science, Business and Practice*. Washington DC: Island Press, p. 384.

Asquith, N.M., M.T. Vargas and S. Wunder (2008). Environmental services: in-kind payments for bird habitat and watershed protection in Los Negros, Bolivia. *Ecological Economics*, **65**, 646–685.

Bennett, M.T. (2008). China's sloping land conversion program: institutional innovation or business as usual? *Ecological Economics*, **65**, 834–852.

Child, B. (1996). The practice and principles of community-based wildlife management in Zimbabwe: the CAMPFIRE programme. *Biodiversity and Conservation*, **5**, 369–398.

Daily, G.C., P.M. Kareiva, S. Polasky, T.H. Ricketts and H. Tallis (2011). Mainstreaming natural capital into decisions. In P. Kareiva, H. Tallis, T.H. Rickets, G.C. Daily and S. Polasky (eds), *Natural Capital Theory and Practice of Mapping Ecosystem Services*. Oxford: Oxford University Press, pp. 3–14.

Engel, S., S. Pagiola and S. Wunder (2008). Designing payments for environmental services in theory and practice: an overview of the issues. *Ecological Economics*, **65**, 663–675.

Ferraro, P.J. (2002). The costs of establishing protected areas in low-income nations: Ranomafana National Park, Madagascar. *Ecological Economics*, **43**, 261–275.

Fisher, B., R.K. Turner and P. Morling (2009). Defining and classifying ecosystem services for decision making. *Ecological Economics*, **68**, 643–653.

Frost, P.G.H. and I. Bond (2008). The CAMPFIRE programme in Zimbabwe: payments for wildlife services. *Ecological Economics*, **65**, 776–787.

Goulder, L.H. and D. Kennedy (2011). Interpreting and estimating the value of ecosystem services. In P. Kareiva, H. Tallis, T.H. Rickets, G.C. Daily and S. Polasky (eds), *Natural Capital Theory and Practice of Mapping Ecosystem Services*. Oxford: Oxford University Press, pp. 15–33.

Grieg-Gran, M., I. Porras and S. Wunder (2005). How can market mechanisms for forest environmental services help the poor? Preliminary lessons from Latin America. *World Development*, **33**, 1511–1527.

Huberman, D. (2008). *A Gateway to PES: Using Payments for Ecosystem Services for Livelihoods and Landscapes*. Markets and Initiatives for Livelihoods and Landscapes Series No. 1, Forest Conservation Programme. Gland, Switzerland: International Union for the Conservation of Nature (IUCN).

Kemkes, R.J., J. Farley and C.J. Koliba (2010). Determining when payments are an effective policy approach to ecosystem service provision. *Ecological Economics*, **69**, 2069–2074.

Kumar, M. and Kumar P. (2008) Valuation of the ecosystem services: A psycho-cultural perspective. *Ecological Economics*, **64**(4), 808–819.

Kumar, P. and R. Muradian (2009). Payments for ecosystem services and valuation. challenges and research gaps. In P. Kumar and R. Muradian (eds), *Payments for Ecosystem Services. Ecological Economics and Human Wellbeing*. Oxford: Oxford University Press, pp. 1–17.

Landell-Mills, N. and I. Porras (2002). *Silver Bullet or Fool's Gold? A Global Review of Markets for Forest Environmental Services and their Impact on the Poor*. London: International Institution for Environment and Development.

Millennium Ecosystem Assessment (2005). *Ecosystems and Human Well-being: Synthesis*. Washington DC: Island Press.

McShane, T.O., P.D. Hirsch, T.C. Trung, A.N. Songorwa, A. Kinzig, B. Monteferri, D. Mutekanga, H.V. Thang, J.L. Dammert, M. Pulgar-Vidal, M. Welch-Devine, J. Peter Brosius, P. Coppolillo and S. O'Connor (2011). Hard choices: making trade-offs between biodiversity conservation and human well-being. *Biological Conservation*, **144**(3), 966–972.

Nelson, E., G. Mendoza, J. Regetz, S. Polasky, H. Tallis, D.R. Cameron, K.M.A. Chan, G.C. Daily, J. Goldstein, P.M. Kareiva, E. Lonsdorf, R. Niadoo, T.H. Ricketts and M.R. Shaw (2009). Modeling multiple ecosystem services, biodiversity conservation, commodity production, and tradeoffs at landscape scales. *Frontiers in Ecology and the Environment*, **7**, 4–11.

Pagiola, S. and G. Platais (2007). *Payments for Environmental Services: From Theory to Practice*. Washington DC: World Bank.

Perrot-Maitre, D. (2006). The Vittel Payments for ecosystem services: a 'perfect' PES case? Project Paper 3, International Institute for Environment and Development, London.

Redford, K.H. and W.M. Adams (2009). Payment for ecosystem services and the challenge of saving nature. *Conservation Biology*, **23**(4), 785–787.

Sharma, E., K. Tse-Ring, N. Chettri and A. Shresta (2008). Biodiversity in the Himalayas: trends, perception and impacts of climate change. International Mountain Biodiversity Conference, Plenary Session 1: Climate Change and its Implications for Mountains, 16–18 November 2008, Kathmandu. Available at: http://www.environmentportalin/files/Biodiversity%20in%20the%20Himalayas.pdf, accessed 2 June 2011.

Spangenberg, J.H. and J. Settele (2010). Precisely incorrect? Monetising the value of ecosystem services. *Ecological Complexity*, **7**, 327–337.

Tallis, D., S. Pagiola, W. Zhang, S. Shaikh, E. Nelson, C. Stanton and P. Shyamsundar (2011). Poverty and the distribution of ecosystem services. In P. Kareiva, H. Tallis, T.H. Rickets, G.C. Daily and S. Polasky (eds), *Natural Capital Theory and Practice of Mapping Ecosystem Services*. Oxford: Oxford University Press, pp. 278–295.

Tallis, H. and P. Kareiva (2005). Ecosystem services. *Current Biology*, **15**(18), 746–748.

Wunder, S. (2005). Payments for environmental services: some nuts and bolts. CIFOR Occasional Paper No. 42, Center for International Forestry Research, Bogor.

Wunder, S. (2008). Payments for environmental services and the poor: concepts and preliminary evidence. *Environment and Development Economics*, **13**, 279–297.

Wunder, S. and M. Alban (2008). Decentralised payments for environmental services: the cases of Pimampiro and PROFAFOR in Ecuador. *Ecological Economics*, **65**(4), 685–698.

Wunder, S., S. Engel and S. Pagiola (2008). Taking stock: a comparative analysis of payments for environmental services in developed and developing countries. *Ecological Economics*, **65**, 834–852.

2. Making payments for ecosystem services work

Rodrigo Arriagada and Charles Perrings

POSING THE QUESTIONS

For over 50 years economists have developed instruments to address the market failures behind the collapse of ecosystem services noted by the Millennium Ecosystem Assessment (2005; MA). Such instruments include taxes, subsidies, user charges, access fees, penalties for non-compliance and the like (Tietenberg 2006). More recently, instruments of this kind have been linked explicitly to the provision of specific ecosystem services through the concept of payments for ecosystem services (PES) (Ferraro and Kiss 2002; Hardner and Rice 2002; Niesten and Rice 2004; Scherr et al. 2004; Wunder 2007; Arriagada and Perrings 2011; Ferraro 2011; Kinzig et al. 2011). PES schemes differ from earlier approaches to the management of ecosystems such as Integrated Conservation and Development Projects or Community-Based Natural Resource Management in three respects: their focus on ecosystem services (the benefits provided by ecosystems), their use of positive financial incentives to achieve the production of additional services and the conditionality of those incentives on some measure of performance (Sanchez-Azofeifa et al. 2007; Swallow et al. 2007; Pagiola 2008; Wunder et al. 2008). Recent attention has focused on PES schemes that connect with climate change, such as the Reduced Emissions from Deforestation and forest Degradation (REDD) scheme in developing countries (Miles and Kapos 2008; O'Connor 2008), but PES schemes have also been developed that offer real financial incentives for local actors to provide a wide range of more localized external, non-market ecosystem services (Engel et al. 2008).

The problem that PES schemes are designed to solve is that approximately 60 per cent of the ecosystem services evaluated in the MA (70 per cent of regulating and cultural services) are being degraded or used unsustainably. The rapid growth in provisioning services – the production of foods, fuels or fibres – in response to the incentives offered by existing markets, has been at the cost of regulating and supporting services such

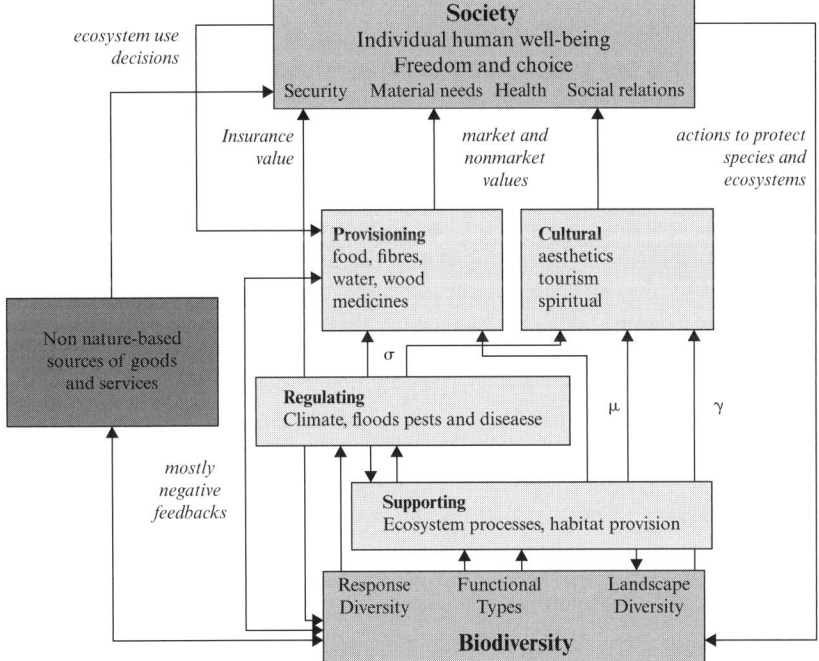

Source: Kinzig et al. (2009).

Figure 2.1 The pathways and processes by which biodiversity influences ecosystem services and ecosystem services influence human well-being. The value of supporting services, most of the value of regulating services and most of the aspects of biodiversity are contained within the value of the directly used provisioning and cultural services. These underlying elements can influence the direct services through altering the mean magnitude of the service (μ) or its variability in time (σ), its variability in space (γ)

as disease and climate regulation, waste processing services, biodiversity and water provision (Millennium Ecosystem Assessment 2005). Yet these services have value that is not reflected in current market prices. Many are ultimately crucial to the sustainability of provisioning services, affecting not just the mean level of output of those services, but also their variability in time and space (Figure 2.1). PES schemes, like other market-based environmental instruments, are designed to signal the importance of these services to the land managers whose decisions determine their supply.

This chapter reviews the factors that make PES schemes work. PES schemes can also offer benefits in terms of improvements in equity and the alleviation of poverty. These aspects are discussed elsewhere.[1] Judging the effects of many PES schemes is challenging, partly because it is not evident what is being paid for compared with more traditional market transactions, and partly because a meaningful appraisal needs to include an estimate of what might have happened without the PES scheme. It is not always possible to calculate the marginal net social benefit of the behaviours induced by the scheme. Given these difficulties, the assessment of current schemes has tended to focus more on 'process' (how the scheme works) rather than 'outcome' (the ecosystem services produced). Since effective design is a necessary, albeit not sufficient, condition for the effectiveness of PES schemes, we follow this approach while emphasizing that it is only part of the story. We pose and answer three questions: What are PES? What can they be expected to achieve? From experiences to date, what lessons can be learned about the efficient design of PES schemes?

WHAT ARE PES?

Like other environmental economic incentives, PES schemes aim to change the behaviours that have led to the degradation of many of the world's most valuable ecosystems. Tropical rainforests disappear due to mining, illegal logging and large-scale slash-and-burn practices, river basins are polluted by agrochemicals, and mountain watersheds are degraded by non-sustainable management practices (Ahlheim and Neef 2006). Over the past 50 years, humans have changed ecosystems more rapidly and extensively than in any comparable period of time in human history, biodiversity is being lost at unprecedented rates, forcing many species to extinction, and many ecosystem services are rapidly deteriorating (Millennium Ecosystem Assessment 2005). Ecosystem services sustain human life. They are the source of food, water, timber, fibre and genetic resources. They contribute to the regulation of climate, floods, disease, pests and water quality, as well as waste treatment. They support recreation, aesthetic enjoyment, knowledge and spiritual fulfilment. They also enable soil formation, pollination and nutrient cycling (Millennium Ecosystem Assessment 2005). They supply food and drinking water, maintain a stock of continuously evolving genetic resources, preserve and regenerate soils, fix nitrogen and carbon, recycle materials, control floods, filter pollutants, pollinate crops and much more (Food and Agriculture Organisation 2007). Many ecosystem services are poorly understood or simply taken for granted by people who cannot see the relation between, for example, milk cartons or medicines and the services of nutrient cycling and biodiversity conservation that make their production possible (Salzman 2005).

PES systems can help address the market failures involved where ecosystem services are 'public goods' or where changes in ecosystem services are 'externalities' of market production. If local land managers do not receive compensation for the production of valuable ecosystem services, they ignore them in their private decision-making, leading to socially sub-optimal land-use decisions. Market failures of this sort may be due to incomplete information (i.e. ignorance and uncertainty regarding ecosystem functioning and conservation), as well as lags in time and space between environmental disturbance and recognition of environmental problems (Mayrand and Paquin 2004; Wertz-Kanounnikoff 2006). PES systems, like other market mechanisms, are intended to induce landowners to incorporate the economic value of ecosystem services into their financial decisions (Rojas and Aylward 2003; Pirard 2012). The benefits of such mechanisms for poverty alleviation and equity lie in the fact that the emergence of market mechanisms for ecosystem services such as carbon sequestration or biodiversity conservation creates new income-generating opportunities for landholders at the same time as generating efficiency gains (Food and Agriculture Organization 2007; Zhang and Pagiola 2011). The principal attraction of market mechanisms is that they enhance efficiency by increasing the supply of socially desirable services and reducing the supply of socially undesirable services (disservices). In fact, in their review of World Bank-funded projects with biodiversity goals, Kareiva et al. (2008) found that the only predictor of the overall success of the project was the development of market mechanisms and new sources of finance for conservation.

In light of this, the search for market-like mechanisms to enhance ecosystem services is gaining attention from policy-makers and private decision-makers (Daily and Matson 2008; Tallis et al. 2008, 2009). Historically, the attempts of governments to correct environmental externality problems have been primarily through the use of command-and-control and other forms of direct interventions, which are easy to implement but can be quite inefficient (Bulte et al. 2008). Yet industrialized nations have used conservation payments for decades to conserve agricultural soil, improve water quality, manage fisheries and protect wilderness on private lands (Hartshorn 2005). The European Union Common Agricultural Policy (CAP) began operating in 1962, and agro-environment schemes have been supported since they were introduced in the CAP reforms of 1992. These schemes encourage farmers to provide ecosystem services that go beyond following good agricultural practice (European Commission 2007).[2]

In less developed countries, projects that have implicitly embraced ecosystem services have historically been categorized as integrated conservation–

BOX 2.1 PAYMENTS FOR AGRICULTURAL BIODIVERSITY CONSERVATION

Many PES schemes currently in operation have their origins in agricultural policies in Organization for Economic Co-operation and Development (OECD) countries, dating from the 1980s (Food and Agriculture Organization 2007). These policies were implemented in response to intensive farming practices. In fact, the best-known conservation payment initiatives are the agricultural land diversion programmes of high-income nations (Ferraro and Simpson 2002). Generally, agro-environmental payments in OECD countries are designed to compensate farmers for forgoing more intensive and more profitable farming practices.

Few programmes in recent US history have had such a large and sweeping effect on farmland use as the Conservation Reserve Program (CRP) (Wu 2000). The CRP was authorized by the Food Security Act of 1985 and re-authorized in subsequent Farm Bills.[3] The CRP offers annual payments for 10–15-year contracts to participants who establish grass, shrub and tree cover on environmentally sensitive lands, with the aim of preventing soil erosion in cropland. The CRP spends about US$1.5 billion annually on contracts for 12–15 million hectares. In Europe, 14 nations spent an estimated US$11 billion between 1993 and 1997 to divert over 20 million hectares into long-term set-aside and forestry contracts (OECD 1998).

In the UK, through the Environmentally Sensitive Areas (ESA) Scheme created in 1987, farmers in eligible areas receive direct payments as compensation for adopting less intensive farming practices that conserve landscape and wildlife values. The total area designated under ESA was estimated in 2003 to be 571 520 hectares (Food and Agriculture Organization 2007). The scheme is voluntary, with farmers being encouraged to adapt their practices so as to enhance or maintain the natural features of the landscape and conserve wildlife habitat. In return, the Department for Environment, Food and Rural Affairs pays the farmer a sum that reflects the financial losses incurred as a result of reconciling conservation with commercial farming. Schemes similar to ESA have been established in Denmark, France, Italy and Spain (Wilson 1996).

> The Australian National Landcare Program was established in 1992 as one of the mechanisms to progress towards sustainable ecosystems, with a primary focus on sustainable agriculture and improved management of the natural resource base – soils, water and vegetation – at farm level.

development projects, community-based natural resource management and, more recently, pro-poor conservation (Adams et al. 2004). The first ecosystem services payment programmes implemented in developing countries formed part of forest conservation initiatives in Latin America, following the limited success of the traditional regulatory approach that emphasized protected areas (Landell-Mills and Porras 2002).

PES schemes have since emerged as a preferred policy solution for realigning the private and social benefits that result from decisions related to the environment. The approach is based on a straightforward proposition, which is to pay individuals or communities to undertake actions that increase levels of desired ecosystem services (Ferraro 2001; Bawa et al. 2004; Berkes 2004; Romero and Andrade 2004; Wunder 2007; Jack et al. 2008).

No formalized definition of PES schemes exists in the literature, which causes some conceptual confusion (Wunder 2007; Pirard 2012). Nevertheless, the basic principle of PES schemes is that those who provide ecosystem services should be compensated for the cost of doing so, whether these are direct costs of specific land-use practices or more indirect opportunity costs of avoiding activities or types of land use (Grieg-Gran and Bann 2003). Others believe that PES schemes should be a first-best (i.e. meeting the maximum number of requirements) direct-payment approach, and an incentive mechanism, used to purchase environmental services from local resource managers who otherwise would not provide the services (Ferraro and Simpson 2002; Wunder 2005, 2007; Engel et al. 2008).[4] In many cases, the term 'PES scheme' seems to be used as a broad umbrella for any kind of market-based mechanism for conservation, including, for example, mechanisms such as eco-certification and charging entrance fees to tourists (Engel et al. 2008). Wunder (2005) defines PES schemes as (1) a voluntary transaction, where (2) a well-defined ecosystem service or land use is likely to secure the service sought, (3) the service is 'bought' by a (minimum one) service buyer, (4) there is a (minimum one) service provider, and (5) if and only if the service provider secures service provision (so-called 'conditionality').

One qualification to this is the requirement that 'ecosystem service

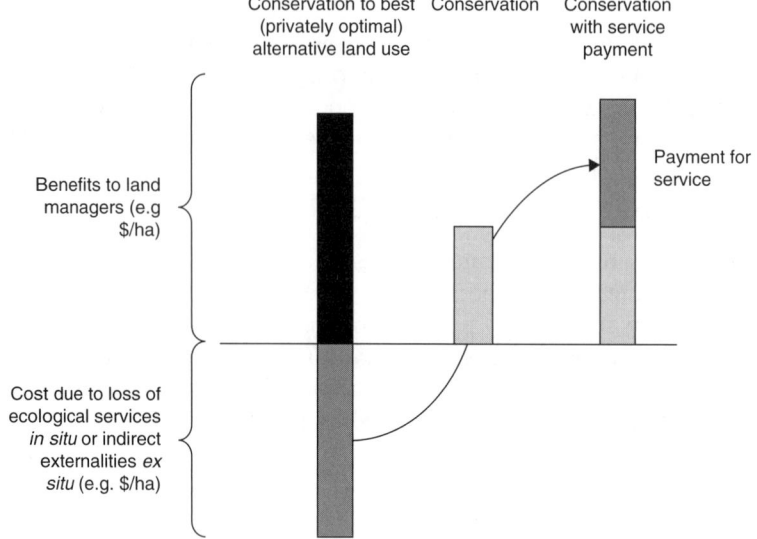

Note: 'Best alternative land use' refers to the land management option that gives the landholder the next best economic return to that obtained from PES.

Source: Adapted from Pagiola (2008).

Figure 2.2 Logic of payments for environmental services

purchasers' should be 'ecosystem service users' (S. Pagiola, personal communication). Thus, rather than having the government or donor agencies financing the provision of ecosystem services, the ultimate beneficiaries should be those paying for the service provision. Payments are then shared by the providers of the ecosystem service under a 'provider gets' principle, which means that those who provide environmental services get paid for doing so (Hodge 2000). A second qualification refers to the requirement that PES schemes should primarily focus on internalizing indirect externalities, i.e. indirect use values obtained from ecosystems that are outside the market, since this is often perceived as the main strength of PES schemes compared with other environmental policy instruments.

The logic of PES schemes illustrated in Figure 2.2 is that beneficiaries are asked or voluntarily decide to pay the landholders for the services provided, thus giving them an incentive to follow land management practices that secure provision of the services. This is achieved through a variety of arrangements that transfer payments from those who benefit from an ecosystem service to those who conserve, restore, and manage the natural ecosystem which provides it. Payments may involve private sector or government financing, and can be made at local, national and global levels (Pagiola and Platais 2007).

Forms of PES scheme are extremely varied (Grieg-Gran and Bishop 2004). The differences reflect differences in the ecosystem services involved, in the social, economic or political context in which they operate, or in the design of the policy instrument (e.g. government-financed vs. user-financed PES schemes). A critical issue concerns the identity of the 'buyers' of the ecosystem services. There is an important distinction between cases in which the buyers are the actual users of the ecosystem services (user-financed programmes) and cases in which the buyers are others (typically the government, an NGO or an international agency) acting on behalf of the users of the ecosystem services (government-financed programmes) (Engel et al. 2008). In the latter case, the funds used to compensate people who suffer lost economic opportunities to protect ecosystem services represent a public investment, and a governmental or other agency is typically responsible for collecting and redistributing the funds (Tallis et al. 2008). In general, the growing role of PES schemes today reflects underlying changes in environmental policy, and especially a greater emphasis on decentralization, flexible mechanisms, the private sector as a provider of public services, corporate self-regulation, consumer sovereignty and civil regulation (Food and Agriculture Organization 2007).

Hundreds of PES schemes are now being implemented around the world.[5] To date, the four main ecosystem services that have been addressed by PES schemes are watershed services, carbon sequestration, landscape amenity and biodiversity conservation. Most current PES schemes are local level arrangements and involve spontaneous, private market-type arrangements. Such schemes tend to be modest in scale, and are very common in nature-based tourism and protection of small watersheds. Large PES schemes tend to be government-driven, working at the state and provincial level (e.g. in Australia, Brazil, China and the United States), or at national level (e.g. Colombia, Costa Rica, China and Mexico) (WWF 2006).

Figures 2.3 and 2.4 indicate the countries in which PES schemes for agro-biodiversity and water, respectively, are currently being implemented. While countries implementing PES schemes for agro-biodiversity are largely developed, countries currently implementing water protection schemes involve a much greater mix of developed and developing countries. This partly reflects the longer history of agro-environmental schemes in developing countries, for example forest conservation initiatives in Latin America began in the 1990s. One of the most notable programmes, initiated in Costa Rica in 1996, was designed to enhance various forest

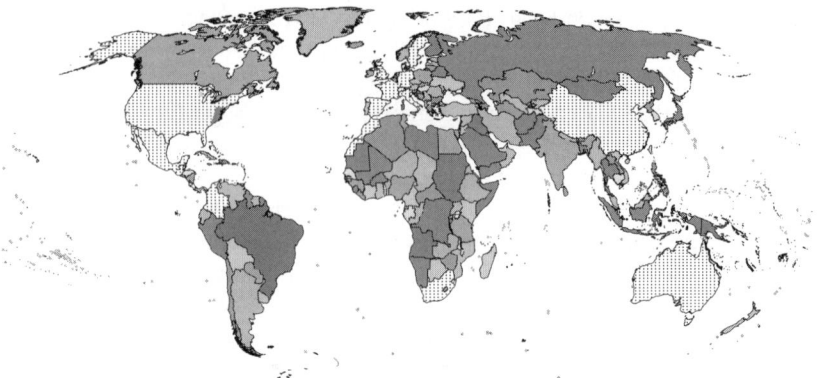

Figure 2.3   Countries implementing agro-biodiversity schemes (pale shading)

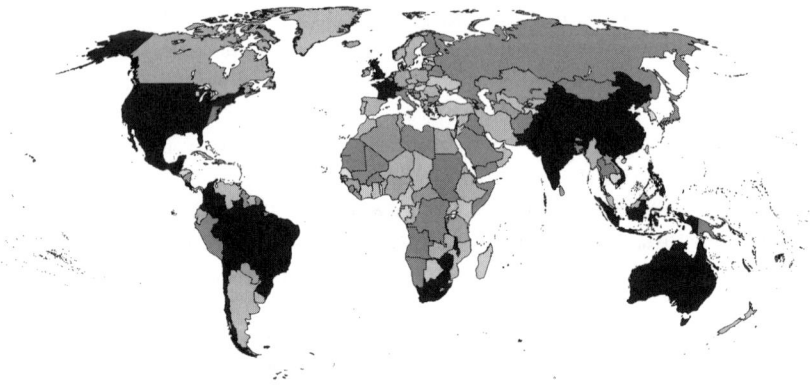

Figure 2.4   Countries implementing PES schemes (dark shading) for water protection

environmental services. Mexico, also, has initiated a national PES programme for forest-based environmental services.

Large PES schemes also exist at the international level (e.g. the European Union) and can involve PES markets created by regulation through an international agreement, such as carbon sequestration markets created by the Kyoto Protocol of the UN Framework Convention on Climate Change (Food and Agriculture Organization 2007). Governments and companies in developed countries may finance tree plantations or forest conservation in developing countries to offset their greenhouse gas emissions. Investors have done so through voluntary carbon offsets (e.g. http://

www.planvivo.org), and less often through the Kyoto Protocol's Clean Development Mechanism (CDM). Currently, only three CDM forestry projects have been approved in the world (one in India, another in China and a third in Moldovia). The characteristics of existing PES schemes are shown in Table 2.1, which reports details of the ecosystem service involved, the buyer, seller, targeting criteria and payment system for a sample of PES schemes. The Chinese 'Grain to Green' programme is illustrative. After severe droughts in 1997 and massive floods in 1998 partially caused by farming on steep slopes and deforestation, China launched the Natural Forest Conservation Program (NFCP, also known as the Natural Forest Protection Program) and the Grain to Green Program (GTGP/ Sloping Land Conversion Program/Farm to Forest Program).

The NFCP and GTGP are the two biggest programmes offering PES schemes in China and worldwide in terms of scale, payment and duration, and their implementation is a milestone of China's forest management because it marks the end of an era dominated by timber production (Liu et al. 2008). To date, PES schemes in these and other programmes have come primarily from China's central government. A total of 96.2 billion Yuan (US$14.1 billion) was designated for NFCP-related activities from 2000 to 2010, of which 81.5 per cent was to come from the central government, with the remainder coming from local governments.

By the end of 2005, more than 90 billion Yuan had been invested in the GTGP, and a planned total investment of 220 billion Yuan by 2010. Currently, the Program covers 1897 counties of 25 provinces, autonomous regions and municipalities. With its completion in 2010, the Program had converted 14.67 million ha of farmland into forestland and grassland, and participants had planted trees on 17.3 million ha of barren mountains, and lands suitable for afforestation (Deng et al. 2012). In 2004, 92 per cent of the accumulated value of the Sloping Lands Program (US$7.6 billion) was provided by the national government (Tallis et al. 2008). As final goals, the NFCP aims to restore natural forests and meet domestic demand for timber in plantation forests, and GTGP aims to reduce environmental degradation, to alleviate poverty and to promote local economic development. The area affected by these schemes is indicated in Figure 2.5.

WHAT ARE PES SCHEMES ABLE TO DO?

The attractiveness of PES schemes can in part be attributed to the interest of governments and civil organizations, especially conservation NGOs, to find new ways of promoting forest conservation while supporting the economic development of rural populations (Corbera et al. 2008). Their

Table 2.1 Characteristics of PES schemes

Case, Country	Environmental Services		Buyer	Seller	Targeting Criteria	Payment Scheme
	Paid for	Non-Paid for				
Government-Financed Programmes						
Grain to Green Program, China (Bennett 2008; Liu et al. 2008)	Cropland retirement; conversion to forest and grasslands, reforestation and afforestation	Carbon sequestration	Chinese government	Chinese farmers	Slope steepness is the main criterion by which plots are chosen for inclusion in the programme	Payment scheme adapts to priorities of participants including: technical assistance, technology transfer and direct payments
Environmental Quality Incentives Program, USA (Claassen et al. 2008)	Watershed protection, biodiversity conservation (benign agriculture and agricultural land retirement)	Landscape beauty	US government	US farmers	Participants are selected based on environmental benefits and cost index	Annual cash payment. A reserved price is based on the rental value of land adjusted for its productive capability
Conservation Reserve Program, USA (Claassen et al. 2008)	Watershed protection, soil conservation, wildlife protection and carbon	Landscape beauty	US government	US farmers	Participants are selected based on environmental benefits and cost index	Annual cash payment. A reserved price is based on the rental value of land adjusted for

Program	Service	Buyer	Seller	Description	Payment
	sequestration (benign agricultural practices and agricultural land retirement)				its productive capability
Environmental Sensitive Area and Countryside Stewardship Scheme, UK (Dobbs and Pretty 2008)	Biodiversity, recreation and watershed protection (benign agriculture and agricultural land retirement)	Carbon sequestration	UK government and European Union	Farmers in targeted areas	The ESA is open to all farmers in targeted areas and CSS selects participants	Cash payments
Australian Bush Tender Program (Department of Sustainability and Environment 2008)	Biodiversity conservation (land management agreements for native vegetation	Watershed protection, carbon sequestration, landscape beauty	Australian government	Private landowners	Private landholders are contracted to improve native vegetation on their land. These contracts are awarded through competitive tendering on a best value for money basis	Landholders establish their own price for the management services they offer to better protect and improve their native vegetation. Successful bids are those that offer the best value for money

Table 2.1 (continued)

Case, Country	Environmental Services		Buyer	Seller	Targeting Criteria	Payment Scheme
	Paid for	Non-Paid for				
Government-Financed Programmes						
Swedish Payments for Wildlife Conservation (Zabel and Holm-Müller 2008)	Number of carnivore reproductions certified on the villages' reindeer grazing grounds		Swedish state	Sami villages	About 20 000 Sami people live in Sweden grouped in 51 villages. All of them are eligible to participate in the programme	Cash payments are determined according to the monetary damage that the offspring are expected to cause throughout their lifetime
Regional Integrated Silvopastoral Approaches to Ecosystem Management Project, Nicaragua (Pagiola et al. 2008)	Biodiversity conservation and carbon sequestration	Wildlife protection, water services	The World Bank (GEF grant). The NGO Nitaplan, is in charge of implementing the programme	Farmers located in the Bulbul and Paiwas micro watersheds	The project developed an 'environmental service index' (ESI) and pays participants for net increases in ESI points	Cash payments are defined based on analyses of the relative profitability of different practices

Working for Water Program, South Africa (Turpie et al. 2008)	Watershed and wetlands protection (clearing invasive alien plants)	Biodiversity conservation	Previously unemployed individuals that tender for contracts to restore public or private lands	Department of Water Affairs and Forestry, water management agencies	The programme prioritize areas using ecological and social rationales	Cash payments to contractors staff that have been previously unemployed
Payments for Hydrological Environ-mental Services, Mexico (Muñoz-Piña et al. 2008)	Watershed protection and aquifer recharge (conservation of pre-existing forest area)	Biodiversity conservation, carbon sequestration, landscape beauty	CONAFOR (state forest agency funded through an earmarked portion of federal fiscal revenues from water fees)	Communal and individual landowners	Applicants selected where severe water problems are linked to deforestation, but where commercial forestry cannot compete against agriculture or ranching	Cash payments are defined according to land value in terms of hydrological services (cloud forest vs. other forested areas)

User-Financed Programmes

The Vittel (Nestlé Waters) watershed protection programme, France	Watershed protection (best practices in dairy farming)		Vittel	Dairy farmers (27 farmers enrolled)		Cash payments are base on new farm investment and the cost of adoption of new farming practices

Table 2.1 (continued)

Case, Country	Environmental Services		Buyer	Seller	Targeting Criteria	Payment Scheme
	Paid for	Non-Paid for				
User-Financed Programmes						
Los Negros, Bolivia	Watershed and biodiversity protection (forest and páramo conservation)		Pampagrande municipality, US Fish and Wildlife Service	Santa Rosa farmers (46 landowners)		In kind plus technical assistance
Pimampiro, Ecuador	Watershed protection	Carbon sequestration, biodiversity conservation, landscape beauty	Metered urban users (20 per cent fee)	Households in Nueva América Cooperative	Participant selection has focused on Nueva América because it is located near the water intake	Three differentiated cash payments according with forest type
PROFAFOR, Ecuador	Carbon sequestration (re- and afforestation)	Water services, biodiversity conservation	FACE (electricity consortium)	Communal and individual landholders	Process of site selection based on biophysical and economic criteria. Trade-offs between ES provision and opportunity costs rule selection	Cash payments plus in-kind subsidies and technical assistance

30

| Scolel Té Project, Mexico | Carbon sequestration | Biodiversity conservation, water services, landscape beauty | Trust fund Fondo Bioclimatico. Purchasers include Int. Automobile Fed, World Economic Forum, Pink Floyd and Future Forests | Individual farmers and communities | Through a management system, Plan Vivo, contacts between the Fondo team and local communities are arranged through farmers' and other organizations in the region | The Fund provides training and support during planning process. |

Notes:
ESA = Environmentally-Sensitive Areas scheme
ESI = Environmental Service Index.

Source: Liu et.al. (2008).

Source: Liu et al. (2008).

Figure 2.5 Current distribution of the NFCP and GTGP in China

effectiveness in meeting conservation goals is, nonetheless, not well understood (Kleijn et al. 2001; OECD 2003). An important feature of incentive systems generally is that since they are voluntary, their outcomes are products of the private decisions of landholders. First, the agency designs a scheme and offers it to landholders. Then the landholders decide whether to participate and, if so, in which areas to enrol. As in any economic (as opposed to management) problem, the agency influences but does not completely control programme outcomes (Siikamaki and Layton 2006).

In principle, one would expect PES mechanisms to be more efficient than other non-market-based policy measures. There is a long standing literature demonstrating the efficiency of market-based instruments of environmental policy relative to traditional command-and-control measures, and the same arguments should apply to PES schemes (S. Pagiola, personal communication; Wertz-Kanounnikoff 2006). It is now widely accepted that the protection and long-term sustainability of diverse ecosystems will be viable only if the full range of services provided by these ecosystems is economically accounted for, thus favouring economic instruments (Liverman 2004; Corbera et al. 2008).

With perfect information, price-based mechanisms (of which PES schemes is an example) and quantity-based mechanisms (such as regulations prescribing particular behaviour) could be equivalent. With incomplete information, the specific circumstances can help define which mechanism is more efficient. The relative efficiency of different mechanisms to address market failures has been the subject of considerable debate in the literature, beginning with the work of Weitzman (1974). Pagiola et al. (2008) found that one of the cases Weitzman examined is particularly relevant to PES, specifically, when there are multiple potential producers of a service (i.e. carbon sequestration) with different marginal costs that are not observable by the buyer, price-based mechanisms are more efficient because they screen out the high cost producers. This encourages them to produce less and the low cost producers to produce more. Ferraro and Simpson (2002) demonstrated that paying for ecosystem protection directly can be far more cost-effective than encouraging activities, such as ecotourism, that indirectly generate ecosystem protection as a joint product. Siikamaki and Layton (2006) showed that schemes that exploit landholders' knowledge about the opportunity costs of ecosystem service provision are more efficient than top-down regulatory schemes.

While PES schemes are motivated by environmental concerns, there is an increasing interest in their potential to deliver development benefits. At the moment it remains unclear to what extent the two objectives of environmental conservation and development can be achieved simultaneously through such market-based mechanisms (Grieg-Gran et al. 2005). Kareiva et al. (2008) did not find that environmental or biodiversity objectives are necessarily consistent with development objectives. Win–win outcomes are not easy to obtain, yet attaining environmental goals without also addressing poverty is equally problematic (Sachs and Reid 2006). Projects that use ecosystem services to simultaneously advance both conservation and human agendas should nevertheless benefit from improved scientific understanding of four overriding issues: sustainable use of ecosystem services; trade-offs among different services; the spatial flows of services; and economic feedbacks in ecosystem services markets (Tallis et al. 2008).

Reactions to PES schemes in conservation and rural development circles have been mixed. Advocates of PES schemes stress that innovation in conservation is needed because current approaches provide too little value for money, that PES schemes can provide new (especially private sector) conservation funding, and that poor communities can improve their livelihoods (Wunder 2007). When buying an ecosystem service, what is being paid for is not self-evident because ecosystem services are provided over time and space. While it is desirable to have an idea about what would hypothetically happen without the PES scheme, rigorous measurement of the counterfactual (i.e. the hypothetical alternative) is non-existent in the conservation literature. PES schemes are being implemented globally in much the same way that previous conservation interventions were implemented, i.e. with an unwavering faith in the connection between interventions and outcomes, and without a plan to judge the effectiveness of such interventions (Ferraro and Pattanayak 2006). Few well-designed empirical analyses assess even the most common biodiversity conservation measures (Millennium Ecosystem Assessment 2005).

There are many instances in which a government-financed programme of PES schemes may be the only option. As the number of ecosystem services buyers increases, transaction costs and incentives for free riding increase as well. When the ecosystem services are public goods, it may be difficult to identify and delimit users, while non-excludability implies that users have strong incentives to free ride. Nevertheless, governments, NGOs or international organizations can play an important role in reducing transaction costs (Engel et al. 2008). User-financed PES schemes are often implemented in situations with single or few local buyers. According to Wunder et al. (2008), user-financed programmes show greater adherence to a pure PES scheme definition, and are more targeted in their effects, compared with the larger, multiple-objective, government-financed programmes that often have broader and less well-defined objectives. Indeed, the latter can sometimes be hard to distinguish from more traditional subsidy programmes, the main differences coming in the conditionality of payments.

As most ecosystems provide not one but a large variety of ecosystem services, efforts are sometimes made to either 'bundle' various services together for sale, or to 'layer' payments from multiple buyers into payments to providers (Engel et al. 2008; Wunder and Wertz-Kanounnikoff 2009). In terms of conservation, it has been suggested that the direct nature of PES transactions induces PES schemes to be both more effective and more cost efficient than indirect tools such as Integrated Conservation and Development Projects or eco-friendly premiums requiring investments in alternative lines of production (Ferraro and Kiss 2002; Ferraro and Simpson 2002, 2005, cited by Wunder 2008).

Several conditions must be met for implementation to be effective, however, including mechanisms for valuing (or at least measuring) a service where none currently exists, identifying how additional amounts of that service can be provided most cost-effectively, deciding which farmers to compensate for providing more of the service, and determining how much to pay them (Food and Agriculture Organization 2007). Whether a PES programme succeeds in generating the desired ecosystem services depends on the successful completion of a series of steps. First, potential

BOX 2.2 ESTIMATING THE IMPACTS OF PES

The most obvious way to assess PES schemes programmes is to quantify the area of ecosystem preserved by participants. However, this does not address an important issue: is the PES schemes programme causing participants to preserve ecosystems that they otherwise would not have preserved?

Answering this question requires estimating a counterfactual (hypothetical) outcome: the area of ecosystems that landowners would have preserved if they had not received payments. However, if we want to estimate the PES schemes counterfactual outcome, we must worry about confounding effects; effects that are contemporaneous with the intervention and could plausibly affect the outcome and thereby mask the intervention's effect (Ferraro and Pattanayak 2006). Historical trends, unrelated programmes or policies, and unobserved environmental and social characteristics are just some examples of these confounding effects. As in all scientific research, confounding effects are addressed through baselines, measures of covariates, and control groups.

One potential confounding effect deserves mention because of its widespread but not well-understood effects on our ability to make inferences about the effectiveness of PES schemes: 'endogenous selection'. In any non-randomized PES scheme, characteristics that influence the outcome also influence the probability of being selected into the scheme. For example, studies of participation in a Costa Rican PES scheme found that participants differed from non-participants in important farm-level characteristics that directly affect land use (Arriagada et al. 2012). Failure to address the issue of endogenous selection can lead to biased estimates of a scheme's effectiveness. Because of the existence of confounding effects, we cannot simply compare the outcome of a participating landowner to that of the average non-participating landowner.

Programme evaluation provides the tools to focus on PES scheme outcomes instead of focusing on 'inputs' (e.g. investment dollars) or 'outputs' (e.g. number of contracts for PES schemes). Evaluation uses randomized experimental policy trials and, when interventions are not randomly assigned, as is usual for PES programmes around the world, appropriate statistical tools are used to evaluate the effects of an intervention. The use of programme evaluation to measure conservation outcomes is almost absent in the conservation literature.

The use of methods to equate, or 'balance' the distribution of characteristics between participants and non-participants ('matching') can address the selection bias and estimate the missing counterfactual without imposing strong distributional assumptions or extrapolating beyond a common support. Propensity score matching, originally proposed by Rosenbaum and Rubin (1983), is perhaps the most popular matching method used in a range of fields. However, there are still proponents of covariate matching, because in cases where there is a good understanding of the determinants of programme participation and the outcomes of interest, matching on these determinants will give unbiased estimates of programme impact (Arriagada 2008). Because matching relies on the assumption that all characteristics that affect participation and outcome are observable and controlled during the matching, results are highly dependent on the quantity and quality of available data. This suggests an important policy recommendation for new programmes of PES schemes. The ideal database for a rigorous empirical evaluation of PES schemes would include observations on land use and characteristics of both participant and non-participant landowners and their properties both before and after the scheme. Collection of these data should be integrated into the scheme operations when rigorous evaluation is a scheme goal but building experiments (random allocation of contracts) into the scheme is not politically feasible. The data should be sufficient to fully characterize the participants and feed this information into the matching process in order to select the most appropriate non-participants for estimating the missing counterfactual. Design of the database should be supported by qualitative studies of participants in PES programmes to identify key determinants of participation and outcomes. Equally important is collecting high-quality time-series data on those programme outcomes.

Finally, targeting payments to areas with high environmental risks (e.g. areas with high deforestation threats) is more appropriate than protecting forest that would be conserved with or without PES. Future evaluations of PES schemes should also concentrate on the impact of the programme on provision of environmental services, given that payments on land with low opportunity costs may be justified if the environmental benefit is high (Arriagada 2008).

BOX 2.3 EFFICIENCY OF THE COSTA RICAN PES PROGRAMME

The Costa Rican Programme of PES (PSA) is currently the longest running programme of PES in the tropics and is mostly devoted to forest conservation (the Costa Rican government assumes that by having forest protected, the provision of ecosystem services will be secured). As it developed, PSA has applied different rules in order to select participants. In early years, contracts were not assigned in a systematic fashion, but in recent years a more targeted selection of applicants has been used. Programme administrators did not design the programme with the intention of empirically evaluating its effectiveness by testing and measuring against a clear baseline or 'control' case. Moreover, forest cover had apparently been increasing in Costa Rica even before the establishment of PSA. To evaluate the effect of PSA on forest cover changes, the analyst must disentangle the effects of PSA from the effects of the elimination of government subsidies that promoted deforestation, incorporate the non-random assignment of contracts, and the economy-wide changes that have made deforestation less appealing. This makes the evaluation of PSA impacts difficult.

Is the Costa Rican programme causing participants to preserve ecosystems that they otherwise would not have preserved? Answering this question requires estimating a counterfactual outcome: the area of ecosystem that landowners would have preserved if they had not received payments. Sills et al. (2007) and Arriagada (2008) in regional parcel-level analyses focusing on the initial years of the Costa Rican programme found that regions with less productive land, fewer roads and lower population density were more likely to have PSA contracts (see Figure 2.6). In their study region, they found that absentee landowners with larger parcels of land that have more steep slopes and that are not used for commercial agriculture are more likely to have enrolled in PSA, as opposed to other landowners who were also eligible but did not enrol. According to these authors, during the initial phase of PSA, the programme did have a statistically significant effect on gross and net deforestation in that the forested area increased. This result significantly extends previous suggestions of relatively low impact from PSA (e.g. Hartshorn et al. 2005; Sierra and Russman 2006; Sánchez-Azofeifa et al. 2007; Pfaff et al. 2008).

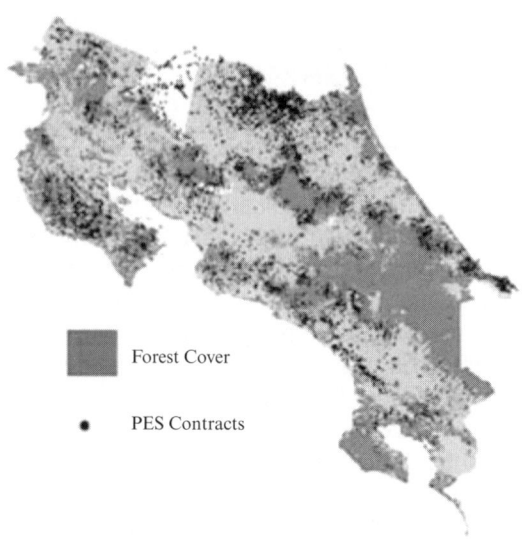

Source: Liu et al. (2008).

Figure 2.6 Distribution of PES schemes contracts in Costa Rica

In a national evaluation of the Costa Rican programme, Arriagada (2008) studied the impact of PSA on three different potential outcomes: forest gain, forest loss and net deforestation. These outcomes are all important dimensions of forest cover in Costa Rica, although they have different implications for the bundle of ecosystem services produced, and consequently are viewed differently by various stakeholder groups (e.g. stopping loss of existing mature natural forests is the priority of many environmental groups, while others interested in climate change and carbon sequestration are most likely to focus on net change in total forest cover). Arriagada (2008) showed that PSA had different impacts on different dimensions of forest cover change. The most robust result is a positive and significant impact on forest gain. PSA has no impact on forest loss, which is not surprising given the deforestation trend in Costa Rica in the last two decades.

service providers must enrol in the programme; second, providers must comply with the terms of their contract; and third, compliance must result in a change in land use compared to what would have happened without the programme (Wunder et al. 2008). The characteristics of a sample of existing PES schemes that correlate most closely with their performance are described in Table 2.2.

Note that conditionality is critical to the implementation of PES. For payments to be conditional, it must be possible to verify the existence of the ecosystem services and to establish a baseline against which additional units provided can be measured. This requires understanding causal pathways (processes), recognizing spatial extent and distribution (patterns), developing proxies or indicators for easy recognition and monitoring, and simplified yet accurate and validated measures of environmental services provided (Tomich et al. 2004). The potential sellers of ecosystem services are those actors who are in a position to secure the delivery of the ecosystem services. As long as participation is voluntary, ecosystem service sellers are unlikely to accept a payment lower than the cost of providing the ecosystem services, while conditionality ensures that they actually comply with their contracts.

Experience to date indicates that PES schemes are frequently inefficient. A number of reasons for this have been noted in the literature: social inefficiency in the adoption of actions where benefits are smaller than their costs (Pagiola 2008); lack of additionality in paying for adoption of practices that would have been adopted anyway (Ferraro and Pattanayak 2006); leakage in the inadvertent displacement of activities damaging environmental service provision to areas outside the geographical zone of PES schemes intervention (Robertson and Wunder 2005); lack of permanence in the construction of PES schemes that are dependent on the continued flow of financing from sources other than the beneficiaries of the services involved. Some authors have noted that this may be a particular problem in government-financed PES programmes, where funding is subject to policy cycles, but that is less likely to be an issue in user-financed programmes as long as the schemes are delivering the ecosystem services for which users are paying (Engel et al. 2008).

WHAT ARE THE LESSONS FOR THE EFFICIENT DESIGN OF PES SCHEMES?

The effectiveness of PES programmes depends on both their design and implementation, given the specific political, socio-economic and environmental context of the programme. Where PES schemes are designed to

Table 2.2 Factors affecting efficiency of PES programmes

Case	Clear definition of participants	Clear definition of service	Additionality	No compliance penalties	Opportunity cost vs. ecosystem services provision cost
Working for Water (WfW) Program, South Africa (Turpie et al. 2008)	Ecosystem services sellers are itinerant service providers in the form of small-scale contractors who perform restoration work on land under any type of ownership	The objective of the programme is to control invasive alien plants to improve water delivery, biodiversity conservation and land productivity	The WfW programme has been hailed as highly successful in terms of its objective of restoring water supply in catchments infested by alien species	The programme is self-supervised by WfW and does not include sanctions	Opportunity costs are low because no land use is displaced and treated land is likely to be more productive. Labour costs are low as the labour employed has few alternative formal sector employment opportunities
Grain to Green (GTGP) Program, China	The programme focuses on farmers in Western China because of its ecological vulnerability, amount of soil erosion, cropland with slope > 25° and poverty	The stated environmental goals include reducing soil erosion and desertification, and increasing China's forest cover and area by retiring steeply sloping and marginal lands from agricultural production	High for land retirement; lower for reforestation. Five years after GTGP implementation, converted plots reduce surface runoff by 75–85 per cent and soil erosion by 85–96 per cent compared with croplands on steep slopes without the GTGP. However, no	Programme compliance is defined in terms of the quality, type and survival rates of the trees/grasses planted. Withholding of subsidies is based on survival rates, but survey results indicate that low survival rates have not generally resulted in significant withholding of subsidies	Farmers of Gansu Province lost 3852–4000 Yuan ha^{-1} partially because of increased prices for agricultural products in 2003. It is possible that NPV of future income from trees and grasses planted under the programme could more than offset farmers for foregone

40

Regional Integrated Silvopastoral Approaches to Ecosystem Management Project, Nicaragua (Pagiola et al. 2008)	Landowners in specific areas can participate. The Matiguás-Rio Blanco site was selected based on its location in a biological corridor	The project developed indices of biodiversity conservation and carbon sequestration under different land uses, then aggregated them into a single ESI	explicit baseline with which to evaluate these gains is presented. The programme pays only for additional ESI score in reference to a baseline. No control groups of non-participants exist to distinguish project-induced land use changes from changes induced by other factors	The project computes changes in ESI over the entire farm – any switch to land uses that reduce service provision would incur negative points, reducing total payment	Based on the relative profitability of different practices, a fixed payment per incremental ESI was established
Pimampiro, Ecuador (Wunder and Albán 2008)	Trade-offs between ecosystem services provision and opportunity costs required selection of participants in high-altitude zones with a minimum contract size to avoid high transaction costs of working with smaller plots	The programme was established for the Palaurco river upper watershed that delivers drinking water for the Municipality of Pimampiro	There is no scientific evidence to assess additionality in terms of water services. A stronger case can be made for land-use additionality, although no formal baseline exists both of service provision and of land-use trends	Range from payment suspension to permanent exclusion depending on amount of forest extraction	Recent work indicates that with a discount rate of around 15–20 per cent, PES schemes yields a higher NPV than incremental deforestation

Table 2.2 (continued)

Case	Clear definition of participants	Clear definition of service	Additionality	No compliance penalties	Opportunity cost vs. ecosystem services provision cost
PROFAFOR, Ecuador (Wunder and Albán 2008)	The process of plantation site selection was based on biophysical conditions and economic criteria	PROFAFOR is an Ecuadorian company acting in extension of the Forests Absorbing Carbon-dioxide Emissions consortium, financed by Dutch electricity companies to offset their carbon emissions	A carbon baseline was built based on vegetation and soil criteria on parcels adjacent to those under contract. In relation to land use, one can safely affirm additionality	For individual owners, contract compliance is pursued by establishing a lien on their lands. For communal lands, members have to reimburse the payments received if they do not fulfil the terms	Average forgone revenues from much more degraded pastureland with predominantly ovine livestock are much smaller than the one for Pimampiro
Scolel Té Project, Mexico (Tipper 2002)	Farmers from six Tzeltal communities and four Tojolobal communities in the municipalities of Chilón and Comitán were selected to participate in the programme	Its point of departure is the land-use activities that communities and individual farmers were seeking to implement and then ask how the carbon benefits could	There is a carbon accounting system where payments are completed only when carbon is generated by the account holder	Annual monitoring is conducted on all sites. The Project's own technical staff checks consistency and accuracy of measurements on 10–20 per cent of sites. Five per cent of the value of timber products will be	An independent economic assessment found that discounted benefits for most participants lie between US$110 and US$1700. These estimates take into account all labour inputs and carbon credit sales, but do

Payments for Hydrological Environmental Services, Mexico (Muñoz-Piña et al. 2008)	2003 almost random 2004 basic grading and regional balance of contract distribution (2005 grading in place)	be packaged and marketed

The programme focuses on forest ecosystems. The hydrological criteria still left a large area of the country to choose from, and other criteria were introduced to narrow the area | Many of the program's payments have been in areas with low deforestation risk | ceded to the Project in the event of non-continuation of the scheme

Annual payments are made after verifying that no land use changes have occurred (the seriousness of the cancellation of payments has not yet been experienced by any forest owner) | not include other possible associated benefits such as soil conservation

Differentiated annual cash payments (higher payments for cloud forests) |

Notes:
ESI = Ecosystem Service Index
NPV = Net Present Value.

meet a particular target level of ecosystem services, then the relevant criterion for programme design is cost-effectiveness (Food and Agriculture Organization 2007). Where they are established to clear the market for ecosystem services, the relevant criterion is efficiency. PES schemes satisfying either criterion typically share a number of the characteristics identified in Table 2.2. The first is a clear identification of the service of interest, and the way it contributes to human well-being. This implies an assessment based on: (1) an understanding of the underlying biophysical science, (2) the ecosystem service supply function (and the price elasticity of supply), and (3) the ecosystem service demand function (and the price elasticity of demand) (Food and Agriculture Organization 2007). From a design perspective, defining the nature of the service for which communities are rewarded and establishing standard methodologies for the evaluation of ecosystem services provision is very important (Corbera et al. 2008).

In addition to understanding the supply and demand functions for ecosystem services, it is important to estimate the transaction costs associated with making an exchange between buyers and sellers. Transaction costs, in this context, include the cost of attracting potential buyers or finding potential providers of ecosystem services, of working with project partners and of ensuring that parties fulfil their obligations. The considerable uncertainties and complexities involved in measuring, monitoring and exchanging services mean that transaction costs can be significant (Food and Agriculture Organization 2007). To be effective, though, any payment to the land manager must make the net benefits derived from maintaining ecosystem services greater than those derived from alternative land uses (WWF 2006).

A further design consideration is whether to pay for the service itself or for some proxy for the service. If ecosystem services can be measured easily, and if cause-and-effect linkages are straightforward, payments will be most effective if made directly for output of the services delivered. In other cases, payments may be linked to observable land-use changes that correlate with provision of the desired ecosystem service. In the vast majority of PES transactions to date, payments have been associated with land-use changes rather than with service provision directly, and the buyers have borne the risk of inadequate service provision. So long as the farmers manage their property in accordance with the terms of the contract, they are paid whether the service is provided or not.

In principle, it should be possible to estimate the marginal benefit of the introduction of a PES scheme. In practice, however, since most PES schemes concentrate on incentives to change land use rather than incentives to change ecosystem service output, there are few effective measures of output. In Costa Rica, for example, direct payments for conservation of existing mature forest has had a statistically significant and positive effect on the establishment of new forest (i.e. forest gain and reduced net deforestation) (Arriagada 2008). More importantly, the programme has had a positive indirect impact on areas not protected by the programme (i.e. beneficial spillover effects) and is meeting some of its non-timber objectives from newly regenerated forests. While these forests may not be equal to the original forest in terms of biodiversity, they do sequester carbon and stabilize soil. Yet even Costa Rica has few direct measures of output.

In the absence of satisfactory measures of output, the only way to evaluate the potential efficiency of PES schemes is through design considerations of the type discussed above. Table 2.2 describes the factors affecting the efficiency of a subset of the PES programmes presented in Table 2.1. In order to evaluate the potential efficiency of these initiatives, we constructed an efficiency index that assigns a score to each of the criteria presented in Table 2.2. The score for an individual criterion extends from one to five where:

- one indicates that the design property is unlikely to enhance the efficiency of the instrument;
- three indicates that the design property may be expected to make an intermediate contribution to the achievement of efficiency;
- five indicates that the design property has a highly significant contribution to the achievement of efficiency.

Table 2.3 reports the resulting efficiency scores of the PES schemes described in Table 2.2. As with environmental governance issues generally, PES schemes should match the scale of the ecosystem service flows at issue. For watershed services, for example, a large share of funding may be secured from local ecosystem service users (Corbera et al. 2008). By contrast, a service like carbon sequestration depends on global sources of funding. In both cases, though, there is likely to be a role for national or local intermediaries. For example, the REDD scheme already referred to (Miles and Kapos 2008; O'Connor 2008) aims to reduce historic rates of deforestation through a system of money transfers to national governments. REDD is part of the land use, land-use change and forestry (LULUCF) provisions of the United Nations Framework Convention on Climate Change (UNFCCC). These provisions impose penalties on regions or countries that are clearing their forests, and reward those which have been reducing rates of deforestation. But this still leaves open the question of how reductions in historic deforestation rates are implemented at the national level. The scheme may introduce unintended consequences

Table 2.3 Efficiency index for PES schemes

Case	Efficiency score					Total efficiency score
	Clear definition of participants	Clear definition of service	Additionality	No compliance penalties	Opportunity cost vs. ecosystem services provision cost	
Scolel Té Project, Mexico (Tipper 2002)	5	5	5	5	3	23
Regional Integrated Silvopastoral Approaches to Ecosystem Mgmt Project, Nicaragua (Pagiola et al. 2008)	5	5	4	5	3	22
PROFAFOR, Ecuador (Wunder and Albán 2008)	5	5	5	5	1	21
Pimampiro, Ecuador (Wunder and Albán 2008)	5	5	3	2	5	20
Grain to Green Program, China	5	3	3	3	1	15
Working for Water Program, South Africa (Turpie et al. 2008)	3	3	5	1	1	13
Payments for Hydrological Environmental Services, Mexico (Muñoz-Piña et al. 2008)	5	3	1	2	2	13

for national decisions, e.g. the potentially perverse incentive it offers either (1) to promote increased emissions and decreased removals at national or sub-national levels in the lead-up to implementation, or (2) to accelerate displacement of deforestation and forest degradation activities from countries that are early entrants into a voluntary REDD mechanism to those that are not (Angelsen 2009). The scheme depends on the development of a system of incentives at the national level that will translate national commitments into the decisions of individual landholders. This implies financial incentives, procedures for setting reference levels, methodologies for monitoring, reporting and verification, and processes to promote the participation of indigenous peoples and local communities. REDD programmes will depend upon effective governance of remote forest regions and an equitable, efficient system of channelling these incentives to the people who control forests (Miles and Kapos 2008).

It is worth highlighting the fact that user-financed schemes score higher than government-funded programmes. The Working for Water Programme (Turpie et al. 2008), which is a PES-like development programme, earned a score similar to that of government-financed programmes. Having buyers and sellers more directly connected assists the efficiency of market transactions through identification of the service at issue, the buyers and sellers, and the marginal benefits and costs of alternative levels of provision. Moreover, whether governments are able to identify the value of environmental public goods depends heavily on the effectiveness of the political process at all levels. Since the provision of ecosystem services that are transboundary environmental public goods requires the involvement of both national governments and international representative bodies, it is extremely important that PES schemes designed to deliver such services properly reflect their value to the different constituencies involved.

PES schemes and their enabling institutions are part of an emerging system of international environmental governance that cuts across scales in novel ways (Tucker and Ostrom 2005; Corbera et al. 2008). This system of governance is rapidly evolving. Indeed, in an increasingly complex world where diverse actors interact across scales, continuous institutional adaptation is important to the long-term effectiveness of governance mechanisms (Biermann 2007). At present, relatively few PES schemes formally recognize off-site environmental benefits. Most are designed to realize conservation benefits and costs that accrue at local scales. Yet increasingly, PES schemes are going to be asked to assure provision of services that provide benefits at much larger scales, to ensure that yield outcomes are clearly verifiable.

Increasingly, PES schemes will have effects on decisions of land use made by those who are not directly involved in the schemes. Such spillovers (also called leakage) can be either negative or positive. Understanding these spillovers is important in understanding the additionality of the scheme. On the positive side, for example, there may be conservation benefits where neighbours of PES schemes are more likely to conserve forest because of the existence of the scheme. The mechanism for these spillovers may be the option value of a future PES scheme's contract, shifts in preferences or increased knowledge about the value of standing forest, or increased enforcement activity (Arriagada 2008). Depending on how the scheme is presented and perceived by landowners, there may also be negative effects: 'a monetary reward to motivate socially desirable behaviour may actually do the opposite because it may crowd out an individual's sense of public-spiritedness' (Cardenas et al. 2000). Scheme benefits may also be reduced by leakages, with recipients of PES investing the revenue in expansion of agriculture or pasture on other properties. According to Robalino and Pfaff (2006), policies that promote agricultural development or forest conservation in a specific area may also affect deforestation rates in non-targeted neighbouring areas. More generally, the additionality of individual PES schemes depends on their impact on wider trends. For deforestation, for example, there is some evidence for the existence of a 'forest transition' analogous to the well-understood 'demographic transition' (Rudel et al. 2005). Derived from historical studies of forests, the idea is that forest cover will change in predictable ways as societies undergo economic development, industrialization and urbanization (Mather 1992). Specifically, a large decline in forest cover occurs, followed by a slow increase in forest cover that takes place as wealth increases (Rudel et al. 2005).

Finally, PES are the latest in a series of mechanisms designed to internalize environmental externalities and enhance the supply of environmental public goods. Their effectiveness ultimately depends on whether they are able to deliver outcomes, measured in terms of the flow of services, that are better than the outcomes in the absence of such schemes. The power of economic incentives to alter behaviour in ways that enhances the efficiency or cost-effectiveness of ecosystem service provision has been demonstrated often enough that the principle of PES is well accepted. But as with many other incentive mechanisms, whether it works as intended depends both on mechanism design and the identification of appropriate measures of performance. Process-based measures are often easier to implement, but are not necessarily the most appropriate. It is important that the incentive to landholders corresponds to the value of the services delivered to all those affected, i.e. that the mechanism captures the interests of all beneficiaries of the service or services produced by land management. It is also important that it reflects the substitutability or complementarity between services, regardless of whether there are 'trade-offs' or 'win–win' associations. While few existing PES schemes do well if evaluated in these terms, some do better than others and are useful guides to the development of new schemes.

CONCLUDING REMARKS

For over 50 years, economists have developed instruments to address the market failures behind the collapse of those ecosystem services noted by the Millennium Ecosystem Assessment (2005). These instruments include taxes, subsidies, user charges, access fees, penalties for non-compliance and, more recently, PES schemes, which have been developed to offer financial incentives for local actors to provide a wide range of ecosystem services that lie outside normal market transactions.

Hundreds of PES schemes are now being implemented around the world, covering four main ecosystem services: watershed services, carbon sequestration, landscape beauty and biodiversity conservation. Most current PES schemes are local level arrangements and involve spontaneous, private market-type arrangements. Such schemes tend to be modest in scale, and are very common in nature-based tourism and protection of small watersheds. Large PES programmes tend to be government-driven, working at the state and provincial level (e.g. in Australia, Brazil, China and the United States) or at national level (e.g. Colombia, Costa Rica, China and Mexico).

Necessary conditions for the implementation of PES schemes to be effective include the creation of mechanisms for valuing (or at least measuring) services that are not currently valued in the market; identifying how additional amounts of that service can be provided most cost-effectively; deciding which farmers to compensate for providing more of the service; and determining how much to pay them. Whether a PES scheme succeeds in generating the desired outcomes (ecosystem services) depends on the successful completion of a series of several steps. Potential service providers must enrol in the scheme. Providers must comply with the terms of their contract, and compliance must result in a change in the provision of the ecosystem service compared with what would have happened without the scheme.

There are several implications for PES design and implementation that follow directly from these observations.

First, user-financed schemes are generally more efficient than government-funded schemes. Having buyers and sellers more directly connected assists the efficiency of market transactions through identification of the service sought, the buyers and sellers involved, and the marginal benefits and costs of alternative levels of provision. Nevertheless, user-financed schemes may ignore services that are public goods.

Second, the provision of ecosystem services at multiple spatial scales should reflect their value to the different constituencies involved. At present, relatively few PES schemes formally recognize off-site environmental benefits. Most are designed to realize the net benefits of conservation at local scales. Yet increasingly, PES schemes are going to be asked to assure provision of services that provide benefits at much larger scales. This requires effective involvement of people whose lives will be affected by the service, as well as an understanding of the substitutability or complementarity between services. In other words whether there are trade-offs or win-win associations between services.

Third, the provision of global ecosystem services through PES schemes (e.g. REDD) requires the involvement of both national governments and international representative bodies. PES schemes designed to provide benefits to people in only a few countries may be negotiated by those countries alone, but schemes that provide global benefits such as carbon sequestration or biodiversity conservation require the involvement of international bodies and coordination with global agreements.

Fourth, PES schemes should avoid negative spillovers (leakage), or provide benefits sufficient to offset unavoidable spillovers. For example, PES schemes frequently have effects on land-use decisions by those who are not directly involved in the scheme. These spillovers should be taken into account in calculating the net benefits (additionality) of a scheme. We note that spillovers might be positive (e.g. where neighbours of PES schemes are more likely to conserve forest because of the existence of the scheme), and these too should be taken into account.

Fifth, the positive incentives offered by PES schemes should be sufficient to 'internalize' the externalities of pre-existing market conditions. The effectiveness of PES schemes depends on whether they are able to deliver outcomes, measured in terms of the flow of services, that are better than the outcomes in the absence of such schemes. As with many other incentive mechanisms, whether they work as intended (in terms of provision of ecosystem services and net economic benefits) depends both on mechanism design and an understanding of the responsiveness of service providers to incentives.

Finally, PES design should be complemented by the measurement of ecosystem services produced through the scheme. Effective PES design is a necessary but not sufficient condition for PES schemes to work. Ultimately, it is necessary to ensure that they deliver additional benefits relative to the status quo. Relatively few existing PES schemes do well if evaluated in these terms, but those that do are useful guides to the design of new schemes.

NOTES

1. See Pascual et al. (2009). PES systems have been closely associated with the target of poverty alleviation and equity. This is consistent with the Convention on Biological Diversity's (CBD) inclusion of the equitable sharing of the benefits of conservation as one goal of the agreement. It also recognizes the importance of incentives to preserve the traditional knowledge, innovation and practices needed to inform conservation efforts (United Nations 2003) and to address the problem of uneven wealth creation (Duraiappah 2004). Although both the Millennium Development Goals (MDG) and the MA recognize the link between poverty, equity and conservation incentives, neither provided the kind of detailed road-map needed by governments to connect them in practical policies (Duraiappah and Roy 2007).
2. Conventional economic theory indicates that input price subsidies or taxes and output price subsidies or taxes will promote intensification or extensification of production processes. Subsidies will increase the use of variable production inputs, such as fertilizer, irrigation water, pesticides and herbicides; they will change the optimal combination or factor proportions with which inputs are used; and output price subsidies will lead farmers to substitute one crop for another or change between crop production and livestock production processes. Associated with this changing behaviour by land users will be different patterns of environmental impacts having both local and wider implications, i.e. the environmental impacts will be felt at local, regional and global levels (Lingard 2000). Literature on the negative environmental consequences of agricultural subsidies has been driven by a perception that the support to farmers neglected important ecosystem services (Pearce 2003; Porter 2003; Summer and Champetier de Ribes 2007).
3. http://www.fsa.usda.gov/FSA/webapp?area=home&subject=copr&topic=crp
4. Following Wunder (2007), the core idea of PES is that external beneficiaries of environmental services make direct contractual quid pro quo payments to local landowners and land users in return for adopting land and resource uses that secure ecosystem conservation and restoration.
5. To date, relatively few PES programmes have targeted farmers and agricultural lands in developing countries. There have also been relatively few examples of private payment mechanisms for the provision of environmental services in agriculture (Food and Agriculture Organization 2007).

REFERENCES

Adams, W., R. Aveling, D. Brockington, B. Dickson, J. Elliot, J. Hutton, D. Roe, B. Vira and W. Wolmer (2004). Biodiversity conservation and the eradication of poverty. *Science*, **306**, 1146–1149.

Ahlheim, M. and A. Neef (2006). Payments for environmental services, tenure security and environmental valuation: concepts and policies towards a better environment. *Quarterly Journal of International Agriculture*, **45**, 303–317.

Angelsen, A. (2009). Realising REDD+ National strategy and policy options. CIFOR, Bogor, Indonesia.

Arriagada, R. (2008). Private provision of public goods: applying matching methods to evaluate payments for ecosystem services in Costa Rica. Unpublished PhD dissertation, College of Natural Resources, North Carolina State University, Raleigh, NC.

Arriagada, R. and C. Perrings (2011). Paying for international environmental public goods. *AMBIO*, **40**, 798–806.

Arriagada, R., P. Ferraro, E. Sills, S.K. Pattanayak and S. Cordero (2012). Do payments for environmental services reduce deforestation? A farm level evaluation from Costa Rica. *Land Economics*, **88**, 382–399.

Bawa, K., R. Seidler and P. Raven (2004). Reconciling conservation paradigms. *Conservation Biology*, **18**, 859–860.

Bennett, M. (2008). China's sloping land conversion program: institutional innovation or business as usual? *Ecological Economics*, **65**, 699–711.

Berkes, E. (2004). Rethinking community-based conservation. *Conservation Biology*, **18**, 621–630.

Biermann, E. (2007). Earth system governance as a cross-cutting theme of global change research. *Global Environmental Change*, **17**, 326–337.

Bulte, E., L. Lipper, R. Stringer and D. Zilberman (2008). Payments for ecosystem services and poverty reductions: concepts, issues, and empirical perspectives. *Environment and Development Economics*, **13**, 245–254.

Cardenas, J., J. Stranlund and C. Willis (2000). Local environmental control and institutional crowding-out. *World Development*, **28**(10), 1719–1733.

Claassen, R., R. Cattaneo and R. Johansson (2008). Cost-effective design of agri-environmental payment programs: US experience in theory and practice. *Ecological Economics*, **65**, 738–753.

Corbera, E., C. González and K. Brown (2008). Institutional dimensions of payments for ecosystem services: an analysis of Mexico's carbon forestry programme. *Ecological Economics*, **68**(3), 743–761.

Daily, G. and P. Matson (2008). Ecosystem services: from theory to implementation. *Proceedings of the National Academy of Sciences of the United States of America*, **105**(8), 9455–9456.

Deng, L., Z. Shangguan and R. Li (2012). Effects of the grain-for-green program on soil erosion in China. *International Journal of Sediment Research*, **27**(1), 120–127.

Department of Sustainability and Environment (2008). *BushTender: Rethinking Investment for Native Vegetation Outcomes. The Application of Auctions for Securing Private Land Management Agreements*. East Melbourne, Australia: Department of Sustainability and Environment.

Dobbs, T.L. and J. Pretty (2008). Case study of agri-environmental payments: the United Kingdom. *Ecological Economics*, **65**, 765–775.

Duraiappah, A. (2004). *Exploring the Links Between Human Well-being, Poverty and Ecosystem Services*. Nairobi: The United Nations Environment Programme and the International Institute for Sustainable Development.

Duraiappah, A. and M. Roy (2007). *Poverty and Ecosystems: Prototype Assessment and Reporting Method. Kenya case study*. Winnipeg, Canada: International Institute for Sustainable Development.

Engel, S., S. Pagiola and S. Wunder (2008). Designing payments for environmental services in theory and practice: an overview of the issues. *Ecological Economics*, **65**, 663–674.

European Commission (2007). *The Common Agricultural Policy Explained*.

Brussels: European Commission Directorate-General for Agriculture and Rural Development.
Food and Agriculture Organization (2007). *The State of Food and Agriculture: Paying Farmers for Environmental Services*. FAO Agricultural Series No. 38. Rome: FAO.
Ferraro, P. (2001). Global habitat protection: limitations of development interventions and a role for conservation performance payments. *Conservation Biology*, **15**, 990–1000.
Ferraro, P. (2011). The future of payments for environmental services. *Conservation Biology*, **25**, 1134–1138.
Ferraro, P. and A. Kiss (2002). Direct payments to conserve biodiversity. *Science*, **298**, 1718–1719.
Ferraro, P.J. and R.D. Simpson (2002). The cost-effectiveness of conservation payments. *Land Economics*, **78**(3), 339–353.
Ferraro, P. and S.K. Pattanayak (2006). Money for nothing? A call for empirical evaluation of biodiversity conservation investments. *PLoS Biology*, **4**(4), 482–488.
Grieg-Gran, M. and C. Bann (2003). A closer look at payments and markets for environmental services. In P. Gutman (ed.), *From Goodwill to Payments for Environmental Services: A Survey of Financing Options for Sustainable Natural Resource Management in Developing Countries*. Washington DC: WWF, Macroeconomics for Sustainable Development Program Office, pp. 27–40.
Grieg-Gran, M. and J. Bishop (2004). How can markets for ecosystem services benefit the poor? In D. Roe (ed.), *The Millennium Development Goals and Conservation: Managing Nature's Wealth for Society's Health*. London: International Institute for Environment and Development, pp. 55–72.
Grieg-Gran, M., I. Porras and S. Wunder (2005). How can market mechanisms for forest environmental services help the poor? Preliminary lessons from Latin America. *World Development*, **33**(9), 1511–1527.
Hardner, J. and R. Rice (2002). Rethinking green consumerism. *Scientific American*, **May**, 89–95.
Hartshorn, G., P. Ferraro and B. Spergel (2005). *Evaluation of the World Bank: GEF Ecomarkets Project in Costa Rica*. Raleigh, NC: North Carolina State University.
Hodge, I. (2000). Agri-environmental relationships and the choice of policy mechanism. *The World Economy*, **23**(2), 257–273.
Jack, K., C. Kousky and K. Sims (2008). Designing payments for ecosystem services: lessons from previous experience with incentive-based mechanisms. *Proceedings of the National Academy of Sciences of the United States of America*, **105**(8), 9465–9470.
Kareiva, P., A. Chang and M. Marvier (2008). Development and conservation goals in World Bank projects. *Science*, **321**, 1638–1639.
Kinzig, A., C. Perrings and R.J. Scholes (2009). Ecosystem services and the economics of biodiversity conservation. ecoSERVICES Group Working Paper, Arizona State University, Phoenix, AZ.
Kinzig, A., C. Perrings, F.S. Chapin III, S. Polasky, V.K. Smith, D. Tilman and B.L. Turner (2011). Paying for ecosystem services: promise and peril. *Science*, **334**, 603–604.
Kleijn, D., F. Berendse, R. Smit and N. Gilissen (2001). Agri-environmental

schemes do not effectively protect biodiversity in Dutch agricultural landscape. *Nature*, **413**, 723–725.
Landell-Mills, N. and I. Porras (2002). *Silver Bullet or Fools' Gold? A Global Review of Markets for Forest Environmental Services and Their Impact on the Poor.* London: Institute for Sustainable Private Sector Forestry.
Lingard, J. (2000). Agricultural subsidies and environmental change. In T. Munn (ed.), *Encyclopedia of Global Environmental Change*. Chichester, UK: John Wiley & Sons Ltd., pp. 168–171.
Liu, J., S. Li, Z. Ouyang, C. Tarn and X. Chen (2008). Ecological and socio-economic effects of China's policies for ecosystem services. *Proceedings of the National Academy of Sciences of the United States of America*, **105**(8), 9477–9482.
Liverman, D. (2004). Who governs, at what scale and at what price? Geography, environmental governance, and the commodification of nature. *Annals of the Association of American Geographers*, **94**(4), 734–738.
Mayrand, K. and M. Paquin (2004). *Payments for Environmental Services: A Survey and Assessment of Current Schemes*. Montreal, Canada: Unisféra International Centre.
Mather, A. (1992). The forest transition. *Area*, **24**, 367–379.
Miles, L. and V. Kapos (2008). Reducing greenhouse gas emissions from deforestation and forest degradation: global land-use implications. *Science*, **320**, 1454–1455.
Millennium Ecosystem Assessment (2005). *Ecosystems and Human Well-being: Synthesis.* Washington DC: Island Press.
Múñoz-Piña, C., A. Guevara, J. Torres and J. Braña (2008). Paying for the hydrological services of Mexico's forests: analysis, negotiations and results. *Ecological Economics*, **65**, 725–736.
Niesten, E. and R. Rice (2004). Sustainable forest management and conservation incentive agreements. *International Forestry Review*, **6**, 56–60.
O'Connor, D. (2008). Governing the global commons: linking carbon sequestration and biodiversity conservation in tropical forests. *Global Environmental Change*, **18**(3), 368–374.
OECD (1998). *Improving the Environment through Reducing Subsidies. Part I: Summary and Conclusions.* Paris: Organisation for Economic Co-operation and Development.
OECD (2003). *Voluntary Approaches for Environmental Policy: Effectiveness, Efficiency and Usage in Policy Mixes.* Paris: Organisation for Economic Co-operation and Development.
Pagiola, S. (2008). Payments for Environmental Services in Costa Rica. *Ecological Economics*, **65**, 712–724.
Pagiola, S. and G. Platais (2007). *Payments for Environmental Services: From Theory to Practice.* Washington DC: World Bank.
Pagiola, S., A.R. Rios and A. Arcenas (2008). Can the poor participate in payments for environmental services? Lessons from the Silvopastoral Project in Nicaragua. *Environment and Development Economics*, **13**, 299–325.
Pascual, U., R. Muradian, L. Rodríguez and A. Duraiappah (2009). *Revisiting the Relationship Between Equity and Efficiency in Payments for Environmental Services.* Nairobi: Ecosystem Services Economics Unit, UNEP.
Pearce, D. (2003). Environmentally harmful subsidies: barriers to sustainable development. In OECD (eds), *Environmentally Harmful Subsidies Policy Issues and Challenges*. Paris: OECD, pp. 9–30.

Pirard, R. (2012). Market-based instruments for biodiversity and ecosystem services: a lexicon. *Environmental Science and Policy*, **19–20**, 59–68.
Pfaff, A., J. Robalino and A. Sánchez-Azofeifa (2008). Payments for environmental services: empirical analysis for Costa Rica. Working Paper Series SAN08-05. Terry Sanford Institute of Public Policy, Duke University, Durham, NC.
Porter, G. (2003). Subsidies and the environment: an overview of the state of knowledge. In OECD (eds), *Environmentally Harmful Subsidies Policy Issues and Challenges*. Paris: OECD, pp. 31–100.
Robalino, J. and Pfaff A. (2006). Contagious development: neighbors' interactions in deforestation. Working Paper. Duke University, Durham. NC.
Robertson, N. and S. Wunder (2005). *Fresh Tracks in the Forest: Assessing Incipient Payments for Environmental Services Initiatives in Bolivia*. Bogor, Indonesia: CIFOR.
Rojas, M. and B. Aylward (2003). *What Are We Learning From Experiences With Markets for Environmental Services in Costa Rica? A Review and Critique of the Literature*. London: International Institute for Environment and Development.
Romero, C. and G.I. Andrade (2004). International conservation organizations and the fate of local tropical forest conservation initiatives. *Conservation Biology*, **18**, 578–580.
Rosenbaum, P. and D. Rubin (1983). The central role of the propensity score in observational studies for causal effects. *Biometrika*, **70**, 41–55.
Rudel, T.K., O.T. Coomes, E. Moran, F. Achard, A. Angelsen, J. Xu and E. Lambin (2005). Forest transitions: towards a global understanding of land use change. *Global Environmental Change*, **15**(1), 23–31.
Sachs, J. and W. Reid (2006). Investments toward sustainable development. *Science*, **312**, 1002.
Salzman, J. (2005). The promise and perils of payments for ecosystem services. *International Journal of Innovation and Sustainable Development*, **1**(1–2), 5–20.
Sánchez-Azofeifa, G.A., A. Pfaff, J. Robalino and J. Boomhowerb (2007). Costa Rica's Payment for Environmental Services Program: intention, implementation, and impact. *Conservation Biology*, **21**(5), 1165–1173.
Scherr, S., A. White and A. Khare (2004). Tropical forests provide the planet with many valuable services. Are beneficiaries prepared to pay for them? *ITTO Tropical Forest Update*, **14**, 11–14.
Sierra, R. and E. Russman (2006). On the efficiency of environmental service payments: a forest conservation assessment in the Osa Peninsula, Costa Rica. *Ecological Economics*, **59**, 131–141.
Siikamaki, J. and D. Layton (2006). Potential cost-effectiveness of incentive payment programs for biological conservation. Resources for the Future, Discussion Paper 06-27, Washington DC.
Sills E., S.K. Pattanayak, P. Ferraro, R. Arriagada, E. Ortiz and S. Cordero (2007). Testing Pigou: private provision of public goods. Presented at the AERE Workshop on Valuation and Incentives for Ecosystem Services, June, 2007 Mystic, CT.
Summer, D. and A. Champetier de Ribes (2007). Role of farm programs in environmental sustainability of agriculture. Briefing paper prepared for the AAAS Meetings, San Francisco, CA.
Swallow, B., M. Kallesoe, U. Ifthikhar, M. van Noordwijk, C. Bracer, S. Scherr, K.V. Raju, S. Poats, A. Duraiappah, B. Ochieng, B. Malle and R. Rumley (2007). Compensation and rewards for environmental services in the developing

world: Framing pan-tropical analysis and comparison. ICRAF working paper. NO.32. World Agroforestry Center, Nairobi.

Tallis, H., P. Kareiva, M. Marvier and A. Chang (2008). An ecosystem services framework to support both practical conservation and economic development. *Proceedings of the National Academy of Sciences*, **105**(28), 9457–9464.

Tallis, H., R. Goldman, M. Uhl and B. Brosi (2009). Integrating conservation and development in the field: implementing ecosystem service projects. *Frontiers in the Ecology and the Environment*, **7**(1), 12–20.

Tietenberg, T. (2006). *Environmental and Natural Resource Economics*, 6th edn. Boston, MA: Addison-Wesley.

Tipper, R. (2002). Helping indigenous farmers to participate in the international market for carbon services: the case of Scolel Té. In S. Pagiola, J. Bishop and N. Landell-Mills (eds), *Selling Forest Environmental Services*. London: Earthscan Publications Ltd., pp. 223–234.

Tomich, T.P., D.E. Thomas and M. van Noordwijk (2004). Environmental services and land use change in Southeast Asia: from recognition to regulation or reward? *Agriculture Ecosystems and Environment*, **104**, 229–244.

Tucker, C. and E. Ostrom (2005). Multidisciplinary research relating institutions and forest transformations. In E. Moran and E. Ostrom (eds), *Seeing the Forest and the Trees: Human–Environment Interactions in Forest Ecosystems*. Cambridge, MA: The MIT Press, pp. 81–104.

Turpie, J.K., C. Marais and J.N. Blignaut (2008). The Working for Water Programme: evolution of a payments for ecosystem services mechanism that addresses both poverty and ecosystem service delivery in South Africa. *Ecological Economics*, **65**, 789–799.

United Nations (2003). Convention on Biological Diversity (with annexes). *Treaty Series*, **1760**, 142–382.

Weitzman, M. (1974). Prices vs. quantities. *Review of Economic Studies*, **41**, 477–491.

Wertz-Kanounnikoff, S. (2006). Payments for environmental services: a solution for biodiversity conservation? Institut du Développement Durable et des Relations Internationales. Idées poor le Débat No 12. Paris.

Wilson, G. (1996). Factors influencing farmer participation in the Environmentally Sensitive Areas Scheme. *Journal of Environmental Management*, **50**, 67–93.

Wu, J. (2000). Slippage effects of the Conservation Reserve Program. *American Journal of Agricultural Economics*, **82**(4), 979–992.

Wunder, S. (2005). Payments for environmental services: some nuts and bolts. CIFOR Occasional Paper 42. Center for International Forestry Research, Bogor, Indonesia.

Wunder, S. (2007). The efficiency of payments for environmental services in tropical conservation. *Conservation Biology*, **21**(1), 48–58.

Wunder, S. (2008). Payments for environmental services and the poor: concepts and preliminary evidence. *Environment and Development Economics*, **13**, 279–297.

Wunder, S. and M. Albán (2008). Decentralized payments for environmental services: the cases of Pimampiro and PROFAFOR in Ecuador. *Ecological Economics*, **65**, 685–698.

Wunder, S. and S. Wertz-Kanounnikoff (2009). Payments for ecosystem services: a new way of conserving biodiversity in forests. *Journal of Sustainable Forestry*, **28**(3–5), 576–596.

Wunder, S., S. Engel and S. Pagiola (2008). Taking stock: a comparative analysis of payments for environmental services programs in developed and developing countries. *Ecological Economics*, **65**, 834–852.

WWF (2006). *Payments for Environmental Services: An Equitable Approach for Reducing Poverty and Conserving Nature*. Gland, Switzerland: World Wildlife Fund.

Zabel, A. and K. Holm-Müller (2008). Conservation performance payments for carnivore conservation in Sweden. *Conservation Biology*, **22**(2), 247–251.

Zhang, W. and S. Pagiola (2011). Assessing the potential for synergies in the implementation of payments for environmental services programmes: an empirical analysis of Costa Rica. *Environmental Conservation*, **38**(4), 406–416.

3. Valuing ecosystem services: benefits, values, space and time

Brendan Fisher, Ian Bateman and R. Kerry Turner

INTRODUCTION

A growing body of evidence suggests that we will continue to face a number of pressing and interrelated problems such as large-scale conversion of ecosystems and the subsequent loss of biodiversity (Millennium Ecosystem Assessment 2005), increasing poverty and water scarcity (Rosegrant et al. 2003), potentially dangerous alteration in the climate system (Schneider 2001; Mastrandrea and Schneider 2004) and global fisheries collapse (Myers and Worm 2003). These problems are occurring on an unprecedented scale and are inherently connected to growing societal demands. The mitigation of these problems requires a deeper comprehension of the environmental infrastructure upon which human existence and welfare depends (Schröter et al. 2005; Sachs and Reid 2006).

The concepts of ecosystem services and 'natural capital' have recently been developed to make explicit this connection between human welfare and ecological sustainability for policy, development and conservation initiatives (Daily 1997; Millennium Ecosystem Assessment 2005). Recent efforts have shown that incorporating ecosystem services into land-use decisions typically favours conservation activities or sustainable management over the conversion of intact ecosystems (Balmford et al. 2002; Turner et al. 2003). Although much ecosystem service research is still in an early stage, systematic approaches to measuring, modelling and mapping of ecosystem services, governance analysis and valuation are needed urgently. In order to make progress in these areas it must be made transparent exactly what is being considered an ecosystem service as opposed to other concepts in the literature such as ecosystem processes, functions, goods and benefits. This delineation is of particular importance to any valuation exercises that might accompany ecosystem service research. Subsequently, there are important economic concepts that need to be made transparent for meaningful estimates to be made. These concepts include the distinction between prices and values, and acknowledging that values are often context specific, which means they may change across space and time. This chapter discusses these issues with the aim of informing valuation exercises from an economic perspective.

SERVICES VERSUS BENEFITS

In 2005, the Millennium Ecosystem Assessment (MA) defined a framework for relating ecosystem services to the larger scientific and policy communities. It proved to be an important development and framework. The MA divided ecosystem services into a few very understandable categories: supporting services, regulating services, provisioning services and cultural services. This in turn makes the classification scheme readily accessible as a framework to decision-makers and non-scientists. The MA delivered a broad definition (by design) of ecosystem services as 'the benefits humans obtain from ecosystems', however, this definition has not been shown to be operational for all research purposes (see Boyd and Banzhaf 2007; Wallace 2007; Fisher and Turner 2008), such as for robust monetary valuation of services. Efforts have been made to classify and understand ecosystem services more carefully, to make the concept more operational for decision-making (see Fisher et al. 2009 for a review).

We have argued elsewhere (Fisher and Turner 2008; Fisher et al. 2009) that a simpler delineation of intermediate services, final services and benefits is more useful than the MA schema for valuation purposes. There are multiple relationships between ecosystem processes and human benefits (see Boyd and Banzhaf (2007) for a description of the benefit dependence aspect of ecosystem services), but what is important for valuation exercises is that you value the endpoints that have a direct affect on human welfare. In economics, these are considered through the use of the term *benefits*. Both intermediate and final services are ecological phenomena (as opposed to things like cultural fulfilment). 'Intermediate services' are equivalent to the MA's 'supporting services'. They combine in complex ways to provide final services, which have direct effects on human welfare. Benefits, which include things like wood, food, cultural aspects and recreation, are related to but different from the services that provide them. For example, regulated water flow and water fit for drinking are not the same thing. Benefits also typically require other forms of capital to affect human welfare. For example, clean drinking water for consumption is a benefit of the combined final services of water provision and water purification. In turn, provision of clean water by an ecosystem is a function of intermediate services including nutrient cycling and soil retention of particles. The end benefit might require some built capital to be realized, perhaps a well or an urban water distribution system.

In this intermediate and final services–benefits scheme we avoid the double counting flaws acknowledged in earlier ecosystem service valuation exercises. This is not the case for the MA classification, which could lead to double counting the value of some ecosystem services. For example, in the MA, nutrient cycling is a supporting service, water flow regulation is a regulating service and recreation is a cultural service. However, if you were a decision-maker contemplating the conversion of a wetland, and utilized a cost–benefit analysis including these three services, you would commit the error of double counting. This is because nutrient cycling and water regulation both help to provide the same service under consideration, providing potable water, and the MA's recreation service is actually a human benefit of that water provision. An analogy is that when buying a live chicken you do not pay for the price of a full chicken plus the price of two legs, two wings, head, neck etc., you simply pay the price of a whole chicken.

Figure 3.1 provides a conceptual example of this schema, where complex ecosystem processes and functions give rise to ecosystem services (final and intermediate), which provide benefits when used by people. Again, some benefits require other forms of capital in order to be realized. For example, hydro-electric power requires water provision and regulation from ecosystems, but also dams, transmission and infrastructure.

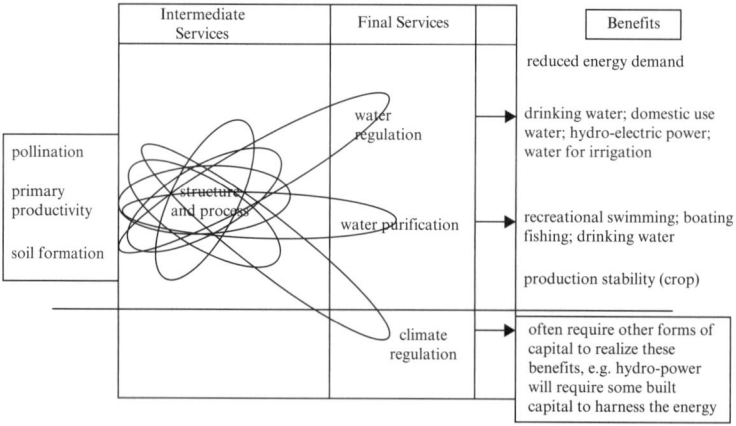

Source: From Fisher et al. (2009).

Figure 3.1 Conceptual delineation between ecosystem services and the benefits derived from them

This line of argument, however, is not meant to imply that intermediate services have no value in themselves. Without a sufficient configuration of structure and processes, ecosystems will function differently and will not provide the diverse range of final services and benefits that they could potentially deliver (Daily 1997; Hooper et al. 2005). Regulating and supporting services that provide the capacity to respond to environmental stresses and shocks are encompassed by the concepts of infrastructure (or primary or glue) value associated with the role that functional diversity can play in certain contexts, providing increased ecological stability and resilience (Hooper et al. 2005). The conservation and protection of ecosystem capacity to provide regulating and supporting services is in some ways a decision about reducing risk and the costs of such a risk-averse strategy.

PRICES VERSUS VALUES

In addition to the services–benefits problem in attempts to value ecosystem services is the confusion over the terms 'value' and 'price', which most people use interchangeably. It is very easy to demonstrate that they are not equivalent. Consider that most basic of all necessities: water. This is the staff of life, without which existence is impossible. Yet the price we pay for water in our household bills is actually very modest. It is clear to see that 'value' and 'price' are not necessarily the same thing. In fact, price is simply that portion of underlying value which is realized within the market place. In many cases, price may be a perfectly acceptable approximation of value, particularly for goods produced in competitive markets where there is no large-scale intervention by governments or other authorities. Indeed even when these latter distortions (i.e. market power and subsidy effects) do arise, economists can often adjust for their influence, to yield what is known as the 'shadow value' of the goods concerned. However, as the water example shows, market price can in some cases be a poor approximation of value. This divergence can often be substantial and is a characteristic of many of the goods produced by the natural environment.

Supply and demand can interact in ways that are highly beneficial to consumers, providing goods at prices which are below the value placed on them by consumers. This excess between price and value is known as the 'consumer surplus'. Decision-makers should be interested in the value provided by different goods, rather than their price. Indeed this constitutes the fundamental difference between accountants and economists; while the former are interested in price, the latter are (or at least should be) interested in value.

The fundamental problem facing any economic analysis is one of measurement, i.e. how do we measure the value or utility provided by any given good? The economist's solution is to use a surrogate measure that is transparent, highly compatible with the decision-making process, and amenable to subsequent adjustment if we wish to allow for different circumstances across individuals. That measure is to assess the amount that individuals are prepared to pay for changes in the provision of goods. Note immediately that we are relating value to willingness to pay (WTP) rather than what actually has to be paid. A simple example serves to illustrate the importance of this difference. Consider the value of walking in woodland. This generates benefits such as exercise, appreciation of nature, perhaps entertainment (e.g. of children), inner calm, etc. If the woodland were publicly owned, the amount paid to enter it is likely to be zero. In this instance, price paid is a highly misleading indicator of value.

Arguably there is no perfect way in which to estimate the value of any good. Nevertheless, several decades of research have resulted in the development of a variety of valuation methods including the following:

Adjusted market prices. For goods that are traded in markets and have prices, we can estimate WTP by examining the reaction of demand to observed variations in prices. Adjustments need to be made for distortions arising from such things as imperfect (non-competitive) markets and policy interventions (e.g. taxes and subsidies). This allows the analyst to estimate consumer surplus and hence values. For example, one can estimate part of the value of improved water quality by examining the increased value of commercial fishing catches. Of course, where either demand or supply changes, so does price. Consider, for example, the long-standing drought in Australia and how this affects water prices.

Productivity methods. Ecosystem services often provide the factors of production required to produce marketed goods. Production functions relating inputs to the output of goods can be estimated and the contributions of individual services assessed. Continuing the water quality improvement example, one could also estimate the value generated by greater agricultural productivity, or the decreased costs of providing clean drinking water.

Revealed preference methods. Many goods that derive from ecological functioning can only be enjoyed through money purchases. For example, individuals may pay extra for homes in quiet neighbourhoods or incur substantial costs to visit areas of natural beauty. By relating behaviour to the characteristics of those goods one can observe the money–environment trade-off and so reveal the values held by individuals for the environmental good.

Stated preference methods. The most direct of all approaches is to ask individuals to state their WTP for some change in the provision of an environmental good. In practice, the costs of conducting new valuation research across the multitude of potential decision situations often means that analysts are forced to rely upon value transfer methods, which transfer existing benefit estimates from studies already completed for another location or issue.

In addition to the various valuation methods described above, many studies adopt simpler 'pricing methods' such as avoided damage or expected damage approaches, which examine the costs of avoiding damages due to lost or at-risk services (e.g. the loss of coastal wetlands and subsequent changes in the impact of storm events).

The damage cost avoided approach is also used by the Intergovernmental Panel on Climate Change (IPCC) to underpin the economic analysis of their climate change assessments (see Pearce et al. 1996). Here the process, in situations as complex as climate regulation, is typically an agglomeration of valuation techniques such as revealed preference approaches for market goods and WTP in hypothetical markets for non-market good and services.

With this approach, the focus is on the benefits or, technically, the avoided costs where damages are the results of climate change impacts on individual and societal welfare. Ecological and atmospheric models are at the cornerstone of this approach because they underpin valuation estimates. Table 3.1 (adapted from Pearce et al. 1996) shows some of the key damages that could be valued as a result of a doubling of CO_2 concentrations in the atmosphere. Here we see that there are both market and non-market effects of CO_2 increases. Major market consequences might be felt in the agricultural sector (Pearce et al. 1996). Climate change will probably affect agricultural production through process effects such as heat stress, soil moisture loss, increases in pests and diseases, a shorter growing season (where temperatures rise too much), a decrease in precipitation and other impacts. Conversely these same effects might produce benefits, such as increased precipitation in some regions, longer growing seasons in higher latitudes and carbon fertilization effects (Godfray et al. 2011). This is a good example of why damage and value estimates are spatially heterogeneous and need to be evaluated at the appropriate scales of the loss or benefit.

In a sector such as forestry, both market and non-market damages can be expected because climate change may negatively affect forest cover and therefore timber values in a region, but also elicit losses in recreational and cultural significance. Several of the impacts in Table 3.1 are easy to

Table 3.1 Potential damages from CO_2 doubling in market and non-market sectors

Market Impacts				Non-market Impacts		
Primary economic sector	Other sectors	Property loss	Extreme event damage	Ecosystem damage	Human impacts	Extreme events
Agriculture Forestry	Water supply Energy demand	Dryland lost Coastal protection	Hurricane damage Damage from drought	Wetland loss Forest loss	Human life Air pollution	Hurricane damage Damage from drought
Fisheries	Leisure activities	Urban infrastructure	Non-tropical storms	Species losses	Water pollution	Non-tropical storms
	Insurance		River floods	Other ecosystem losses	Migration	River floods
	Construction Transport Energy Supply		Hot or cold spells		Morbidity Physical comfort Political stability Human hardship	Hot or cold spells

Source: Adapted from Pearce et al. (1996).

estimate while several are rather difficult. Examples of the latter include the values of species loss, and damages due to increases in hot and cold spells, which will be both spatially and temporally heterogeneous.

We can see that estimating these damages is heavily reliant on the ecological and atmospheric models that predict how changes in greenhouse gas emissions will affect land cover, seascapes and ecosystem functions and responses. Even with sophisticated, spatially explicit models there are still numerous caveats and assumptions that need to be highlighted. On the biophysical side, damages will be a function of the rate of change as well as the degree to which different ecosystems are linked. For example, how exactly does the evapotranspiration from the Amazon affect agricultural productivity in North America? That is, how tightly linked are regions and ecosystems?

Another approach has been to use replacement costs as a proxy for the loss of existing ecosystem services. This is not a true valuation method as it is not based on WTP but can be an effective approach to demonstrate the importance of ecosystem services to policy-makers. In this context the 'cost versus value' distinction raises similar concerns to the 'price versus value' distinction explained above. It is easy to make the error of assuming that the replacement cost is the true value (benefit) of the service under assessment. It is also the case that the method can result in unrealistically high estimates.

'HERE AND NOW' VERSUS 'THERE AND THEN'

As indicated above, the value of an ecosystem service is dependent on where the service is delivered, and the time at which the value is being assessed. This temporal and spatial context of valuation is what we mean by 'here and now' versus 'there and then'.

There are several spatial relationships between where an ecosystem service is produced and where the benefit is felt. Some ecosystem services are produced in the same area as where the benefit is realized. For example, soil formation occurs in a given spot and may be utilized as an ecosystem service when a farmer plants a crop in the same place. Some ecosystem services are produced in one place, while the benefits are felt elsewhere. A good example can be found in water regulation, where an upslope vegetated landscape may conduct runoff from heavy rain to the water table which will return to the surface as part of a river flow elsewhere. Here a downstream user benefits from the upstream landscape. Another such relationship is where a service is produced in a particular spot, but the entire world may receive benefits from it. Carbon storage is one example since it does not matter where the carbon is being stored; all of humanity benefits from it (if we desire our current relatively stable climate regime). In essence, ecosystem services often 'flow' from a point of production to a point of use.

This 'flow' of services changes through space in at least three ways:

1. The biophysical process itself varies across the landscape or seascape. This is obvious in the above example if one considers the way that carbon storage or net primary productivity will vary with slope, aspect, elevation, species and structural diversity.
2. The benefits and beneficiaries will change across a landscape. Water regulation might be an important service for providing irrigation to farmers abutting a forest or woodland. The same service might provide the benefit of hydro-electric power to beneficiaries far downstream. User groups will hold different values and preferences for this water regulation, report different WTP and hold different information about how the system works. All of these will affect an aggregate valuation assessment.
3. The costs of provision of the ecosystem service are likely to vary spatially. Consider forest protection for the sake of regulating water flows. To local people who use forest resources for non-timber forest products (NTFP) collection, the opportunity cost is equivalent to losing the ability to collect resources. To local livestock keepers, the cost might be the increased risks of predators or disease transmission to livestock. To urban water users, the cost might be only a small additional fee on water bills.

The fact that ecosystem service provision changes in terms of ecological processes, magnitude, beneficiaries and costs across space is critical for any valuation process. This demands that spatially explicit ecological models, a detailed understanding of benefit stakeholders, and knowledge of all costs (including opportunity costs) should be incorporated for ecosystem service valuation exercises to be robust.

The *now* and *then* of ecosystem service research implies that just as services and benefits change across space, they also change over time. We consider three reasons why ecosystem services or the value of their benefits change across time, and why this is important for valuation.

First, ecological conditions or processes themselves change over time. For example, a restored wetland attenuates larger stormwater surges, assimilates more heavy metals, and houses more breeding waders than a degraded wetland might. Conversely, a woodland reduced in size will probably produce less NTFPs, store less carbon and house fewer pollinating species than it did before it was degraded. Any ecosystem service assessment must occur at a point in time. Future changes in ecosystem condition or function can be modelled based on past changes, forecasted based on predicted future drivers of change, or perhaps instigated through scenario-building and analysis. The very fact that ecosystem service research has risen to such prominence in science and policy circles is based on an acknowledgement that over the past few decades we have seen precipitous declines in the provision of some services in certain places.

Second, over time society's preferences change. For example, wetlands were once more commonly and derisorily termed 'bogs' and 'swamps'. They were often considered to be wastelands to be improved by drainage. Today, in many places wetlands are highly valued for their ability to provide wildlife habitat, store carbon, and treat or store pollutants such as nitrogen and heavy metals. To some degree this change in preferences can be explained by the dwindling area of remaining natural resources (it is noticeable that in countries where such resources are still common they are often less prized). Increasing real incomes and leisure time together with better transportation and a growing appreciation of the services of such areas all play a part in the transformation of societal preferences, and this causes considerable difficulty for economic analyses because it is difficult to assess changes in future preferences.

The sustainability literature offers a strategy based on the maintenance of a set of 'opportunities' to be bequeathed across intergenerational timescales. The core idea is that future generations should possess at least an equivalent set of economic, social and cultural opportunities as previous generations. Natural capital (ecosystems and their relationships) should be conserved as a store of wealth and wealth creation opportunities. Taking a weak sustainability position (Neumayer 2003), natural capital can be extensively substituted for by human and physical capital. From a strong sustainability (Neumayer 2003) perspective, important components of natural capital are critical to life support and other services and cannot be substituted, their loss being effectively irreversible. So called 'social capital' also needs to be nurtured in a world of strong sustainability policy (Neumayer 2003).

Third, and linked to the second point, is the complexity that individuals tend to prefer benefits to be provided sooner rather than later (and the opposite for costs). For example, people typically prefer $100 today rather than $105 a year from today. This seemingly innocuous aspect of preferences leads to the problem of discounting, i.e. the present day value of a future benefit (or cost) will fall over time, and will fall to a greater extent the further into the future that value is considered. Determining the nature and rate of this decline is important as it can radically alter present day assessments of the value of different options. For ecosystem service assessments it is becoming more accepted that the economist should not adopt the discount preferences of the individual but rather use a social discount function (Turner 2007). It is also becoming more obvious from a strong sustainability viewpoint that the discount function (particularly for non-market benefits, e.g. from ecosystem services) should be declining in nature (i.e. the discount is less) for large-scale societal decisions (Turner 2007). This means that in each time period the rate at which a benefit (or cost) is discounted should itself decline. This reflects the longevity of society and the greater weight placed upon delayed benefits and costs relative to the preferences of individuals. That said, the choice of discount function can have a major impact upon the economic assessment of long-term concerns such as ecosystem services. Sensitivity analyses of the impact of different discounting strategies are advisable. In the end, ethics plays as important a part as economics in the discounting debate.

CONCLUSION

The importance of providing policy-makers with timely and robust estimates of the value and benefits of well-functioning ecosystems has never been more critical. While there is still much ignorance about how ecosystems function to provide benefits to human society, how humans behave and value such benefits, and how these two interact in the face of diminishing natural capital, we are beginning to make some progress conceptually and analytically so that we can deliver estimates and recommendations to decision-makers. We have discussed casually a few of the conclusions that natural scientists and economists are arriving at regarding this literature, including the understanding that market prices may serve as a poor proxy for individual or societal values, and that ecosystem service assessment needs to include spatial and temporal aspects to be truly policy relevant. While these are just a few small conceptual steps in the typical long journey of an ecosystem service assessment, they are critical steps for each journey.

REFERENCES

Balmford, A., A. Bruner, P. Cooper, R. Costanza, S. Farber, R.E. Green, M. Jenkins, P. Jefferiss, V. Jessamy, J. Madden, K. Munro, N. Myers, S. Naeem, J. Paavola, M. Rayment, S. Rosendo, J. Roughgarden, K. Trumper and R.K. Turner (2002). Economic reasons for conserving wild nature. *Science*, **297**, 950–953.

Boyd, J. and Banzhaf S. (2007). What are ecosystem services? *Ecological Economics*, **63**, 616–626.

Daily, G.C. (1997). *Nature's Services: Societal Dependence on Natural Ecosystems*. Washington DC: Island Press.
Fisher, B. and R.K. Turner (2008). Ecosystem services: classification for valuation. *Biological Conservation*, **141**, 1167–1169.
Fisher, B., R.K. Turner and P. Morling (2009). Defining and classifying ecosystem services for decision making. *Ecological Economics*, **68**, 643–653.
Godfray, H.C.J., J. Pretty, S.M. Thomas, E.J. Warham and J.R. Beddington (2011). Linking Policy on Climate and Food, *Science*, **331**, 1013–1014.
Hooper, D.U., F.S. Chapin, J.J. Ewel, A. Hector, P. Inchausti, S. Lavorel, J.H. Lawton, D.M. Lodge, M. Loreau, S. Naeem, B. Schmid, H. Setälä, A.J. Symstad, J. Vandermeer and D.A. Wardle (2005). Effects of biodiversity on ecosystem functioning: a consensus of current knowledge, *Ecological Monographs*, **75**(1), 3–35.
Mastrandrea, M.D. and S.H. Schneider (2004). Probabilistic integrated assessment of 'dangerous' climate change. *Science*, **304**, 571–575.
Millennium Ecosystem Assessment (2005). *Ecosystems and Human Well-being: Synthesis*. Washington DC: Island Press.
Myers, R.A. and B. Worm (2003). Rapid worldwide depletion of predatory fish communities. *Nature*, **423**, 280–283.
Nuemayer, E. (2003). *Weak Versus Strong Sustainability*. Cheltenham, UK and Northampton, MA: Edward Elgar.
Pearce, D., W. Cline, A. Achanta, S. Fankhauser, R. Pachauri, R.S.J. Tol and P. Vellinga (1996). The social costs of climate change: greenhouse damage and the benefits of control. In R.T. Watson, M.C. Zinyowera and R.H. Moss (eds), *Climate Change 1995. Economic and Social Dimensions of Climate Change*. Cambridge: Cambridge University Press.
Rosegrant, M.W., X.M. Cai and S.A. Cline (2003). Will the world run dry? Global water and food security. *Environment*, **45**, 24–36.
Sachs, J.D. and W.V. Reid (2006). Environment: investments toward sustainable development. *Science*, **312**, 1002–1002.
Schneider, S.H. (2001). What is 'dangerous' climate change? *Nature*, **411**, 17–19.
Schröter, D., W. Cramer, R. Leemans, C. Prentice, M.B. Araújo, N.W. Arnell, A. Bondeau, H. Bugmann, T.R. Carter, C.A. Gracia, A.C. de la Vega-Leinert, M. Erhard, F. Ewert, M. Glendining, J.I. House, S. Kankaanpää, R.J.T. Klein, S. Lavore, M. Lindner, M.J. Metzger, J. Meyer, T.D. Mitchell, I. Reginster, M. Rounsville, S. Sabaté, S. Sitch, B. Smith, J. Smith, P. Smith, M.T. Sykes, K. Thonicke, W. Thuiller, G. Tucj, S. Zaehle and B. Zier (2005). Ecosystem service supply and vulnerability to global change in Europe. *Science*, **310**(5752), 1333–1337.
Turner, R.K. (2007). Limits to CBA in the UK and European environmental policy: retrospects and future prospects. *Environmental and Resource Economics*, **37**, 253–269.
Turner, R.K., J. Paavola, P. Cooper, S. Farber, V. Jessamy and S. Georgiou (2003). Valuing nature: lessons learned and future research directions. *Ecological Economics*, **46**, 493–510.
Wallace, K.J. (2007). Classification of ecosystem services: problems and solutions. *Biological Conservation*, **139**(3–4), 235–246.

4. Managing trade-offs in ecosystem services

Thomas Elmqvist, Magnus Tuvendal, Jagdish Krishnaswamy and Kristoffer Hylander

INTRODUCTION

The concept of ecosystem services was successfully introduced into the global policy arena by the Millennium Ecosystem Assessment (Millennium Ecosystem Assessment 2005) and has been welcomed by both the conservation and development communities as a potential bridge between the biodiversity and sustainable development discourses (e.g. Tallis et al. 2008). Ecosystem services are here defined following the recent perspectives developed within the *The Economics of Ecosystems and Biodiversity* (TEEB 2010) study as 'the direct and indirect contributions of ecosystems to human well-being'. Despite the apparent success of the concept of ecosystem services, progress in practical applications in land-use planning and local decision-making has been slow (e.g. Naidoo et al. 2008; Daily et al. 2009). This stems not only from failures of markets to capture values of ecosystem services, but also from our limited understanding of:

1. how different services are interlinked with each other and to various components of biodiversity;
2. the influence of differences in temporal and spatial scales of demand and supply of services, and;
3. the potential trade-offs among services and in particular, the lack of knowledge of the relationship between provisioning and regulating services (TEEB 2010).

We define provisioning services as 'fluxes of nutrients, soil, water and biomass (food, wood, fibre, medicine, etc.) that are generated by ecosystems and harvested and used by people' and regulating services as 'eco-physiological functions and ecosystem processes necessary for maintaining functioning ecosystems, which directly underpin the production of provisioning services'.

Functioning ecosystems produce multiple services that interact in complex ways, different services being interlinked and therefore affected negatively or positively as one service (e.g. food provision) increases. Some ecosystem services co-vary positively (more of one means more of another). For example, maintaining soil quality may promote nutrient cycling and primary production, enhance carbon storage and hence climate regulation, help regulate water flows and water quality, and improve most provisioning services, notably production of food, fibre and other chemicals. Other services co-vary negatively (more of one means less of another), such as when increasing provisioning services, for example increased provision of agricultural crops, may reduce many regulating services, for example soil quality, climate regulation and water regulation.

Most studies so far have focused on one or a few services such as pollination, or food versus water quality and quantity. Attempts to characterize multiple ecosystem services across regions have only recently emerged (e.g. Schroter et al. 2005), and the little quantitative evidence available to date has led to mixed conclusions (e.g. Bohensky et al. 2006). The spatial correlation among different services varies widely (Naidoo et al. 2008), with quantification and mapping of services being challenging (Meyerson et al. 2005). Assessing how multiple ecosystem services are scaled and coupled in 'bundles' represent a major research gap (Carpenter et al. 2009). Attempts at quantifying spatial aspects of multiple services include that of the service-providing unit (SPU), defined by Luck et al. (2003) as ecosystem structures and processes that provide specific services at a particular spatial scale. For example, an SPU might comprise all those organisms contributing to pollination of a single orchard, or all those organisms contributing to water purification in a given catchment area (Luck et al. 2003, 2009). One of the major challenges in applying the SPU concept is to translate the unit into tangible and ideally mappable units of ecosystems and landscape or seascape, but the concept potentially offers an approach that focuses on multiple services and where changes to key species or population characteristics have direct implications for service provision.

Focusing on single provisioning ecosystem services in isolation from regulating services has frequently resulted in policy failures. Perhaps the best recent example of this is the new European biofuels policy (Directive 2003/30/EC). The European target of 10 per cent of motor fuels derived from biofuels has resulted in management of ecosystems for a single provisioning service, but other services of importance for climate regulation such as carbon storage and trace gas regulation, performed by the same or other organisms in the same system, have been completely ignored. By ignoring regulating services, the capacity to fulfil long-term goals of sustainable landscape management, e.g. maintaining agricultural productivity, conservation of biodiversity and reducing the rate of climate change, may be seriously jeopardized.

Today 35 per cent of the Earth's land surface is used for agriculture: growing crops or rearing livestock (Millennium Ecosystem Assessment 2005). Grazing land alone accounts for 26 per cent of the Earth's surface, and animal feed crops account for a third of all cultivated land (Steinfel et al. 2006). The extensive use of the Earth's surface for agriculture severely affects the generation of many regulating ecosystem services that underlie human well-being (Millennium Ecosystem Assessment 2005). Adding to the existing pressure, global food production will need to increase more than 50 per cent within the next four decades to meet the demands of a growing human population. In addition, the development of biofuels is placing massive and rapid demands on land. Biofuels accounted for almost half of the increase in consumption of major food crops in 2006–07 (TEEB 2008). Given these rapid global trends, we need to understand how trade-offs among services can be addressed and to what extent new insights in ecology and innovations in institutions and governance may help to reduce some of the most undesirable trade-offs. Here we outline the possibility of moving towards 'winning more and losing less' by developing a framework that could deepen our understanding of how losses of regulating services may be reduced under various scenarios of development of provisioning services.

TRADE-OFFS AMONG PROVISIONING AND REGULATING SERVICES

It has been suggested that major ecosystem degradation tends to occur as simultaneous failures in multiple ecosystem services (Carpenter et al. 2006). The dry lands of sub-Saharan Africa provide one of the clearest examples of these multiple failures, causing a combination of failing crops and grazing, declining quality and quantity of freshwater and loss of tree cover. On the other hand, a synthesis of over 200 cases of investments in organic agriculture in developing countries around the world (both dry lands and non-dry lands) showed that the implementation of various novel agricultural techniques and practices could result in a reduction of ecosystem service trade-offs, even as crop yields increased (Pretty et al. 2006). In other words, multiple failures can be avoided with the appropriate knowledge, incentives and institutions at hand. In Figure 4.1, we illustrate three possible trade-off patterns between provisioning and regulating ecosystem services. In type A, there is a steep decline in regulating services even with a moderate increase in provisioning services production. In type B, there

Managing trade-offs in ecosystem services 73

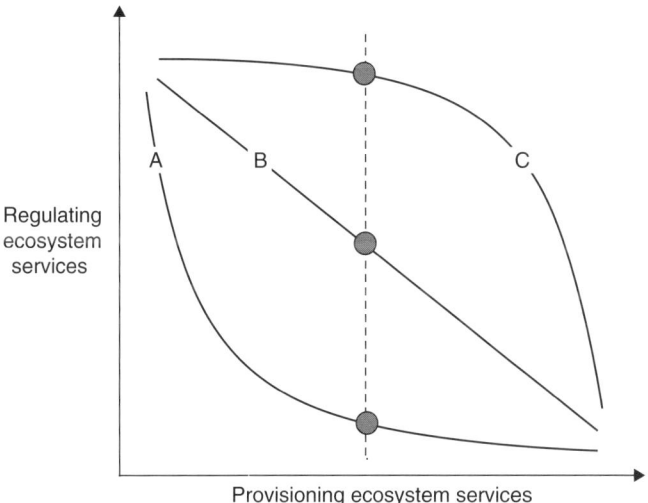

Note: Depending on the type of trade-off (type A, B or C), the supply of regulating services can be low, intermediate or high for similar levels of provisioning services.

Figure 4.1 The potential trade-offs between provisioning services (e.g. food, timber) and regulating services (e.g. soil quality maintenance, pollination, biological control and water regulation)

is a linear relationship between the two service categories, and in type C, levels of provisioning services may increase to very high levels before there is a decline in regulating services. Depending on the type of response the supply of regulating services can be low, intermediate or high for a similar level of provisioning services.

A major task for forestry and agricultural research during the next decade is to design studies which generate information on how current production, which may have strongly negative effects on regulating services (a type A response), can be transformed into type B or even type C.

REGULATING ECOSYSTEM SERVICES

Regulating services represent the role ecosystems have in climate regulation, water regulation, biological control, pollination, maintenance of soil quality, erosion prevention and hazard control (see classification in TEEB 2010). These services are based on complex processes and interactions in ecosystems, and may exhibit different thresholds in response to environmental drivers, for example land cover change. Current knowledge about the dynamics of these services is summarized below (based on Elmqvist et al. 2010):

Climate regulation. All soils store carbon, but to widely varying extents. Forests are the only major ecosystems where the amount of carbon stored in the biomass of the plants exceeds that in the soil; deforestation therefore also affects climate regulation. Agricultural ecosystems currently have low soil carbon stores owing to intensive production methods, and there is scope for enhancing those stores. There are major uncertainties associated with this service, particularly related to large time lags in the feedbacks between changes in ecosystem processes and the atmosphere. The complex interactions and long time lags make it very difficult to forecast eventual outcomes or whether important thresholds have already been passed (Rockström et al. 2009).

Water regulation. Vegetation is a major determinant of water flows and quality, and micro-organisms play an important role in the quality of groundwater. Ecosystems such as forests and wetlands with intact groundcover and root systems are considered very effective at regulating water flow and improving water quality. Water reaches freshwater stores (lakes, rivers, aquifers) by a variety of routes, including direct precipitation, surface and subsurface flows and human intervention. In all but the first case, the water quality is altered by the addition and removal of organisms and substances. Ecosystems therefore play a major role in determining water quality. In particular, the passage of water through soil has a profound impact, both through the dissolution of inorganic (for example nitrate, phosphate) and organic (dissolved organic carbon compounds, pesticides) compounds and the modification of many of these by soil organisms. Most changes to the capacity of ecosystems to regulate and provide freshwater seem to derive from, and be generally proportional to, land-use change. There are, however, situations in which a relatively small additional change may trigger a disproportionate, and sometimes difficult to reverse, response from an ecosystem's hydrological function (Gordon et al. 2008). For example, human-induced eutrophication can lead to sudden shifts in water quality from clear to turbid conditions (Scheffer et al. 2001), which affect freshwater fisheries and recreational use of water bodies. Reduction of nutrient concentrations is usually insufficient to restore the original state, with restoration necessitating very substantially lower nutrient levels than those at which the regime shift occurred. In addition, climate change can potentially trigger sudden changes, particularly in regions where ecosystems are already highly water-stressed.

Biological control. The relationship between densities of natural enemies and the biological control services they provide is unlikely to be linear (Losey and Vaughan 2006) and biological control functions may decline disproportionately when a tipping point in natural enemy diversity is passed. The importance of natural enemy assemblage composition in some instances of biological control (Shennan 2008) indicates that changes in composition can lead to disproportionately large, irreversible and often negative shifts in ecosystem services (Diaz et al. 2006).

Pollination. It is possible that a threshold in pollinator species or functional diversity exists below which pollination services will become too scarce or too unstable for efficient pollination of a field of crops (Klein et al. 2007). Such a tipping point might occur when, in a landscape context, sufficient habitat is destroyed that the next marginal change causes a population crash in multiple pollinators. Alternatively, a threshold in habitat loss may lead to the collapse of particularly important pollinators, leading to a broader collapse in pollination services. Larsen et al. (2005) found that large-bodied pollinators tended to be both more extinction-prone and more functionally efficient compared with small-bodied pollinators, which supports the concept of rapid loss of function with loss of habitat.

Maintenance of soil quality. The process of soil formation is governed by the nature of the parent materials, biological processes, topography and climate. The progressive accumulation of organic materials is characteristic of the development of most soils and depends on the activity of a wide range of microbes, plants and associated organisms (Brussaard et al. 1997; Lavelle and Spain 2001). Soil quality is underpinned by nutrient cycling, which occurs in all ecosystems and is strongly related to productivity. A key element is nitrogen, which occurs in enormous quantities in the atmosphere and is converted to a biologically useable form (ammonium) by bacteria. Nitrogen fertilizer is becoming ever more expensive, with about 90 per cent of the cost directly related to use of energy in production (typically from gas), therefore supplies are not sustainable. Nitrogen fixation by organisms accounts for around half of all nitrogen fixation worldwide, and sustainable agricultural systems will have to rely on this process increasingly in future. Soil formation is a continuous process in all terrestrial ecosystems, but is particularly important and active in the early stages after rock surfaces are exposed. Agricultural expansion into new areas often occupies terrains that are not particularly suitable for agriculture, and soil fertility may decline very quickly as crops sequester soil nutrients (Carr et al. 2006).

Erosion prevention and hazard control. Vegetation cover is the key factor preventing soil erosion. Landslide frequency seems to be increasing, and it has been suggested that land-use change, particularly deforestation, is one of the causes. In steep terrain, forests protect against landslides by modifying the soil moisture regime (Sidle et al. 2006). The ecosystem service is generally not species-specific nor dependent on biodiversity in general, though in areas of high rainfall or in extreme runoff events, forests may be more effective than grassland or herb-dominated communities. The effect of ecosystems on natural hazard mitigation is still poorly understood and it is uncertain to what extent abrupt changes in this service may be associated with abrupt changes in ecosystem extension and condition, for example from the degradation of forests due to climate change.

We may conclude that, even though uncertainty is high, to varying degrees these regulating services may respond along the A, B or C response curves in Figure 4.1 as provisioning services increase (corresponding to a specific land-use or land-cover change) depending on the specific spatial and temporal context. The question is, can strategies be found for designing production of provisioning services and landscape management that can reduce the likelihood of type A responses and increase the likelihood of type B or even type C responses? We will illustrate the potential of such management with an example of one important provisioning service: coffee.

Coffee Production and Regulating Services

The montane rainforests of southwestern Ethiopia are widely considered to be the centre of origin of arabica coffee *Coffea arabica* (Sylvain 1955; Meyer 1965). Still, more or less wild coffee can be found throughout these forests (Woldemariam 2003; Schmitt 2006). People living in agricultural areas surrounding these forests and forest fragments utilize this coffee by picking the berries from scattered shrubs but also by managing certain areas within the forests to increase coffee productivity. A widespread activity is to clear some forest understory and increase coffee density below the canopy of indigenous trees by planting or by allowing natural regeneration. There is thus a gradient in coffee density within the forest ecosystem from true forest coffee to semi-forest coffee systems where most small trees and shrubs have been removed in favour of coffee (Senbeta and Denich 2006). Besides these systems with continuous canopy cover, farmers also cultivate coffee in home gardens below single shade trees (Hylander and Nemomissa 2008). This practice promotes a tree-rich matrix and decreases the dependency on forests for many forest plants and animals (Gove et al. 2008; Hylander and Nemomissa 2008).

Arabica coffee is nowadays grown in similar climatic contexts (rather humid climates at intermediate elevations) throughout the tropics (Klein et al. 2008). Moguel and Toledo (1999) have summarized the large variety in coffee systems found in Latin America as a gradient from coffee grown without any shade (sun coffee) to coffee planted in the understory of indigenous forests (rustic systems) with several intermediate steps that vary in density and diversity of tree species. In India, there is a clear distinction between systems that are shaded by monocultures of exotic species, e.g. *Grevilliea* and systems that utilize indigenous species. Shade coffee in the Western Ghats biodiversity hotspot in India is grown in close proximity to remaining (protected) forests and has replaced medium elevation evergreen and deciduous forests. Shade coffee's intrinsic biodiversity values are a complex function of both local shade practices and landscape factors, and lie either in proximity to biodiversity-rich protected forests or within a corridor that may be a major driver of patterns of biodiversity (mammals, birds and butterflies) in coffee plantations (Bali et al. 2007; Anand et al. 2008; Dolia et al. 2008).

In the Western Ghats, coffee is regarded as one of the most biodiversity-friendly agro-ecosystems (Das et al. 2006). Similarly, many studies from Latin America have shown the superiority of various shaded coffee systems compared with sun coffee in terms of providing habitats and enhancing biodiversity (e.g. Moguel and Toledo 1999; Gordon et al. 2007; Anand et al. 2008). Less is known regarding the capacity of various coffee systems to regulate water balance (infiltration and ground water production), erosion control, local climate, pollination, soil productivity and other services, but the general view is that high tree cover with a variety of tree species would increase the capacity to generate these functions (cf. Steffan-Dewenter et al. 2007). For example, in coffee agroecosystems, ants, birds and bats can control important coffee pests (Borkhataria et al. 2006; Philpott and Armbrecht 2006), and pollinator activity could increase close to natural forest fragments (Rickets 2004; Klein et al. 2008). Coffee in the Western Ghats and elsewhere straddles a gradient of carbon storage, sequestration and evapotranspiration levels depending on shade practices and management. In general, areas with shade coffee are expected to have similar hydrologic functions and carbon sequestration services as the native forest types that they have replaced (Kumar and Nair 2006; Olchev et al. 2008; Krishnaswamy et al. 2009). However, it may have severe detrimental effects on water quality because of discharge of coffee pulp effluent into streams. In addition, biodiversity values are often more strongly a function of proximity to remnant forest rather than management practices or shade type (Anand et al. 2008; Dolia et al. 2008; Bali et al. 2007).

From these examples it is clear that the various coffee production land uses vary in terms of both provisioning ecosystem services and regulating ecosystem services ranging from low yield per hectare in forested

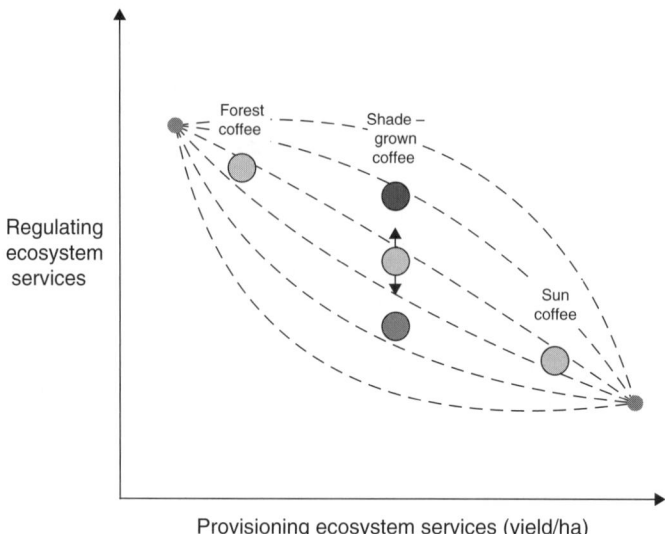

Note: Forest coffee represents more or less wild-harvested coffee with rather low yield per hectare, and maintenance of regulating ecosystem services in the landscape. Sun coffee is high yielding under intense input of pesticides and fertilizers, but with much reduced biodiversity and regulating services in the landscape. Shade-grown coffee represents an intermediate form where a different management approach may result in the same yield per hectare but with very different impacts on regulating services (type A, B and C responses). In this specific case, density and diversity of trees may be used as a proxy for levels of regulating services.

Figure 4.2 A model of trade-offs in coffee production

landscapes to very high yields in areas of intense inputs of fertilizers and pesticides, and with regulating services showing responses from type A to type C. (Figure 4.2).

In Figure 4.2, we analyse production using yield per hectare and assume a strong relationship between yield and intensity of cultivation (sun coffee produces more per hectare than forest and shade coffee) (Perfecto et al. 2005). Sun-grown coffee is often dependent on intensive use of pesticides and fertilizers (cf. Soto-Pinto et al. 2002). A shift towards a less intensive way of cultivating could be expected if the price for pesticides, fertilizers and energy (i.e. fossil fuels) increases. Then, regulating ecosystem services such as biological pest control and natural fertilizing through nitrogen fixing organisms may become more attractive.

In the centre of Figure 4.2 we illustrate that for shade-grown coffee, the level of regulating services may be high, low or intermediate at the same yield per hectare. This represents an opportunity to introduce the

Managing trade-offs in ecosystem services 79

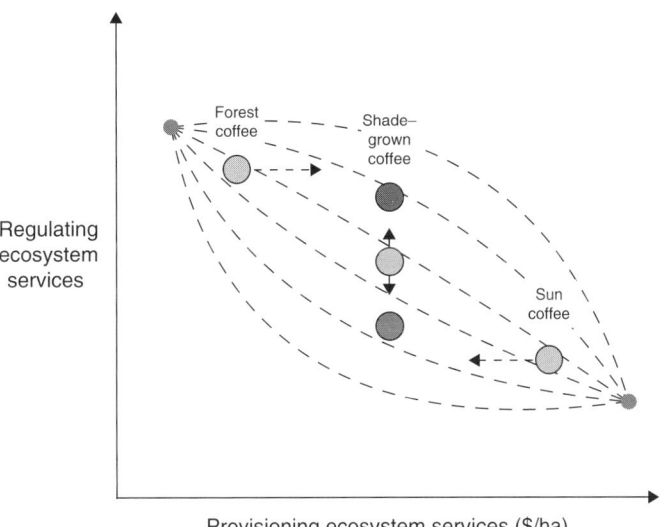

Note: Note that all three production systems may generate approximately the same income per hectare but with very different levels of impact on regulating services.

Source: http://www.accessmylibrary.com/coms2/summary_0286-7587403_ITM (accessed 28 September 2009).

Figure 4.3 A model of trade-offs in coffee production with income per hectare as a variable

right incentives for farmers to move from the lower to the higher levels of regulating service without sacrificing yield (see further discussion below in relation to Figure 4.3).

Regulating ecosystem services are complex, however, and the specific relationship will look different for different services. The difference in regulating services provision response to a conversion of one ecosystem to another can be illustrated by the different responses of different taxa along the gradient from rustic to sun coffee, where butterflies may have a type A response (i.e. the lower curve in Figure 4.1) and ants a type C response (i.e. the upper curve in Figure 4.1) (Perfecto et al. 2003). If instead, income per hectare is used to analyse production of the provisioning service (Figure 4.3) the pattern is slightly different since the price per volume can vary widely depending on quality. Wild-harvested coffee can have a very high price on the global market (hatched arrow to the right) and the high costs of input in sun coffee often reduce net revenues (hatched arrow to the left). For example, in a recent study from Mexico, Gordon et al. (2007)

showed that there was small or no difference in net profitability between 'rustic' coffee plantations and more intensively cultivated farms.

Another mechanism that can increase revenues of lower intensity production is represented by certification schemes with price premiums for certain cultivation techniques (Philpott et al. 2007). Figure 4.3 illustrates that there are potentially many levels of regulatory ecosystem services for the same income level, especially at intermediate incomes. There is some empirical evidence of this. Perfecto et al. (2005) predicted a range of relationships between coffee production and biodiversity combined with shade complexity and proposed that certain small changes, in for example species composition of trees which do not affect shade or productivity, may significantly enhance regulating ecosystem services such as pest control, pollination, and functional groups of taxa. Benefits may also include the possibility of utilizing other provisioning services than coffee from the same area. In Ethiopia, both timber and non-timber products are collected from the same areas that coffee is cultivated, e.g. harvesting of honey from beehives placed in shade trees in forest, semi-forest and home garden coffee systems (Woldemariam 2003; Hylander and Nemomissa 2008).

A FRAMEWORK FOR MANAGING TRADE-OFFS BETWEEN REGULATING AND PROVISIONING SERVICES

We propose a general framework for handling undesired trade-offs between provisioning and regulating services. The framework is based on a typology of landscapes, (1) intensive agricultural landscapes, (2) conservation landscapes and (3) degraded landscapes, each with a distinct configuration of levels of provisioning and regulating services (Figure 4.4). We define landscape at a spatial scale that is sufficiently large to incorporate important ecological process (e.g. large-scale disturbances) and thus on a scale of hundreds of square kilometres. Although most landscapes would represent a mix of the three types in Figure 4.4, it is useful to discuss them as distinct points of departure and trajectories, since transitions to a more desirable state involve different types of management and governance strategies. None of the landscape typologies in Figure 4.4 are likely to be stationary; they are subject to changing biophysical, social and economic drivers. We need to analyse the conditions under which these systems tend to move along the more undesirable trajectories indicated by hatched arrows in Figure 4.4. Moving along the more desirable alternative trajectories, indicated by filled arrows in Figure 4.4, is only partly

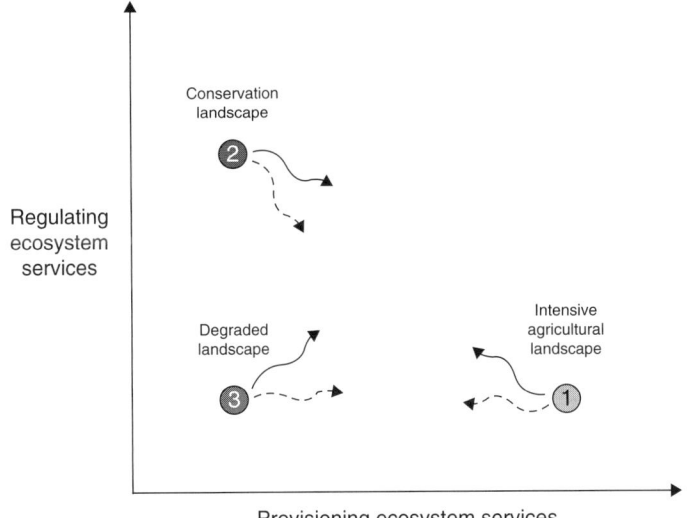

Notes: Intensive agricultural landscapes (1), conservation landscapes (2) and degraded landscapes (3). Future trends may include transitions where regulating services are lost or stay at a low level (type A responses in Figure 4.1, hatched lines) or where they may increase or are maintained at a high level (type B or C response in Figure 4.1, solid lines).

Figure 4.4 Different trajectories of change for intensively used agricultural landscapes

dependent on technology and new innovations. Even more important are incentive schemes, institutions and governance structures that may either encourage or provide barriers to such a development (cf. Folke et al. 2005).

1. *Intensively used agricultural landscapes* are located at the far right, where high generation of provisioning services is maintained at least temporarily at the expense of regulating services. In the future, increasing fossil fuel prices, climate change and water shortage may drive many such areas to the left. For example, the Intergovernmental Panel on Climate Change (IPCC) predicts that even slight global warming will substantially decrease agricultural productivity in many tropical and subtropical countries, mainly due to water shortage for irrigation (IPCC 2007). Further, as stated earlier, nitrogen fertilizer is increasingly expensive and supplies are therefore not sustainable, and we may already have passed a critical threshold likely to threaten global environmental stability and human well-being (Rockström et al. 2009).

Sustainable agricultural systems will therefore increasingly have to rely on nitrogen fixation by organisms.

The transformation of intensive agriculture under this scenario may occur along at least two trajectories, one representing a monotonic decrease in productivity (type A response in Figure 4.1), another representing a simultaneous increase in regulating services (type B or C response in Figure 4.1) such as biological control, nitrogen fixing and climate regulation. We still lack financial incentives and institutions for encouraging the type B or C response, although in some cases these may be triggered by new financial mechanisms where for example conversion of cropland to forests has a very large effect on carbon sequestration (Righelato and Spracklen 2007).

2. *Conservation landscapes* are characterized by conservation and protected area policies, where high levels of regulating services also mean low generation of provisioning services. Demographic and market pressures will, in most parts of the world, lead to intensification of land use, and thus drive such areas down and right in Figure 4.4, i.e. losing regulating and cultural services while gaining provisioning services. For example, natural areas will continue to be converted to agricultural land, and land currently under extensive (low-impact) forms of agriculture will be increasingly converted to intensive agricultural use with 5–7 million km^2 expected to be converted by 2050 (Braat and ten Brink 2008). Two trajectories are also possible in this case, one leading to an increase in provisioning services and a sharp decline in regulating services (a type A response), and the other to maintaining higher levels of regulating services (a type B or C response).

What institutions and incentives should be put in place for the type B or C trajectories? One important and yet unresolved question in this context is to what extent the net production of regulating and provisioning services is greater when a portion of the landscape is converted to very intensive agriculture and the rest turned to conservation, or when most of the landscape is managed in ways that aim to balance agricultural production with biodiversity conservation (e.g. Bengtsson et al. 2003; Fischer et al. 2008).

3. *Degraded landscapes* may generate few services as a result of extremely low productivity and often the absence of institutions to regulate boom-and-bust economic drivers. However, there are currently incentives emerging to transform such areas by production of biomass energy. Global terrestrial annual plant growth is more than five times the 8 billion tonnes of carbon released to the atmosphere in fossil fuel combustion. In principle, diverting a small fraction of

total plant growth into biomass energy could satisfy the majority of global energy needs. The large potential for producing biomass energy without negative effects on climate and food production lies in using degraded and abandoned agricultural lands (Field et al. 2007). Houghton (1991) estimated the area of degraded abandoned land to be around 500 million hectares globally, with 100 million hectares in Asia, 100 million hectares in Latin America and 300 million hectares in Africa, while Field et al. (2007) estimated the global area to be 386 million hectares. Metzger and Hiittermann (2009) argued that the world energy needs could be provided sustainably and economically primarily from lignocellulosic biomass grown on areas which have been degraded by human activities in historical times. With the right incentives additional advantages such as CO_2 sequestration, soil formation, water conservation and desertification control may be achieved. Increased demand for energy offers an opportunity to convert these degraded landscapes. If the conversion is profitable and novel in character there is a window of opportunity for influencing this shift so that many regulating services are enhanced at the same time, corresponding to a type B or C response rather than an type A response (Figure 4.4).

CONCLUSION

In a simplified example with one manager (a farmer) and one agricultural system in a landscape, the space for possible trade-offs in Figure 4.3 highlights that even though many options are available, most do not currently provide an improved economy for the farmer, and hence do not present an economic incentive. However, when a farmer reduces yield by making room for biodiversity and promotes regulating services, and then compensates the revenue loss with higher prices thanks to a certification scheme, the situation could be a 'zero-sum game' for the farmer but a real improvement for the landscape as a whole. In this context, two important and urgent questions need to be addressed:

1. What institutions are required to obtain more sustainable landscape management by 'mainstreaming' ecosystem services and, for example, promoting choices that might be zero-sum for the individual manager but of great value to the larger landscape?
2. What are the impediments to incorporating ecosystem services into everyday land-use decision-making by individuals, companies and governments (cf. Polasky et al. 2008)?

One solution suggested by Goldman et al. (2007) is the development of incentives within 'Ecosystem Services Districts' where the full range of services generated within a landscape is monitored, valued and made part of responsible decision-making processes. The district could constitute a combination of regulation and incentives as well as voluntary and non-voluntary approaches. A potential advantage is that such districts might facilitate cooperation amongst landowners who are then more likely to manage their communal resources sustainably (Ostrom 1990). Cooperation is likely to reduce individual direct costs and transaction costs (Wagner and Kreuter 2004). An Ecosystem Service District would allow landowners to bundle services together and offer the provision of multiple services within one defined boundary, enabling them to address type B and C responses in land management decisions. Novel solutions like this need to be developed, tested and evaluated in real landscapes (e.g. Cowling et al. 2008).

ACKNOWLEDGEMENTS

The authors would like to thank the NRM-group at the Department of Systems Ecology for a stimulating discussion around the theme of this chapter. This work has been supported by a grant from Formas to Thomas Elmqvist. The authors would like to also thank SIDA for providing support for J. Krishnaswamy and exchange between SRC and ATREE, and support of research and field-work in Ethiopia for K. Hylander.

REFERENCES

Anand, M.O., J. Krishnaswamy and A. Das (2008). Proximity to forests drives bird conservation value of shade-coffee plantations: Implications for certification. *Ecological Applications*, **18**(7), 1754–1763.

Bali, A., A. Kumar and J. Krishnaswamy (2007). The mammalian communities in coffee plantations around a protected area in the Western Ghats, India. *Biological Conservation*, **139**, 93–102.

Bengtsson J., P. Angelstam, T. Elmqvist, U. Emanuelsson, C. Folke, M. Ihse, F. Moberg and M. Nyström (2003). Reserves, resilience and dynamic landscapes. *Ambio*, **32**(6), 389–396.

Bohensky, E.L., B. Reyers and A.S. Van Jaarsveld (2006). Future ecosystem services in a Southern African river basin: A scenario planning approach to uncertainty. *Conservation Biology*, **20**, 1051–1061.

Borkhataria, R.R., J.A. Collazo and M.J. Groom (2006). Additive effects of vertebrate predators on insects in a Puerto Rican coffee plantation. *Ecological Applications*, **16**, 696–703.

Braat, L. and P. ten Brink (eds) (2008). The cost of policy inaction (COPI): the case of not meeting the 2010 biodiversity target. DG Environment. Study EC 2007/S95-116033 for the European Commission. Alterra, IEEP, FEEM, GHK, UNEP-WCMC, Wageningen, the Netherlands.

Brussaard, L., P.C. de Ruiter and G.G. Brown (2007). Soil biodiversity for agricultural sustainability. *Agriculture, Ecosystems and Environment*, **121**(3), 233–244.

Carr, D., A. Barbieri, W. Pan and H. Iranavi (2006). Agricultural change and limits to deforestation in Central America. In F. Brouwer and B.A. McCarl (eds), *Agriculture and Climate Beyond 2015*. Dordrecht, the Netherlands: Springer, pp. 91–107.

Carpenter, S.R., R. DeFries, T. Dietz, H.A. Mooney, S. Polasky, W.V. Reid and R.J. Scholes (2006). Millennium Ecosystem Assessment: research needs. *Science*, **314**, 257–258.

Carpenter, S.R., H.A. Mooney, J. Agard, D. Capistrano, R.S. DeFries, S. Diaz, T. Dietz, A.K. Duraiappah, A. Oteng-Yeboah, H.M. Pereira, C. Perrings, W.V. Reid, J. Sarukhan, R.J. Scholes and A. Whyte (2009). Science for managing ecosystem services: beyond the Millennium Ecosystem Assessment. *Proceedings of the National Academy of Sciences of the United States of America*, **106**(5), 1305–1312.

Cowling, R., B. Egoh, A.T. Knight, P.J. O'Farrell, B. Reyers, M. Rouget, D.J. Roux, A. Welz and A. Wilhelm-Rechman (2008). An operational model for mainstreaming ecosystem services for implementation. *Proceedings of the National Academy of Sciences of the United States of America*, **105**, 9483–9488.

Daily, G., S. Polasky, J. Goldstein, P.M. Kareiva, H.A. Mooney, L. Pejchar, T.H. Ricketts, J. Salzman and R. Shallenberger (2009). Ecosystem services in decision making: time to deliver. *Frontiers in Ecology and Environment*, **7**(1) 21–28.

Das, A., J. Krishnaswamy, K.S. Bawa, M.C. Kiran, V. Srinivas, N. Samba Kumar and K.U. Karanth (2006). Prioritization of conservation areas in the Western Ghats, India. *Biological Conservation*, **133**, 16–31.

Diaz, S., J. Fargione, F.S. Chapin III and D. Tilman (2006). Biodiversity loss threatens human well-being. *PLoS Biology* **4**(8) e277.

Dolia, J., M.S. Devy, N.A. Aravind and A. Kumar (2008). Adult butterfly communities in coffee plantations around a protected area in the Western Ghats, India. *Animal Conservation*, **11**, 26–34.

Elmqvist, T., E. Maltby, T. Barker, M. Mortimer, C. Perrings, J. Aronson, R. DeGroot, A. Fitter, G. Mace, J. Norberg, I. Sousa Pinto and I. Ring (2010). Biodiversity, ecosystems and ecosystem services. In P. Kumar (ed.), *The Economics of Ecosystems and Biodiversity: Ecological and Economic Foundations*. London and Washington DC: TEEB Earthscan, Chapter 2.

Field, C.B., D.B. Lobell and H.A. Peters (2007). Feedbacks of terrestrial ecosystems to climate change. *Annual Review of Environment and Resources*, **32**, 1–29.

Fischer, J., B. Brosi, G.C. Daily, P.R. Ehrlich, R. Goldman, J. Goldstein, D.V. Lindenmayer, A.D. Manning, H.A. Mooney, L. Pejchar, J. Ranganathan and H. Tallis (2008). Should agricultural policies encourage land sparing or wildlife-friendly farming? *Frontiers in Ecology and the Environment*, **6**(7), 380.

Folke, C., T. Hahn, P. Olsson and J. Norberg (2005). Adaptive governance of social–ecological systems. *Annual Review of Environment and Resources*, **30**, 441–473.

Goldman, R.L., B.H. Thompson and G.C. Daily (2007). Institutional incentives for managing the landscape: Inducing cooperation for the production of ecosystem services. *Ecological Economics*, **64**, 333–343.

Gordon, C., R. Manson, J. Sundberg and A.Cruz-Angon (2007). Biodiversity, profitability, and vegetation structure in a Mexican coffee agroecosystem. *Agriculture Ecosystems & Environment*, **118**, 256–266.

Gordon, L.J., G.D. Peterson and E. Bennett (2008). Agricultural modifications of hydrological flows create ecological surprises. *Trends in Ecology and Evolution*, **23**, 211–219.

Gove, A.D., K. Hylander, S. Nemomissa and A. Shimelis (2008). Ethiopian coffee cultivation: implications for bird conservation and environmental certification. *Conservation Letters*, **1**, 208–216.

Houghton, R.A. (1991). Tropical deforestation and atmospheric carbon dioxide. *Climatic Change*, **19**(1–2), 99–118.

Hylander, K. and S. Nemomissa (2008). Home garden coffee as a repository of epiphyte biodiversity in Ethiopia. *Frontiers in Ecology and the Environment*, **6**, 524–528.

IPCC (2007). *Climate Change 2007: The Physical Science Basis*. Geneva: IPCC Secretariat.

Klein, A., B.E. Vaissiere, J.H. Cane, I. Steffan-Dewenter, S.A. Cunningham, C. Kremen and T. Tscharntke (2007). Importance of pollinators in changing landscapes for world crops. *Proceedings of the Royal Society of London B*, **274**, 303–313.

Klein, A.-M., S.A. Cunningham, M. Bos and I. Steffan-Dewenter (2008). Advances in pollination ecology from tropical plantation crops. *Ecology*, **89**, 935–943.

Krishnaswamy, J., K.S. Bawa, K.N. Ganeshaiah, and M.C. Kiran (2009). Quantifying and mapping biodiversity and ecosystem services: utility of a multi-season NDVI-based *Mahalanobis* distance surrogate. *Remote Sensing of Environment*, **113**, 857–867.

Kumar, B.M and P.K.R. Nair (eds) (2006). *Tropical Homegardens: A Time-Tested Example of Sustainable Agroforestry*. Dordrecht, the Netherlands: Springer.

Larsen, T.H., N.M. Williams and C. Kremen (2005). Extinction order and altered community structure rapidly disrupt ecosystem functioning. *Ecology Letters*, **8**, 538–547.

Lavelle, P. and A.V. Spain (2001). *Soil Ecology*. Dordrecht, the Netherlands: Kluwer Academic Publishers.

Losey, J.E. and M. Vaughan (2006). The economic value of ecological services provided by insects. *Bioscience*, **56**, 331–323.

Luck, G.W., G.C. Daily and P.R. Ehrlich (2003). Population diversity and ecosystem services. *Trends in Ecology and Evolution*, **18**(7), 331–336.

Luck, G.W., R. Harrington, P.A. Harrison, C. Kremen, P.M. Berry, R. Bugter, T.R. Dawson, F. de Bello, S. Díaz, C.K. Feld, J.R. Haslett, D. Hering, A. Kontogianni, S. Lavorel, M. Rounsevell, M.J. Samways, L. Sandin, J. Settele, M.T. Sykes, S. Van Den Hove, M. Vandewalle and M. Zobel (2009). Quantifying the contribution of organisms to the provision of ecosystem services. *Bioscience*, **59**, 223–235.

Metzger, J.O. and A. Hiittermann (2009). Sustainable global energy supply based on lignocellulosic biomass from afforestation of degraded areas. *Naturwissenschaften*, **96**, 279–288.

Millennium Ecosystem Assessment (2005). *Ecosystems and Human Well-being: Synthesis*. Washington DC: Island Press.

Meyer, F.G. (1965). Notes on wild *Coffea arabica* from southwestern Ethiopia, with some historical considerations. *Economic Botany*, **19**, 136–151.

Meyerson, L.A., J. Baron, J.M. Melillo, R.J. Naiman, R.I. O'Malley, G. Orians, M.A. Palmer, A.S.P. Pfaff, S.W. Running and O.E. Sala (2005). Aggregate measures of ecosystem services: can we take the pulse of nature? *Frontiers in Ecology and Environment*, **3**(1), 56–59.

Moguel, P. and V.M. Toledo (1999). Biodiversity conservation in traditional coffee systems of Mexico. *Conservation Biology*, **13**, 11–21.

Naidoo, R., A. Balmford, R. Costanza, B. Fisher, R.E. Green, B. Lehner, T.R. Malcolm and T.H. Ricketts (2008). Global mapping of ecosystem services and conservation priorities. *Proceedings of the National Academy of Sciences of the United States of America*, **105**(28), 9495–9500.

Olchev, A., A. Ibrom, J. Priess, S. Erasmi, C. Leemhuis, A. Twele, K. Radler, H. Kreilein, O. Panferov and G. Gravenhorst (2008). Effects of land-use changes on evapotranspiration of tropical rain forest margin area in Central Sulawesi (Indonesia): modelling study with a regional SVAT model. *Ecological Modelling*, **212**(1–2), 131–137.

Ostrom, E. (1990). *Governing the Commons: The Evolution of Institutions for Collective Action*. New York: Cambridge University Press.

Perfecto, I., A. Mas, T. Dietsch and J. Vandermeer (2003). Conservation of biodiversity in coffee agroecosystems: a tri-taxa comparison in southern Mexico. *Biodiversity and Conservation*, **12**, 1239–1252.

Perfecto, I., J. Vandermeer, A. Mas and L.S. Pinto (2005). Biodiversity, yield, and shade coffee certification. *Ecological Economics*, **54**, 435–446.

Philpott, S.M. and I. Armbrecht (2006). Biodiversity in tropical agroforests and the ecological role of ants and ant diversity in predatory function. *Ecological Entomology*, **31**, 369–377.

Philpott, S.M., P. Bichier, R. Rice and R. Greenberg (2007). Field-testing ecological and economic benefits of coffee certification programs. *Conservation Biology*, **21**, 975–985.

Polasky, S., E. Nelson, J. Camm, B. Csuti, P. Fackler, E. Lonsdorf, C. Montgomery, D. White, J. Arthur, B. Garber-Yonts, R. Haight, J. Kagan, A. Starfield and C. Tobalske (2008). Where to put things? Spatial land management to sustain biodiversity and economic returns. *Biological Conservation*, **141**(6), 1505–1524.

Pretty, J.N., A.D. Noble, D. Bossio, J. Dixon, R.E. Hine, F.W.T.P. de Vries and J.I.L. Morison (2006). Resource-conserving agriculture increases yields in developing countries. *Environmental Science & Technology*, **40**, 1114–1119.

Ricketts, T.H. (2004). Tropical forest fragments enhance pollinator activity in nearby coffee crops. *Conservation Biology*, **18**, 1262–1271.

Righelato, R. and D.V. Spracklen (2007). The carbon benefits of fuels and forests: response. *Science*, **318**(5853), 1066–1068.

Rockström, J., W. Steffen, K. Noone, A. Persson, S. Chapin, E.F. Lambin, T.M. Lenton, M. Scheffer, C. Folke, H.J. Schellnhuber, B. Nykvist, C.A. de Wit, T. Hughes, S. van der Leeuw, H. Rodhe, S. Sörlin, P.K. Snyder, R. Costanza,

U. Svedin, M. Falkenmark, L. Karlberg, R.W. Corell, V.J. Fabry, J. Hansen, B. Walker, D. Liverman, K. Richardson, P. Crutzen and J.A. Foley (2009). A safe operating space for humanity. *Nature*, **461**, 472–475.

Scheffer, M., S.R. Carpenter, J.A. Foley, C. Folke and B. Walker (2001). Catastrophic shifts in ecosystems. *Nature*, **413**, 591–596.

Schmitt, C.B. (2006). Montane rainforest with wild *Coffea arabica* in the Bonga region (SW Ethiopia): plant diversity, wild coffee management and implications for conservation. PhD thesis. University of Bonn, Bonn, Germany.

Schroter, D., W. Cramer, R. Leemans, L.C. Prentice, M.G. Araújo, N.W. Arnell, A. Bondeau, H. Bugmann, T.R. Carter, C.A. Gracia, A.C. de la Vega-Leinert, M. Erhard, F. Ewert, M. Glendining, J.I. House, S. Kankaanpää, R.J.T. Klein, S. Lavorel, M. Lindner, M.J. Metzger, J. Meyer, T.D. Mitchell, I. Reginster, M. Rounsevell, S. Sabaté, S. Sitch, B. Smith, J. Smith, P. Smith, M.T. Sykes, K. Thonicke, W. Thuiller, G. Tuck, S. Zaehle and B. Zierl (2005). Ecosystem service supply and vulnerability to global change in Europe. *Science*, **310**, 1333–1337.

Senbeta, F. and M. Denich (2006). Effects of wild coffee management on species diversity in the Afromontane rainforests of Ethiopia. *Forest Ecology and Management*, **232**, 68–74.

Shennan, C. (2008). Biotic interactions, ecological knowledge and agriculture. *Philosophical Transactions of the Royal Society B*, **363**(1492), 717–739.

Sidle, R.C., A.D. Ziegler, J.N. Negishi, A.R. Nik, R. Siew and F. Turkelboom (2006). Erosion processes in steep terrain – Truths, myths, and uncertainties related to forest management in Southeast Asia. *Forest Ecology and Management*, **224**(1–2), 199–225.

Soto-Pinto, L., I. Perfecto and J. Caballero-Nieto (2002). Shade over coffee: its effects on berry borer, leaf rust and spontaneous herbs in Chiapas, Mexico. *Agroforestry Systems*, **55**, 37–45.

Steffan-Dewenter, I., M. Kessler, J. Barkmann, M.M. Bos, D. Buchori, S. Erasmi, H. Faust, G. Gerold, K. Glenk, S.R. Gradstein, E. Guhardja, M. Harteveld, D. Hertel, P. Höhn, M. Kappas, S. Köhler, C. Leuschner, M. Maertens, R. Marggraf, S. Migge-Kleian, J. Mogea, R. Pitopang, M. Schaefer, S. Schwarze, S.G. Sporn, A. Steingrebe, S.S. Tjitrosoedirdjo, S. Tjitrosoemito, A. Twele, R. Weber, L. Woltmann, M. Zeller and T. Tscharntke (2007). Tradeoffs between income, biodiversity, and ecosystem functioning during tropical rainforest conversion and agroforestry intensification. *Proceedings of the National Academy of Sciences of the United States of America*, **104**, 4973–4978.

Steinfel, H., P. Gerber, T. Wassenaar, V. Castel, M. Rosales and C. de Haan (2006). *Livestock's Long Shadow: Environmental Issues and Options*. Rome: FAO.

Sylvain P.G. (1955). Some observations on *Coffea arabica* L. in Ethiopia. *Turrialba*, **5**, 37–53.

Tallis, H., P. Kareiva, M. Marvier and A. Chang (2008). An ecosystem services framework to support both practical conservation and economic development. *Proceedings of the National Academy of Sciences of the United States of America*, **105**(28), 9457–9464.

TEEB (2008). The economics of ecosystems and biodiversity: an interim report. European Commission, Brussels. Available at www.teebweb.org (accessed 1 September 2009).

TEEB (2010) *The Economics of Ecosystems and Biodiversity: Ecological and Economic Foundations*. London and Washington: Earthscan.

Wagner, M.W. and U.P. Kreuter (2004). Groundwater supply in Texas: private land considerations in a rule-of-capture state. *Society and Natural Resources*, **17**, 359–367.

Woldemariam, T. (2003). Vegetation of the Yayu Forest in SW Ethiopia: Impacts of human use and implications for in situ conservation of wild *Coffea arabica* L. populations. PhD thesis. University of Bonn, Bonn, Germany.

5. Revisiting the relationship between equity and efficiency in payments for ecosystem services*

Unai Pascual, Roldan Muradian, Luis C. Rodriguez and Anantha K. Duraiappah

INTRODUCTION

This chapter addresses the relationships between the economic efficiency and distributional implications of market-based incentive mechanisms, and more specifically of payments for ecosystem services (PES). We point out that the mainstream interpretation of PES as a mere market-based instrument to internalize environmental externalities overlooks the complex interactions, including power relations, that take place between agents, and determine both equity and efficiency outcomes. We propose an approach that gives emphasis to the roles of (1) uncertainty, as perceived by agents in setting up PES schemes, (2) the institutional context required to set up the payments and services, and (3) the notions of fairness held by different agents, e.g. providers, beneficiaries of the ecosystem services and intermediaries. Some of the most important questions to be answered are the following: What are the key factors conditioning the relationship between equity and efficiency in PES schemes? What is the role of social perceptions about fairness in the performance of PES? What is the contribution of social science, and economics in particular, to our understanding of the relationship between efficiency and equity in PES?

Hitherto, the standard economics paradigm (internalization of the costs of externalities through locally agreed compensation agreements) has dominated the conceptualization of PES schemes (Pagiola et al. 2005; Wunder 2007; Engel et al. 2008). The mainstream policy thinking about PES goes as follows. The core justification for the design and implementation of PES schemes is that a lack of markets for ecosystem services (ES) results in market failures; in other words, in a failure to account for the negative externalities of agricultural or industrial activity. Private incentives (in the form of a payment to ES providers) are meant to generate market signals to encourage private agents to take into account the costs and benefits of providing ecosystem services when making decisions about land use, and thus solve potential environmental externality problems. It is generally assumed in the PES literature that as long as transaction costs are low enough, individuals would trade their rights away until an efficient allocation is achieved.

From this mainstream viewpoint, there are at least two general and three more specific requirements for internalizing externalities through voluntary exchanges between stakeholders. The general requirements are that (1) property rights are clearly specified, and (2) there is freedom for their exchange (i.e. a market is created). The more specific conditions are that (1) each party must be able to provide something the other wants at a lower cost (including transaction costs) than the others could provide on their own, (2) there must be a way for the parties to agree on a formal or informal contract mechanism within the existing legal and institutional structure, i.e. an institutional enabling environment is in place, and finally (3) the parties involved must be able to capture at least some of the gains from trade, or at least not to be harmed by it. Within this theoretical framework, what is at stake is overall net efficiency. That is, the objective of the policy-maker is to maximize the difference between the aggregate gains and losses of welfare across individuals. This would render the so-called most socially efficient outcome. Such an outcome may in principle be achieved by a variety of interventions (command and control, markets, taxes, subsidies, etc.). Hence, the economic efficiency of implementing a PES scheme ought to be evaluated relative to the opportunity cost of achieving the same result by other policy means (Ferraro and Simpson 2002).

The distributional aspects of PES schemes have been largely overlooked; however, there is a new focus on assessing their income effects on poor people engaged in PES: so-called 'pro-poor' effects. For the most part, attention is paid to analysing the factors conditioning the eligibility, ability and willingness of the poor to participate in PES schemes, and to answering the question of whether low-income stakeholders participate as providers in PES programmes (Grieg-Gran et al. 2005; Pagiola et al. 2005; Proctor et al. 2008; Wunder 2008). Thus, efficiency and equity issues have been addressed in the mainstream literature in a dissociated way. We argue that this piecemeal approach cannot help in understanding the intertwined relationship between efficiency and equity in PES schemes. The following are what we consider some of the limitations of the mainstream approaches to the analysis of PES.

It is well known that the main contribution of the mainstream economics literature to the internalization of environmental externalities is the proposition that a socially efficient solution may be achieved through market transactions, independent of the initial wealth endowment or allocation of property rights. From this point of view, the key issue is the identification of property rights. This is a main concern of environmental economics, since from a neoclassical perspective environmental problems tend to be seen as the result of market failures due to poorly defined property rights. However, stressing aggregated efficiency gains independently of the allocation of property rights, the so-called Coasean approach, implies a neglect of distributional effects of market transactions and thus of the role of social values and power relations that are key considerations in the distribution of costs and benefits.

This mismatch can be avoided if both efficiency and equity are considered as structural features of PES schemes. Practitioners, particularly in developing countries, are usually confronted with trade-offs between these criteria and are not able to detach efficiency from equity considerations. Thus, in reality, PES schemes tend to be part of broader rural development interventions (not only designed to solve environmental externalities).

Furthermore, in many cases the efficiency of PES schemes can hardly be demonstrated in practice. The contexts in which PES operate are normally characterized by high uncertainty about efficiency gains. The two main sources of uncertainty are structural and informational. First, due to ecological complexities, the structural relationships between land-use changes and the provision of ES are often context-dependent and hard to evaluate (particularly in the case of water-related ES). Second, the level of uncertainty is determined by the type and level of information available. Gathering further information can reduce this uncertainty but raises the transaction costs of the scheme. Thus, typically there exists a trade-off between (1) the need to improve confidence in the actual costs and gains (level of efficiency) that potentially will follow from implementing the scheme, and (2) the economic feasibility of it, which is closely associated with the level of transaction costs, including set up and initial implementation costs, subsequent monitoring, enforcement, etc.

High uncertainty and incomplete information constitute a problem for the Coasean approach because two of the specific requirements cannot be met with confidence, namely: both parties must be able to reap benefits from trade; and providers must be able to supply a service that buyers cannot procure at a lower cost. Even from a neoclassical economics perspective, most PES schemes can only be seen as second best instruments, since the benefits created from the traded ES cannot be known with enough certainty, given the complexity of ecosystem functioning and interdependencies between ecosystem flows and the subjective value people attach to them over time.

This chapter also argues that the pro-poor approach, though pertinent, is limited in scope, particularly for addressing the relationship between equity and efficiency. This is because it does not tackle the equity and economic fairness dimension of PES in a comprehensive way. Such a dimension surpasses the typical deciding factors concerning the eligibility, ability and willingness of the poor to participate in PES programmes. The distribution of benefits is not only a matter of who participates, but also of the conditions of participation (including the distribution of bargaining power among involved and excluded stakeholders). This calls for efforts to identify who holds the power of decision about the 'rules of the game' (so-called institutional setting), and the distribution of costs and benefits. As Corbera et al. (2007) have pointed out, PES schemes are likely to reinforce existing power structures and inequalities in access to resources, which have significant equity implications.

Given the importance of distributional conflicts in the practical feasibility of PES schemes, we argue that equity concerns should be considered as important as efficiency matters in the design of a scheme. We have developed a conceptual framework to address both concerns (equity and efficiency), which overcomes some of the previously mentioned piecemeal limitations of Coasean and the pro-poor approaches in shaping the relationships between these two concerns. Special attention is paid to the role of the institutional setting and considerations of political economy.

In the next section we introduce some definitions of key categories needed to construct the conceptual model. Then we focus on the main economic fairness criteria and the methods that can be used to analyse efficiency and equity in PES schemes before discussing in more detail the relevance of the economic fairness criteria, the role of the intermediary and the equity and efficiency trade-offs. We then analyse the political economy and institutional factors that set the shape of the dual (efficiency and equity) policy trade-off curve using some heuristic models and hypothetical examples.

MODEL SET UP: SOME KEY DEFINITIONS

We first introduce a few concepts that will be used throughout the chapter. Here 'environmental additionality' is understood as the net positive impact of the provision of ES in comparison with the baseline scenario or a hypothetical situation where the scheme is not in place. By definition, environmental additionality is a necessary (but not sufficient) condition for increasing efficiency in a PES scheme. If a scheme is unable to lead to an environmental net benefit, it would be inefficient because operational costs would already have been incurred without making some individuals better off (in the Paretian sense).

We define efficiency as the difference between the total welfare benefits to the target population and the total cost incurred. Efficiency may be evaluated using an array of indicators, including monetary. Additionally, equity is understood as the net impact of payments on inequality, among a reference set of agents, compared with a baseline scenario. The assessment of equity, however, depends on two factors: the first relates to the choice of fairness criterion (chosen during the design of the scheme) for the distribution of net benefits that will accrue from implementation of the payments; the second factor is related to the method chosen to evaluate the scheme when all is in place.

Given the two-fold objectives (efficiency and equity), the term 'policy trade-off curve' (PTOC) is central to our analysis. The PTOC describes the potential outcomes of combinations of levels of efficiency and equity, given the institutional context and the set of economic and ecological factors that constitute a PES scheme. In other words, it identifies the feasible set of potential combinations. A PES scheme may lie at a specific point on the PTOC at a given moment. The position along the curve may change over time depending on underlying institutional factors (we discuss this in some detail below). Of course, the PTOC concept is heuristic, but is useful in so far as it depicts the interdependency of both efficiency and equity.

The idea of the PTOC is illustrated in Figure 5.1. Four quadrants are associated with positive and negative representations of efficiency and equity. It should be noted that whether the equity of a payments scheme is positive or negative depends on the adopted values.[1] Note also that Figure 5.1 depicts a PTOC located in quadrant I, although it could theoretically be positioned in any of the four quadrants (depending on a set of factors explained below).

For the sake of illustration, as the indicator of equity we have chosen the inverse of the income Gini coefficient to measure the evenness of the distribution of payments across individuals receiving the payment in exchange for providing the ecosystem service. In addition, following the standard welfare economics tradition, the efficiency gain of a given PES scheme could be measured by the aggregate change in the consumer and producer surplus, net of transaction costs. In other words, net welfare gain is a measure of efficiency.

The intercept in the vertical axis of Figure 5.1 represents maximum achievable equity, since this point indicates the highest willingness to pay (WTP) by buyers of the ecosystem service. This may be achieved at the expense of efficiency, but it is unlikely because providers of ES are seldom associated with having market power (normally buyers can exert more market power than providers because the former are well-off consumers in urban areas). In addition, the intercept in the horizontal axis represents a hypothetical scenario either where buyers and providers interact freely to find economic compensation for the costs of externalities, or where better-off buyers exert maximum market pressure and hence pay providers their minimum willingness to accept (WTA). This amount should represent the providers' opportunity cost for each level of the ES provision, i.e. perfect price discrimination imposed on buyers. This scenario is unlikely to exist in reality if the property rights to the land are vested in the ecosystem service providers and are not totally subordinated to the urban buyers. Since the intermediary usually plays an important role in mediating between the conflicting interests of buyers (aiming to maximize efficiency) and providers (aiming to maximize equity), PES schemes are most likely to fall somewhere between the two extreme ends of the PTOC.

The combination of efficiency and equity outcomes would place a PES scheme at a given point on the PTOC in Figure 5.1. External pressures that are themselves potentially subject to policy measures can influence the scheme's position on the curve. For instance, a diverse set of factors associated with the bargaining power of individual stakeholders and the role of intermediaries can influence the performance of the PES scheme. The

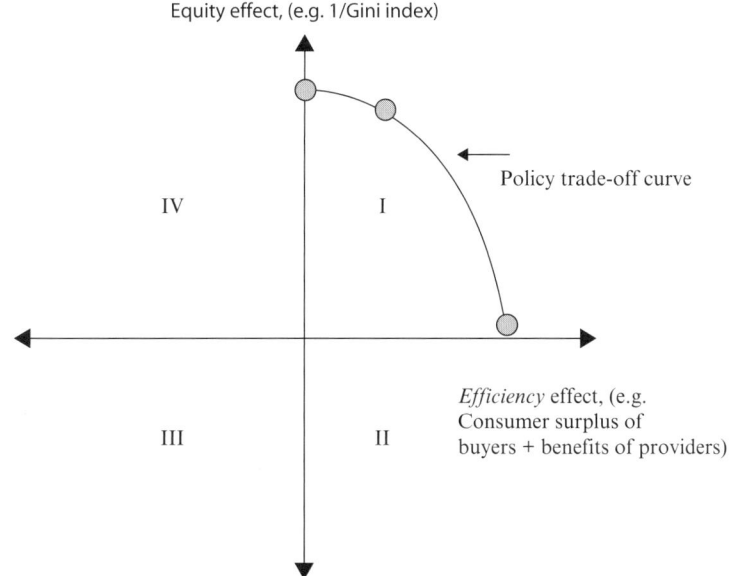

Figure 5.1 A possible policy (efficiency–equity) trade-off curve in a policy intervention space

factors affecting the bargaining power of stakeholders may include the levels of past association within the group, their socio-economic status or political influence, their level of empowerment and their relationships with the intermediary. Other factors that can influence the position of a PES scheme under a given efficiency–equity trade-off curve are also associated with the role of the intermediary. These include the intermediary's ability to manage information and uncertainty, strategies to tackle free-riding, ability to negotiate, reputation (relative to trust) and the magnitude of the transaction costs resulting from the work.

ECONOMIC FAIRNESS AND WELFARE EVALUATIVE CRITERIA IN PES

Alternative perceptions of what is a fair PES scheme may significantly alter decisions about the allocation of resources within the scheme and hence its outcomes in terms of social welfare and the social legitimacy of the rules of the schemes. In this vein, we refer to economic fairness as the shared view by a given social group of the embedded distributive justice of the payments. It is associated with the social notion of what distributive rule (procedure for allocating costs and benefits) and outcomes are deemed as fair according to different stakeholders. Of course, this is a social construct that is context-dependent, hence multiple criteria may exist. The key point is that whatever local fairness criterion is used at the design stage of the PES scheme will influence the perception by stakeholders of the equity of the scheme. Ultimately, it would influence the feasibility of the scheme. The equity criteria in any PES design can play an important role in legitimizing the process by generating alliances or conflicts between individuals, and encouraging or discouraging the participation of the target population. Although the selection of an explicit fairness criterion is rarely a condition considered in the design of PES, every scheme has an implicit criterion with regard to the way the payments are designed and implemented.

A comprehensive notion of fairness should take into account both fairness as a procedural rule and fairness at the level of distribution of outcomes. For the sake of the conceptual model, we focus on fairness in outcomes.[2] Further, it should be noted that each culture and social group has its own local conceptions of what is fair, equitable or just. The notion of economic fairness is a highly subjective phenomenon that varies widely in social, cultural and institutional contexts (Konow 1996, 2003; Schokkaert 1998; Schokkaert and Devooght 2003). Furthermore, different social groups participating in a given PES scheme may differ about the fairness criteria that they support. Which criteria prevail in any given PES

Table 5.1 Fairness criteria for PES schemes

Fairness criterion	Design implications
Compensation	Payments should compensate landholders for the forgone benefits related to the provision of ecosystem services. Payments are differentiated according to the cost of provision.
Consensus	Design should promote group decision-making processes to distribute the available funds in a consensual basis. The criteria for payment differentiation is decided by consensus.
Egalitarian	Design should distribute the fund equally among all the providers (per unit of land area, for example) independently of the level and cost of ecosystem service provision. Payments are not differentiated.
Maxi-min	Payments aim to maximize the net benefit to the poorest landholders, even at a cost of efficiency loss. Payments are differentiated according to the income of providers.
Actual provision	The allocation of funds among landowners corresponds to the measured level of ecosystem services provided. Payments are differentiated according to the actual provision of ecosystem services.
Expected provision	Payments to landholders depend on the expected level of provision of services for a given land use. Payments are differentiated according to the expected provision of ecosystem services. These payments compensate landholders for particular land use changes or practices expected to enhance the provision of ecosystem services.
Status quo	Payments should maintain the previous level of relative distribution of income among providers. Payments are differentiated according to impact on income.

scheme depends on power relations and the critical role of the intermediary, as we discuss below.

What Fairness Criteria for what PES?

There is no simple way to prefer, a priori, one fairness criterion over another because all of them can be equally justifiable both in ethical and operative terms by different stakeholders. Here we summarize some of the economic fairness criteria that could be applied to PES schemes and illustrate with examples of schemes that have been implemented in developing countries. In addition, we discuss how the adoption of any given criteria have implications for equity.[3] Table 5.1 summarizes some of the economic fairness criteria that could be applied or are implicit in PES schemes.

As Table 5.1 illustrates, each economic fairness criterion is associated with specific economic objectives. Some criteria may face serious limitations when applied to PES schemes, e.g. because of practical difficulties in estimating the provision of ES associated with a particular land use. This would be the case for example in the *compensation* and *actual provision* criteria. Clarity in these criteria suffers from uncertainties about the links between land-management decisions and the provision of ES, and problems in estimating the adequate level of compensation for landholders where information is incomplete (e.g. about costs). Other criteria may be hard to implement because of a power imbalance among the social groups involved, as in the case of the *maxi–min* criterion, or owing to potential conflicts between societal goals and the overall objective of the PES scheme, e.g. the *egalitarian* criterion.

Some of these criteria have already been applied in well-known PES schemes in developing countries. For example, in the nationwide PES programme for carbon sequestration in Mexico, the *consensus* criterion was used to allocate the transferred money to members of communities involved in the programme. In this case, the communal nature of the land facilitated the application of the criterion because the government distributes the funds to the communities directly, rather than to individuals.

By contrast, in the well-known nationwide forest conservation PES programme in Costa Rica run by FONAFIFO, an equal payment per hectare was implemented independently of spatial variability in the provision of ES or the costs of the required management practices by different landholders (Pagiola 2008). This implies the explicit adoption of the *egalitarian* fairness criterion. Yet in the watershed-protection PES programme implemented in Jesus de Otoro, Honduras, the *expected provision* fairness criterion was used. Payments were differentiated according to the type of forest and the type and number of water-quality improvement practices adopted by farmers, and were thus ultimately based on the expected level of provision of ES and their contribution to the protection of the watershed (Kosoy et al. 2007). Another illuminating example is the PES programme run by the PROFAFOR project in Ecuador, where the *actual provision* criterion was preferred given that the amount of funds allocated to each landholder was estimated based on the evaluation of the quantity of carbon sequestrated in their land (Wunder and Alban 2008).

EVALUATING WELFARE OUTCOMES

In addition to the type of fairness criteria used in the design of PES schemes, it is also essential for us to make clear the method of evaluation that will be used to judge the final welfare outcomes of the respective stakeholders, both in terms of efficiency and equity.

It is well established that evaluating equity using only income as the unit of measure can be misleading (Millennium Ecosystem Assessment 2005), an observation equally valid for PES schemes. While the efficiency of PES schemes may be assessed in terms of the average increase in income, clearly this does not translate automatically into an increase in well-being. An increase in income generated by a PES scheme might not be sufficient to compensate for increases in the cost of other ES that may have declined. For example, putting aside land for restoring water-regulating services could cause a decrease in the supply of food-provisioning services. This might increase prices for food in the market, resulting in reduced mean food intake and nutritional value for the community. There is thus a need for a more holistic approach to measuring well-being, and subsequently the equity changes that may result from PES schemes.

'Capability' is a term denoting an individual's capacity, through personal attributes combined with local constraints and opportunities, to live in the way they have chosen to live. Since evaluating people's capabilities goes beyond using income alone to evaluate individual welfare changes and the distribution of such changes, the concepts of fairness and capabilities are intertwined. A person's life can be seen as comprising various 'doings and beings' (called by Sen (2006) 'functionings'). People can achieve several functionings simultaneously, e.g. having self-respect and being well nourished. Every individual thus can be seen as possessing a bundle of functionings, i.e. several linked functionings that are possessed in concert, at any given time. Of course, some functionings cannot be achieved simultaneously and, moreover, there are limits to how much any one person can do and be at any one time. A person may have a variety of incompatible bundles of functionings that could be achieved in the context of any PES scheme. The set of all such bundles in operation simultaneously is the capability set. Basic capabilities include the ability to avoid preventable diseases, the ability to acquire an education, the ability to be adequately nourished and the ability to live in a safe environment.

The Millennium Ecosystem Assessment (Millennium Ecosystem Assessment 2003) documented links between ES and the capabilities of different groups. Trade-offs and how they differ between stakeholders are important considerations. For example, a farmer in Africa may decide to make a trade-off in favour of food production over conserving biodiversity. A pharmaceutical company interested in establishing a PES scheme for biodiversity conservation might choose an option that calls for setting aside large parcels of land. Although both services can be complementary, there will be a need to establish land-use options to maximize the conservation of biodiversity while at the same time minimizing impacts on farm production.

By representing a person's meaningful opportunities for quality of life, capabilities may constitute an appropriate metric for evaluating human well-being. Any PES scheme is likely to affect the extent of capabilities within the population and therefore evaluations of efficiency (as the increase in well-being of the target populations) and fairness, and consequently equity. When evaluating the collective capabilities of stakeholders, the indicators used to assess efficiency and equity should also be appropriate for evaluation of the impacts of the PES scheme in enhancing the capabilities needed for a decent quality of life in the target population.

If we depart from a measure of income for evaluating efficiency and equity, and instead rely on capability, the notion of agency becomes an important point. Agency is defined by Sen (1992) as the freedom of individuals to take part in an activity they have reason to value. Therefore, although the outcome is an important element, the act of being an active stakeholder in delivering the outcome adds to an individual's overall well-being. This is often ignored in the design and implementation stages of PES schemes, as these stages are evaluated (in either efficiency or pro-poor effects) in income terms only.

The prevailing fairness criterion in a given PES scheme arises during the planning and implementation process. It is conditioned by practical considerations related to the feasibility of the scheme and governed by the capacity for involvement by stakeholders and the power relations of the involved parties (e.g., governments, NGOs, ecosystem service providers, beneficiaries and intermediaries). The role of the intermediary in facilitating the implementation of the scheme should not be overlooked. Intermediaries try to reconcile the interests and expectations of buyers and providers of ES by finding equitable solutions acceptable to all parties, while ensuring the economic feasibility of the scheme. As noted above, economic feasibility might be jeopardized by the adoption of a particular fairness criterion. For instance, the application of a *compensation* criterion would require detailed information about the opportunity costs from each individual provider, which might be very costly to obtain, and therefore could negatively affect the practical feasibility of the scheme. These transaction-cost-related constraints have a considerable capacity to reduce the freedom to choose such criteria.

The intermediary has the main role of avoiding conflicts between social groups. These often derive from the previous history of interaction between groups, current distributive conditions and incompatible notions of fairness. For example, among the relevant participants in a PES scheme, latent disputes caused by perceived unfairness in current or past land tenure, or distributional conflicts triggered by differing notions of economic fairness may lead to a low level of commitment by providers of ES to fulfil the conditions of the contract. The role of the intermediary is thus challenged.

An example might be the application of a *maxi–min* criterion that is focused mainly on poverty reduction, which may negatively affect better-off groups since they tend to bear the largest share of the cost. This criterion would then provide better-off landholders with little incentive to comply with the contracts. Conversely, if the *compensation* criterion is applied, the initial allocation of land rights plays a critical role. If the land ownership is not evenly distributed, as is the case in many developing countries, then this criterion might pose social problems, especially if the ES in question relate to public goods that many other stakeholders, in particular the socially disadvantaged and vulnerable groups, are dependent on for their well-being. This is especially the case for regulating services like water management, disaster risk reduction and maintenance of microclimate, among others. In addition, the importance of the means of evaluation cannot be ignored. Even if monetary compensation is provided to socially disadvantaged groups, it might not be sufficient to cover the extra costs incurred from, for example, the higher frequency of disasters. Therefore evaluation of income might suggest an improvement of well-being while an evaluation of capabilities might suggest otherwise.

When disadvantaged groups participate in PES schemes they tend to be the weakest party, while more powerful groups have greater influence on the decision-making process, exerting leverage by, e.g. promoting a particular criterion favouring their own interests and marginalizing the poor from market benefits (Grieg-Gran and Bann 2003). Additionally, through choice of fairness criterion, an intermediary can exert power on the efficiency performance of PES and the distributional impacts across stakeholders engaged in PES. Even if, in general, PES schemes are easier to implement where powerful stakeholders are supportive, the intermediary should avoid supporting particular groups seeking to take control of the project and thus shape the equity considerations to the detriment of disadvantaged groups.

The implicit fairness criteria of PES schemes both reflect and determine the relative weights given to equity and efficiency concerns during the design of the programme. For example, at one extreme, a payment scheme adopting a *maxi–min* criterion would primarily be designed to increase the net benefit to the poorest eligible landholders even if efficiency is negligible. However, this kind of pro-poor PES design might find itself competing with other poverty relief programmes (Turpie et al. 2008) or

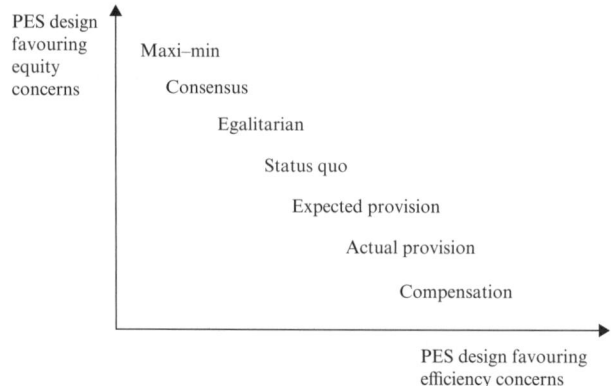

Figure 5.2   Economic fairness criteria and the emphasis of equity and efficiency concerns in PES design

'conditional cash transfers', which are not linked with conservation efforts or environmental considerations (Wunder 2008).

By contrast, a PES scheme designed under a status quo criterion would display positive efficiency, but negative equity (regardless of how evaluated). This situation will be more evident in circumstances where the individuals providing services in the most cost-efficient and environmentally effective way are not necessarily disadvantaged landholders (see Scheer et al. 2004). Other criteria included in Table 5.1, like the *egalitarian* and *consensus* criteria, are more difficult to evaluate in terms of their impact on trade-offs between efficiency and equity. However, the *compensation, actual provision* and *expected provision* criteria place more emphasis on environmental additionality (hence, on efficiency) than promoting equity, given that these criteria favour the allocation of payments based on environmental merit.

Obviously, each specific PES scheme would have a context-dependent effect on equity and efficiency, mediated by the economic fairness criterion and method of evaluation. Figure 5.2 draws on a tentative proposition regarding the positions of the different economic fairness criteria identified in Table 5.1 in relation to their relative emphasis on equity and efficiency concerns in the design stage of a PES scheme. We have assumed for Figure 5.2 that better-off landowners have more access and control over larger and better quality land areas (in terms of ecosystem service provision). This is supported by empirical evidence from developing countries (e.g. Umali-Deininger and Shapouri 2002). We do not think, however, that this assumption should hold for all PES schemes and thus this model should be adapted in scheme-specific ways (both in terms of evaluation and assumptions of underlying political economy and power relations).

Apart from the prevailing criteria governing fairness and evaluation, multiple institutional factors influence the final outcome regarding equity and efficiency. Here, two sets of factors can be distinguished. One set includes those that can be said to set the shape of the PTOC, thus constraining the potential combinations of equity and efficiency. The other includes those that influence the actual combination of equity and efficiency effects (and possible paths for changes) within the PTOC. The following sections identify and address the importance of these factors.

FACTORS INFLUENCING EQUITY AND EFFICIENCY IN PES

There are three main factors that determine the equity outcome in a PES scheme. First, it is influenced by the method chosen to evaluate the scheme, which can involve for instance the distribution of either income or capabilities. Second, the existing wealth of individuals can be used to compare equity; this would set the origin point (0,0) in Figure 5.1. Third, it is crucial to know which are the relevant stakeholders or reference set of participants of the PES scheme, because equity concerns have to be compared with potential changes in the well-being of the different social groups.

The efficiency of the PES scheme depends on at least three other key factors: environmental additionality; the value of the ES provided; and the cost of implementation of the scheme. Environmental additionality changes the economic efficiency of the PES scheme compared with a 'business-as-usual' scenario but needs a comparative reference state, which may change over time. Second, the value of the service provided must be determined, which is difficult and costly in practice. Third, the costs of implementation need to be identified and valued. These include all types of transaction and opportunity costs, the sum of all payments made by buyers of the services and all additional costs (covered, for example, by external agents), as well as the cost of providing the service by other means. In principle, calculations of benefits and costs need to consider the net present value of the PES scheme, that is, taking into consideration variations over time. This involves the additional challenge of identifying a proper discount rate. Implicit in this are fairness criteria about the distribution of costs and benefits in the future, reflected in the use of discount rates. However, as discussed earlier, if the capability approach is used, then efficiency may change or even be reversed.

These points can be illustrated with the example of a PES scheme for the water-related services of forests implemented in Mexico (Muñoz Piña et al. 2008). To begin with, there were doubts about the net additional effect on forest conservation (since most forests under the scheme are not highly threatened). However, the scheme is effective as an income redistribution mechanism since most of the participant communities are classified as very poor. For this example, we assume that while the scheme contributes to equity, since it transfers income from donors and tax payers to rural indigenous communities, there is no environmental additionality. Thus, there is no net gain in efficiency, which means that the PTOC can only be located in quadrant IV in Figure 5.1, i.e. no or negative efficiency in addition to positive equity outcome.

The theoretical PTOC in Figure 5.1 also illustrates the outcome of any potential PES design in relation to efficiency and equity. The shape of the PTOC depends on factors affecting the size and distribution of the transfer of resources (affecting income or capabilities) between buyers and providers of the ES. Based on the more simple evaluation of income, at least three factors determine the size of transfers: (1) The ES providers' WTA compensation for the level of ES they provide, (2) the WTP of buyers of the ecosystem services, and (3), the implementation cost of the PES scheme. We briefly address these three factors in turn.

The willingness of ecosystem service providers to accept compensation depends on a variety of factors, which in some instances can overlap. These include among others: (1) providers' perceptions about foregone revenues, i.e. the flow of the opportunity cost of ES supply, (2) the potential on-site benefits of the ecosystem services, (3) the nature of land use, e.g. ancestral land-use patterns, (4) the degree of livelihood dependency on the land, (5) the size and quality of their land, and (6) social perceptions about liability, property rights or reputation of other individuals associated with the scheme.

By contrast, the WTP of buyers depends mainly on (1) the cost of alternative policy options, (2) their disposable income or ability to pay, (3) social perceptions about liability, property rights, or the reputation of others, (4) confidence in the relationship between land use and provision of ES, which in turn is associated with (5) the degree of uncertainty about the actual provision of those services.

Lastly, the implementation costs depend largely on (1) the amount and quality of the information that needs to be gathered and managed, (2) the difficulty of the negotiation and contracting process, (3) the structure and organization of the intermediary organization(s), (4) the costs of enforcement, monitoring (post-contractual activities), and the way the funds are managed, and also critically on (5) the number and heterogeneity of stakeholders in the PES programme.

CONCLUDING REMARKS

We have argued that a comprehensive theoretical framework is needed for understanding the interactions that shape relationships between equity and efficiency in PES schemes. We think that such a framework should rest on institutional and political economy insights, and should aim to analyse, among others, the role of the adopted notions of economic fairness criteria, possible efficiency–equity trade-offs and power relations between stakeholders.

The relationship between equity and efficiency in PES schemes might be very complex and context-dependent, owing to the great number of factors that influence the shape of the PTOC. We argue that efficiency and equity in PES schemes are intertwined, since they are mutually dependent, and we argue that trade-offs between them should be considered structural features of PES, since the conflicting interests of buyers and providers are pervasive.

The allocation of property rights is not neutral from the point of view of fairness or social values, and is to a large extent influenced by power relations. The analysis of who has the power to set the rules of the game, including the size of the transfer and the prevailing economic fairness criteria, is key for understanding the outcomes of a given PES scheme in terms of equity and efficiency. PES tend to be built from the demand side, so they share a propensity to have a demand-driven institutional structure. That is, the interests of buyers are usually better represented than those of providers, and buyers normally hold greater bargaining power.

The level of uncertainty has considerable implications for the relationship between efficiency and equity in PES schemes. The relationship between land use and the provision of ES is often hard to demonstrate, not only because the necessary studies are costly but also because of ecological complexity. This fact usually makes it difficult to meet one of the key neoclassical conditions for ensuring efficiency, namely the identification and estimation of the provision of an environmental service for a given amount of payment. Uncertainty is amplified by the fact that monitoring systems (checking the actual provision of the service) are often loosely implemented on the ground in developing countries, owing to their relatively high cost. In addition, the costs of alternative policy options often are not fully known, which creates difficulties for the evaluation of the cost-effectiveness of PES. A major implication of high uncertainty is that practitioners and stakeholders have to take 'precautionary' decisions without having full information about their efficiency implications in economic terms.

Uncertainty increases the importance of information asymmetry, and therefore it confers larger power to intermediaries. Uncertainty also affects the size of the transfer, since we may infer that the WTP of buyers is negatively correlated with the degree of uncertainty about the benefits derived from the transaction. Thus, uncertainty reduces the capacity of intermediaries to introduce sizable fees among buyers.

The role of intermediary agents in steering the performance of PES has been rather neglected in the literature on PES. However, owing to the need to deal with uncertainty and to coordinate actions between stakeholders, the intermediary plays a key role. Having control over information, and holding an important relational role confer to the intermediary significant ability to steer the transfer of resources, select the beneficiaries and set the 'rules of the game', which determine to a large extent the efficiency and equity performance of this kind of scheme.

NOTES

* This article was published in *Ecological Economics*, **69**(6), Unai Pascual, Roldan Muradian, Luis C. Rodriguez and Anantha K. Duraiappah, Exploring the links between equity and efficiency in payments for environmental services: a conceptual approach, 1237–1244, Copyright Elsevier BV (2009).
1. This simple example adopts a monetary income measure, although more encompassing indicators may be used.
2. We analyse the effect of fairness based on the consequences of the approach rather than on the morality of actions leading to it.
3. Here we consider the distributional effects among providers of ES for the sake of exposing the argument. The analysis can be extended to a wider set of agents (buyers, providers, potential providers, etc.). However for illustrative purposes this chapter focuses only on the providers of ES.

REFERENCES

Corbera, E., K. Brown and W.N. Adger (2007). The equity and legitimacy of markets for ecosystem services. *Development and Change*, **38**(4), 587–613.

Engel, S., S. Pagiola and S. Wunder (2008). Designing payments for environmental services in theory and practice: An overview of the issues. *Ecological Economics*, **62**, 663–674.

Ferraro, P. and R.D. Simpson (2002). The cost-effectiveness of conservation payments. *Land Economics*, **78**(3), 339–353.

Grieg-Gran, M. and C. Bann (2003). A closer look at payments and markets for environmental services. In P. Gutman (ed.) *From Goodwill to Payments for Environmental Services: A Survey of Financing Options for Sustainable Natural Resource Management in Developing Countries*. Washington DC: WWF Macroeconomics for Sustainable Development Program Office M/P/O, pp. 27–40.

Grieg-Gran, M., I.T. Porras and S. Wunder (2005). How can market mechanisms for forest environmental services help the poor? Preliminary lessons from Latin America. *World Development*, **33**, 1511–1527.

Konow, J. (1996). A positive theory of economic fairness. *Journal of Economic Behaviour and Organization*, **31**, 13–35.

Konow, J. (2003). Which is the fairest one of all? A positive analysis of justice theories. *Journal of Economic Literature*, **41**(4), 1186–1237.

Kosoy, N., M. Martinez-Tuna, R. Muradian and J. Martinez-Alier (2007). Payments for environmental services in watersheds: insights from a comparative study of three cases in Central America. *Ecological Economics*, **61**, 446–455.

Millennium Ecosystem Assessment (2003). *Ecosystems and Human Well-being: A Framework for Assessment*. Washington DC: Millennium Ecosystem Assessment, WRI.

Millennium Ecosystem Assessment (2005). *Our Human Planet: Summary for Decision Makers*. Washington DC: Millennium Ecosystem Assessment, WRI.

Muñoz-Piña, C., A. Guevara, J.M. Torres and J. Braña (2008). Paying for the hydrological services of Mexico's forests: Analysis, negotiations and results. *Ecological Economics*, **65**, 725–736.

Pagiola, S. (2008). Payments for environmental services in Costa Rica. *Ecological Economics*, **65**, 712–724.

Pagiola, S., A. Arcenas and G. Platais (2005). Can payments for environmental services help reduce poverty? An exploration of the issues and the evidence to date. *World Development*, **33**, 237–253.

Proctor, W., T. Köllner and A. Lukasiewicz (2008). Equity considerations and payments for ecosystem services. Environmental Economy and Policy Research Working Papers 31.2008, Department of Land Economics, University of Cambridge, Cambridge.

Scheer, S.J., A. White and A. Khare (2004). *For Service Rendered. Current Status and Future Potential of Markets for Ecosystem Services of Tropical Forests: An Overview*. Technical Series 21. Yokohama, Japan: International Tropical Timber Organization.

Schokkaert, E. (1998). Mr. Fairmind is post-welfarist: Opinions on distributive justice. Discussion paper series. 98.09. Center for Economic Studies. Department of Economics, Catholic University of Leuven, Leuven.

Schokkaert, E. and K. Devooght (2003). Responsibility-sensitive fair compensation in different cultures. *Social Choice and Welfare*, **21**, 207–242.

Sen, A. (1992). *Inequality Reexamined*. Oxford: Clarendon Press.

Sen, A. (2006). Conceptualizing and measuring poverty. In D.B. Grusky and R. Kanbur (eds), *Poverty and Inequality*. Stanford, CA: Stanford University Press, pp. 30–46.

Turpie, J.K., C. Marais and J.N. Blignaut (2008). The working for water programme: evolution of a payments for ecosystem services mechanism that address both poverty and ecosystem service delivery in South Africa. *Ecological Economics*, **65**, 788–798.

Umali-Deininger, D. and S. Shapouri (2002). Food security: is India at risk? In R. Lal, D. Hansen, N. Uphoff and S.A. Slack (eds), *Food Security and Environmental Quality in the Developing World*. Boca Raton, FL: CRC Press, pp. 31–52.

Wunder, S. (2007). The efficiency of payments for environmental services in tropical conservation. *Conservation Biology*, **21**, 48–58.

Wunder, S. (2008). Payments for environmental services and the poor: concepts and preliminary evidence. *Environment and Development Economics*, **13**, 279–297.

Wunder, S. and M. Alban (2008). Decentralized payments for environmental services: the case of PIMAMPIRO and PROFAFOR in Ecuador. *Ecological Economics*, **65**, 685–698.

6. Are the amounts of payments for environmental services enough to contribute to poverty alleviation efforts in developing countries?

Luis C. Rodriguez, Unai Pascual and Roldan Muradian

INTRODUCTION

Environmental conservation and poverty reduction are two areas receiving increasing attention from donors, governments, research organizations and development agencies around the world (Sachs and Reid 2006). The use of market-based instruments, such as payments for ecosystem services (PES) or conditional cash transfers (CCT), is considered to be among the most cost-effective instruments to achieve these environmental (Ferraro and Simpson 2002) and social (Rawlings 2005) goals. Both approaches involve payments that are transferred to a number of selected beneficiaries, previously identified in a targeting exercise as fulfilling a number of conditions designed to internalize an environmental externality in the case of PES, or a social externality in the case of CCT (Wunder 2007; World Bank 2009). Thus, PES are aimed at compensating landholders for the cost of providing positive environmental services to stakeholders that can directly pay (by market transaction) or indirectly pay (by using public funds from taxes or donors) for the services they expect to receive.

These payments, when properly designed and implemented, are linked to a requirement to fulfil a set of environmental conditions and land management practices that promote the provision of ecosystem services (Pascual et al. 2010). Therefore, while PES are conservation instruments, they indirectly affect people's livelihoods (Smith and Scherr 2002; Niesten and Rice 2004). Accordingly, they have been the starting point for the exploration of their potential as a poverty reduction tool (Grieg-Gran et al. 2005; Suyanto et al. 2007). Owing to this aspect of PES schemes, they have attracted the attention of donors for their potential pro-poor effects rather than for their positive environmental impacts (Wunder 2008). However, despite the increasing body of literature about the capacity of the poor to participate and benefit from PES schemes (Zilberman et al. 2008; Pagiola et al. 2005, 2008a), there is no information on whether the magnitude of payments under PES schemes is enough to have significant impacts on poverty reduction efforts.

The aim of this chapter is to fill that knowledge gap, providing evidence about the potential of PES to contribute to poverty alleviation in developing countries. We first present CCT in the context of poverty reduction initiatives and then analyse the amounts of PES in relation to poverty reduction efforts in selected countries of Latin America, Africa and Asia, before putting these in context as tools for poverty alleviation.

REDUCING POVERTY

Poverty is a complex concept that goes beyond economic constraints related to income or job access, to include other dimensions such as human health, education, political empowerment or exclusion, socio-cultural issues of status and dignity, and protection of individuals and communities in terms of risk and vulnerability. These dimensions are interrelated and their relative importance changes over time. Thus, poverty reduction might have different meanings in different circumstances, meaning that a different set of strategies should be implemented to alleviate poverty (see Asselin 2009).

Traditionally, these strategies have included promoting economic growth, attracting development aid, supporting good governance and debt relief, improving markets and improving the social environment and abilities of the poor. More recently, the use of CCT has become one of the preferred instruments for poverty alleviation around the world. CCT are innovative approaches to promote immediate poverty reduction and break intergenerational poverty traps. They do this by providing money to a defined target group (usually poor and vulnerable households), and generally are linked to compliance with a set of conditions aimed at promoting investment in human capital such as sending children to school or bringing them to health centres on a regular basis (Das et al. 2004).

The amount of payments needed to have an impact on poverty is estimated during the early stages of the design of each CCT programme based on in-depth poverty analysis. As a general rule of thumb, the payment should represent between 20% and 40% of the minimum per capita income considered adequate (total poverty line), in order to be meaningful to the beneficiaries (Handa and Davis 2006).

Are PES enough to contribute to poverty alleviation? 111

The amounts of CCT represent an adequate estimate of the level of external support required to have a significant impact on poverty reduction. Therefore it can be argued that by comparing the amounts of CCT and PES implemented in the same countries, it is possible to estimate whether PES amounts are of the order of magnitude required to make a significant contribution to poverty reduction initiatives.

COMPARING PES AND CCT AMOUNTS

CCT and PES schemes are relatively novel approaches for conservation and poverty alleviation and, despite their increased use worldwide, there are still few developing countries where both types of programme have been in place (Figure 6.1). The use of CCT in Asia and Africa is a very recent phenomenon (World Bank 2009). We studied ten countries where both PES and CCT have been implemented: five from Latin America, three from Asia and two from Africa.

Table 6.1 presents information on environmental services associated with PES schemes and annual per household payment amounts from PES and CCT for each country. It should be noted that some CCT and PES payments are expressed as a range, since they are designed as minimum and maximum amounts per beneficiary household, depending on their particular circumstances (World Bank 2009).

Paired comparisons between the average annual amount received per

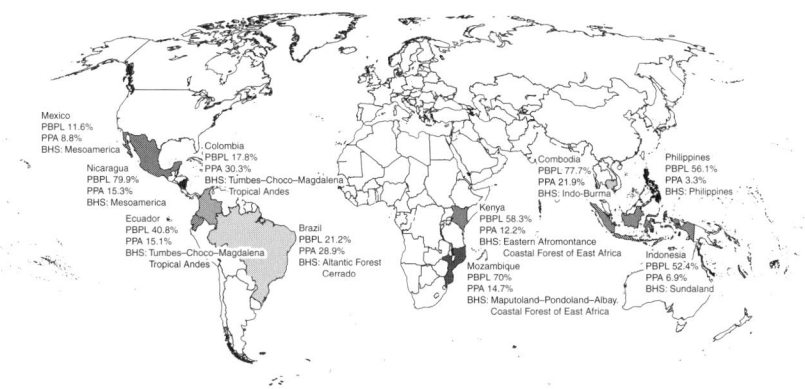

Note: The map indicates countries with populations below the US$2/day poverty line.

Figure 6.1 Selected countries where both PES and CCT have been implemented

Table 6.1 Household benefits per year from CCT and PES

	Programme name	Household benefit US$ per year*	Sources
Latin America			
Ecuador			
CCT	Bono de Desarrollo Humano	360	World Bank (2009)
PES (Carbon sequestration)	PROFAFOR	252	Wunder and Alban (2008)
Colombia			
CCT	Familias en Accion	552–768	World Bank (2009)
PES (Carbon sequestration– Biodiversity)	Silvopastoral Project	607	Rios and Pagiola (2009)
Brasil			
CCT	Bolsa Familia	506–703	World Bank (2009)
PES (Carbon sequestration)	Bolsa Floresta	517	Viana (2008)
Mexico			
CCT	Oportunidades	371–1068	World Bank (2009)
PES (Carbon sequestration)	PSA–CABSA Niños Heroes	545	Corbera et al. (2009)
Nicaragua			
CCT	Programme Atencion a Crisis	325–378	World Bank (2009)
PES (Carbon sequestration– biodiversity)	Silvopastoral Project	592	Rios and Pagiola (2009)
Asia			
Cambodia			
CCT	Cambodia Education Sector Support Project	60	World Bank (2009)
PES (Biodiversity)	Biodiversity Conservation Payments	120–160	Clements et al. (2010)
Indonesia			
CCT	Programme Keluarga Harapan	65–152	World Bank (2009)
PES (Watershed protection)	Tlekung and Cidanau Watershed payments	42–132	Munawir and Vermeulen (2007)
Philippines			
CCT	Pantawid Pamilyang Pilipino Programme	202–342	World Bank (2009)

Table 6.1 (continued)

	Programme name	Household benefit US$ per year*	Sources
PES (Watershed protection)	Bakun Watershed Protection	31	Beria et al. (2009)
Africa			
Kenya			
CCT	Cash Transfer for Orphan and Vulnerable Children	252	Pearson and Alviar (2009)
PES (Biodiversity)	Wildlife Conservation Lease Programme	306	Reto-o-Reto (2006), World Bank (2008)
Mozambique			
CCT	Bolsa Escola	21	Massingarela and Nhate (2006)
PES (Carbon sequestration)	Nhambita Community Carbon	34	Jindal et al. 2008

Notes: *Authors' calculations based on information from the sources.

beneficiary household from PES schemes and the annual payment to beneficiary households of CCT programmes implemented in the same countries are made in Figure 6.2. These show that in most cases, the PES amounts are within the range of CCT payments, and the difference between PES and CCT amounts for the selected countries is not significant. However, government-funded and public-funded PES schemes pay larger amounts to beneficiaries than that paid by private PES schemes. Carbon sequestration schemes generally provide larger payments to beneficiaries than biodiversity and watershed protection programmes, while payments for jointly supplied services might be substantially higher (see Figure 6.3).

THE USE OF PES FOR POVERTY ALLEVIATION

There is evidence suggesting that poor people who provide environmental services are able to get access to and benefit from PES programmes (Wunder et al. 2008). Participation in both PES and CCT is voluntary, but in order to increase programme efficiency, participants are selected based on either environmental quality in the case of PES or household poverty

114   *Values, payments and institutions for ecosystem management*

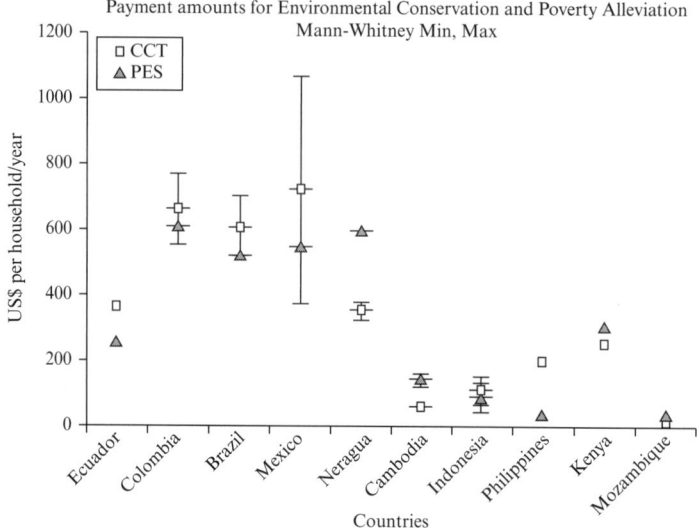

Notes: Payments for CCT and PES are statistically similar (Wilcoxon matched pairs test Z = 0.764, p < 0.44). Publicly funded schemes provided greater amounts to beneficiaries than did private schemes (Mann-Whitney U test Z = 2.61, p < 0.01).

Figure 6.2   Paired comparison of PES and CCT implemented in selected developing countries

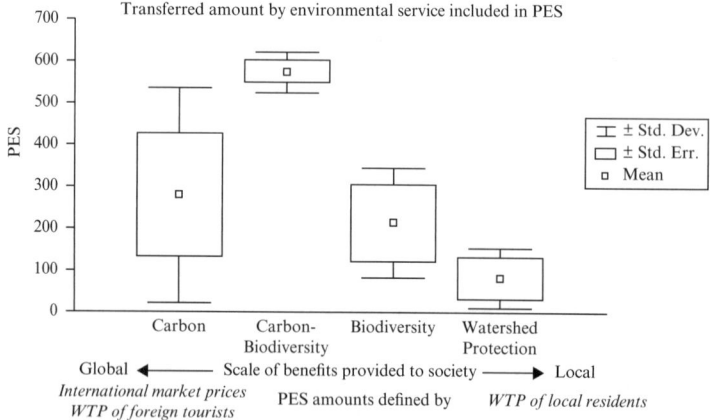

Figure 6.3   Annual per household payment amounts by type of environmental service

level in the case of CCT. Funds are distributed among beneficiaries in a way that generates environmental additionality or promotes poverty reduction. Participation in PES or CCT involves costs for the beneficiaries, however, the incurred private costs necessary to generate a behavioural change towards the desired actions are not revealed in either scheme. The payments are just enough to cover the estimated costs associated with the required conservation or social actions, i.e. neither PES nor CCT designs offer a surplus. Participation in the schemes is voluntary, and the incentive to beneficiaries to join the scheme and fulfil the associated conditions will be present only if the amount offered covers the costs incurred.

For CCT programmes, the payments are designed to cover the direct costs of sending children to school, which include school fees, supplies and transport, and the opportunity cost of their time, i.e. foregone income of children (PIDS 2007), but not a surplus in case it generates an incentive to parents not to participate in the labour market (Medeiros et al. 2008). For PES, the payment should compensate the landholder for foregone revenue and for the opportunity cost of land management to promote the provision of ecosystem services (Pascual et al. 2010). However, a surplus, i.e. a payment higher than the value of the services provided, is not offered since it leads to an inefficient outcome (Engel et al. 2008). Since for each programme the level of benefits ought to match the costs incurred by the participants, by comparing just the level of benefits provided by PES and CCT, it is possible to infer whether the magnitude of PES is sufficient to have an impact on poverty. The results indicate that for the selected countries, there are no significant differences between the amounts of PES and CCT, suggesting that properly designed and targeted PES schemes could make a significant contribution to poverty alleviation.

PES are conservation instruments that do not explicitly target poor landholders. In practical terms, however, many PES schemes have been coincidently implemented in poor areas, making poor people over-represented among the beneficiaries (Wunder et al. 2008). These PES schemes did not include specific measures to target poverty at the household level, but the fact that environmentally-rich areas are frequently inhabited by poor communities make them equivalent to geographic target areas. This geographic targeting is a common first-stage practice in selecting areas and beneficiaries for poverty reduction initiatives (Grosh et al. 2008). In areas of high conservation value chosen for implementing PES schemes, a household targeting exercise might increase the social benefits of the intervention by preferential selection of poor households able to provide the desired level of environmental services. However, targeting is expensive and an appropriate mechanism for it should be selected to avoid the costs from exceeding any gain in efficiency (Coady et al. 2004).

For beneficiary households, regular and predictable PES payments might represent a significant portion of household income (Miranda et al. 2003), and become part of their risk management strategies (Pagiola 2008). Moreover, there is evidence that households can invest money from PES in areas such as health and education, which are usually covered by CCT programmes. For example, for some biodiversity payment projects in Kenya, payments intended for environmental purposes were allocated to school fees and medicines (World Bank 2008).

This example highlights the linkages between environmental and socio-economic systems and their potential contribution to well-being at the household level. However, at the level of society, the same linkages are more likely to be influenced by the scale of benefits provided by nature and the institutional arrangements to ensure their provision. Thus, in the selected schemes, PES based on global benefits such as carbon sequestration or biodiversity conservation usually granted higher payments to landholders than smaller schemes. This is because they are linked by intermediaries to the prices of international carbon markets or to wealthy foreign visitors through international tourism. On the other hand, PES schemes that provide local benefits, like watershed protection, in developing countries usually received lower payments than large schemes because they are constrained by the willingness to pay of the beneficiaries of the service i.e. local people and industries. In contrast, the level of payments for the same service in a developed country might be substantially higher (Perrot-Maitre 2006). Within our sample, PES schemes funded by government or public sources generally made larger payments to beneficiaries than those in private agreements, but in most cases they were associated with services that provide global rather than local benefits. Around the world, despite the increasing number of PES schemes funded by international sources, there are still no defined global mechanisms available for their finance and implementation (Farley et al. 2010). However, it is becoming clearer that improving the targeting of PES towards bundled ecosystem services (e.g. biodiversity and carbon sequestration together) might be cost-effective (Venter et al. 2009; Wendland et al. 2010).

The linkages between environmental and socio-economic systems are also becoming more evident for policy-makers and decision-makers. It is now recognized that environmental degradation is a major barrier to poverty reduction. At the same time, reaching environmental conservation goals requires progress in the eradication of poverty. Thus, in order to be successful, poverty reduction and environmental initiatives should be linked and implemented together (see Sachs and Reid 2006). With regard to this, converting PES schemes into poverty reduction programmes might be inefficient from a conservation perspective if there is no correlation between incidence of poverty and improvements in environmental quality (Pagiola et al. 2008b). However, in regions where such a correlation exists, the level of benefits provided to landholders by PES schemes seem to have the economic potential to contribute to poverty alleviation efforts if properly targeted and designed.

REFERENCES

Asselin, L.M. (2009). *Analysis of Multidimensional Poverty: Theory and Case Studies. Economic Studies in Inequality, Social Exclusion and Wellbeing*. London and Ottawa: Springer and IDRC.
Beria, L., J. Laxman and M. van Noordwijk (2009). Can rewards for environmental services benefit the poor? Lessons from Asia. *International Journal of the Commons*, **3**, 82–107.
Clements, T., J. Ashish, K. Nielsen, A. Dara, T. Seth and E.J. Milner-Gulland (2010). Payments for biodiversity conservation in the context of weak institutions: Comparison of three programs from Cambodia. *Ecological Economics*, **69**(6), 1283–1291.
Coady, D., M. Grosh and J. Hoddinott (2004). *Targeting of Transfers in Developing Countries: Review of Lessons and Experience*. Washington DC: World Bank, Regional and Sectoral Studies.
Corbera, E., C. González Soberanis and K. Brown (2009). Institutional dimensions of payments for ecosystem services: an analysis of Mexico's carbon forestry programme. *Ecological Economics*, **68**, 743–761.
Das, J., Q.T. Do and B. Ozler (2004). Conditional cash transfers and the equity–efficiency debate. World Bank Policy Research Working Paper No. 3280. The World Bank, Washington DC.
Engel, S., S. Pagiola and S. Wunder (2008). Designing payments for environmental services in theory and practice: an overview of the issues. *Ecological Economics*, **62**, 663–674.
Farley, J., A. Aquino, A. Daniels, A. Moulaert, D. Lee and A. Krause (2010). Global mechanisms for sustaining and enhancing PES schemes. *Ecological Economics*, **69**, 2075–2084.
Ferraro, P.J. and R.D. Simpson (2002). The cost effectiveness of conservation payments. *Land Economics*, **78**, 339–353.
Grieg-Gran, M., I.T. Porras and S. Wunder (2005). How can market mechanisms for forest environmental services help the poor? Preliminary lessons from Latin America. *World Development*, **33**, 1511–1527.
Grosh, M., C. del Ninno, E. Tesliuc and A. Ouerghi (2008). *For Protection and Promotion: The Design and Implementation of Effective Safety Nets*. Washington DC: World Bank.
Handa, S., and B. Davis (2006). The experience of conditional cash transfers in Latin America and the Caribbean. *Development Policy Review*, **24**, 513–536.
Jindal, R., B. Swallow and J. Kerr (2008). Forestry-based carbon sequestration projects in Africa: Potential benefits and challenges. *Natural Resources Forum*, **32**, 116–130.
Massingarela, C. and V. Nhate (2006). Background paper for the Chronic Poverty

Report 2008–09 the politics of what works: a case study of food subsidies and the Bolsa-Escola in Mozambique. Chronic Poverty Research Centre, Brazil.

Medeiros, M., T. Britto and F. Veras Soares (2008). Targeted cash transfer programmes in Brazil: BPC and the Bolsa Familia, Working Paper 46, International Poverty Centre, United Nations Development Programme, Brazil.

Miranda, M., T. Porras and M. Moreno (2003). *The Social Impacts of Payments for Environmental Services in Costa Rica. A Quantitative Field Survey and Analysis of the Virilla Watershed*. London: International Institute for Environment and Development.

Munawir, S. and S. Vermeulen (2007). *Fair Deals for Watershed Services in Indonesia*. Natural Resource Issues 9. London: International Institute for Environment and Development (IIED).

Niesten, E. and R. Rice (2004). Sustainable forest management and conservation incentive agreements. *International Forestry Review*, **6**, 56–60.

Pagiola, S. (2008). Payments for environmental services in Costa Rica. *Ecological Economics*, **65**, 712–724.

Pagiola, S., A. Arcenas and G. Platais (2005). Can payments for environmental services help reduce poverty? An exploration of the issues and the evidence to date. *World Development*, **33**, 237–253.

Pagiola, S., A. Rios and A. Arcenas (2008a). Can the poor participate in payments for environmental services? Lessons from the Silvopastoral Project in Nicaragua. *Environment and Development Economics*, **13**, 299–325.

Pagiola, S., W. Zhang and A. Colom (2008b). Assessing the potential for payments for watershed services to reduce poverty in Guatemala. Presented at the Annual Meeting of the American Agricultural Economics Association, Orlando, Florida, July 27–29, 2008.

Pascual, U., R. Muradian, L.C. Rodriguez and A. Duraiappah (2010). Exploring the links between equity and efficiency in payments for environmental services: a conceptual approach. *Ecological Economics*, **69**, 1237–1244.

Pearson, R. and C. Alviar (2009). *Cash Transfers for Vulnerable Children in Kenya: From Political Choice to Scale-up*. New York: United Nations Children's Fund (UNICEF), Policy, Advocacy and Knowledge Management, Division of Policy and Practice.

Perrot-Maitre, D. (2006). The Vittel payments for ecosystem services: a 'perfect' PES case? Project Paper 3, International Institute for Environment and Development, London.

PIDS (2007). Conditional cash transfers: social assistance and human development combined. Philippine Institute for Development Studies. *The Economic Issue of the Day*, **7**(3).

Rawlings, L. (2005). A new approach to social assistance: Latin America's experience with conditional cash transfer programs. *International Social Security Review*, **58**, 133–161.

Reto-o-Reto (2006). Conservation in Kitengela: keeping land open for people, livestock and wildlife. Policy Brief number 1. ILRI, Nairobi.

Rios, A.R. and S. Pagiola (2009). Poor household participation in Payments for Environmental Services in Nicaragua and Colombia. MPRA Paper No. 13727. World Bank, Washington DC.

Sachs, J.D. and W.V. Reid (2006). Investments towards sustainable development. *Science*, **312**, 1002.

Smith, J. and S. Scherr (2002). Forest carbon and local livelihood: assessment of

opportunities and policy recommendations. Occasional Paper 37. Center for International Forestry Research, Bogor, Indonesia.
Suyanto, S., N. Khususiyah and B. Leimona (2007). Poverty and environmental services: case study in Way Besai watershed, Lampung Province, Indonesia. *Ecology and Society*, **12**(2), 13.
Venter, O., E. Meijaard, H. Possingham, R. Dennis, D. Sheil, S. Wich, L. Hovani and K. Wilson (2009). Carbon payments as a safeguard for threatened tropical mammals. *Conservation Letters*, **2**(3), 123–129.
Viana, V.M. (2008). Bolsa Floresta (Forest Conservation Allowance): an innovative mechanism to promote health in traditional communities in the Amazon. *Estudos Avancados*, **22**, 64.
Wendland, J.K., M. Honzak, R. Portela, B. Vitale, S. Rubinoff and J. Randrianarisoa (2010). Targeting and implementing payments for ecosystem services: opportunities for bundling biodiversity conservation with carbon and water services in Madagascar. *Ecological Economics*, **69**, 2093–2107.
World Bank (2008). World Bank Kenya Wildlife Conservation Leasing Demonstration Project. GEF Project Brief Report 37471. Washington DC.
World Bank (2009). Conditional Cash Transfers. Reducing Present and Future Poverty. A World Bank Policy Research Report. Washington DC.
Wunder, S. (2007). The efficiency of payments for environmental services in tropical conservation. *Conservation*, **21**, 48–58.
Wunder, S. (2008). Payments for environmental services and the poor: concepts and preliminary evidence. *Environment and Development Economics*, **13**, 279–297.
Wunder, S. and M. Alban (2008). Decentralized payments for environmental services: the case of Pimampiro and PROFAFOR in Ecuador. *Ecological Economics*, **65**, 685–698.
Wunder, S., S. Engel and S. Pagiola (2008). Taking stock: a comparative analysis of payments for environmental services in developed and developing countries. *Ecological Economics*, **65**, 834–852.
Zilberman, D., L. Lipper and N. McCarthy (2008). When could payments for environmental services benefit the poor? *Environment and Development Economics*, **13**, 255–278.

7. Unifying environmental and social protection: learning from PES and CCT in developing countries*

Luis C. Rodriguez, Unai Pascual, Roldan Muradian, Nathalie Pazmino and Stuart Whitten

INTRODUCTION

Environmental conservation and poverty alleviation are the two key policy areas that continue to receive increasing attention from governments, donors and non-governmental organizations (NGOs) in the developing world. The complex links between poverty and environmental degradation continue to be the focus of much research and debate and it is unlikely to be closed (Dasgupta 2003; Ruijs et al. 2008). However, there is increasing consensus about the policy instruments to be used to tackle both problems, albeit in a rather piecemeal approach. The current thrust is mostly on payments for ecosystem services (PES) to correct market failures that lead to excessive levels of environmental degradation (Pagiola et al. 2005; Wunder 2006, 2008; Bulte and Zilberman 2008; Engel et al. 2008; Pascual et al. 2010). On the other hand, conditional cash transfer (CCT) programmes are being implemented to correct market failures that lead to underinvestment in social protection in developing countries (Schubert 2005; Chapman 2006; Farrington and Slater 2006).

PES constitute economic transfers (in the form of monetary payments or in-kind rewards) aiming to compensate a target group (normally landholders) for the opportunity cost of providing positive environmental externalities (ecosystem services) to a group of stakeholders that can pay directly (by market creation or other informally devised institutional mechanisms) or indirectly (public funds) for the environmental services they expect to receive. These transfers, when properly designed and implemented, link the payment with a set of environmental conditions and land management practices that promote the provision of socially valuable environmental services (Bulte and Zilberman 2008; Engel et al. 2008; Muradian et al. 2010).

While PES programmes are primarily cost-effective conservation strategies (Ferraro and Simpson 2002), there is increasing interest in the potential positive effect of PES on the livelihoods of poor landholders (Niesten and Rice 2004; Grieg-Gran et al. 2005; Pagiola et al. 2005; Wunder 2008; Zilberman et al. 2008). However, PES programmes have also received criticisms due to their potential negative effects in changing the balance of local political or economic power, and generating changes in the behaviour of participants towards a vision that links conservation with exploitation of tenants over rent (Karsenty 2004; Kosoy and Corbera 2010; Pascual et al. 2010).

CCT programmes in developing countries, on the other hand, are primarily designed as a means for poverty alleviation or, more generally, as social protection mechanisms. CCT programmes are an innovative approach that promotes immediate and intergenerational poverty reduction based on the provision of monetary transfers to socially vulnerable households (Winters and Davis 2009). Similarly to PES, they are linked to compliance with a set of conditions aimed at promoting investment in human capital. A typical application is offering payments to households so that children attend school or are brought regularly to health centres (Das et al. 2004). However, it has also been argued that the direct provision of cash may generate disincentives to participate in the labour market and might not be a satisfactory solution to the poverty problem if the transfer promotes dependency on external aid (Medeiros et al. 2008). Cash transfers can be applied in versatile ways and have been adapted to different urban and rural contexts and circumstances (Skoufias 2005; Schubert and Slater 2006). There is emerging evidence of the positive impact of CCT on rural poverty alleviation (Rawlings 2005; Veras Soares et al. 2008) and indirectly on environmental quality because of investments of the transferred money in land rehabilitation activities (Standing 2007).

The positive effect of some PES schemes on the livelihoods of beneficiaries has been the starting point to explore the potential of PES as a poverty reduction tool (Grieg-Gran et al. 2005; Suyanto et al. 2007), a concept that has attracted the attention of donors interested in the pro-poor effects of the payments, not just in their positive environmental impacts (Wunder 2008). In a similar way, due to the versatility of CCT programmes, there is a call for revisiting them to take advantage of their positive effects on poverty reduction as well as their ability to help with more than human capital alone, such as the purchase of livestock and seeds, which can have important environmental consequences (Handa and Davis 2006; Cuesta

2007). Table 7.1 outlines the main characteristics of PES and CCT in terms of these defining aspects.

This chapter seeks to review the similarities of both approaches in order to provide a unifying framework for targeting efficient economic transfers to deal with both environmental protection and poverty alleviation. In the following section, we review the conceptual similarities between PES and CCT. We then present a framework for unifying both approaches. Some important considerations for the design of money transfers are discussed, and some recommendations are proposed for designing a new breed of joint PES–CCT programmes in rural areas of developing countries.

PES AND CCT: FROM CONCEPT TO PRACTICE

Although both have different objectives, both PES and CCT schemes have similarities in terms of their respective conceptual designs. We argue that, from a broad perspective, both programmes can be considered market-based interventions aimed at internalizing an externality in which a payment or reward-in-kind is transferred to a number of beneficiaries previously identified in a targeting exercise as fulfilling a set of conditions. Table 7.2 presents a comparison of transfer amounts, target mechanisms, conditions and monitoring of compliance of PES and CCT programmes in countries of Latin America, Asia and Africa.

PES AND CCT AS MARKET-BASED INTERVENTIONS

Both of these policy interventions aim at correcting market failures (underprovision or underdemand of a valuable environmental or social service, respectively) by adjusting the behaviour of individuals through economic incentives. In CCT programmes, interventions targeted at poor populations are designed to be compatible with individual household decisions about investment in human capital with wider social preferences, mainly in the areas of education and health (World Bank 2009). It is argued that an increase in education or health by an individual generates positive benefits for others. These positive externalities are not rewarded by market forces, therefore it is expected that poor households underinvest in human capital from a social optimum perspective (Das et al. 2004; de Janvry and Sadoulet 2004). The core justification for the design and implementation of PES schemes is similar. It is aimed at internalizing environmental externalities generated by individual landowners due to their underprovision

Table 7.1 Comparison of PES and CCT schemes from a conceptual perspective

	PES	CCT
Market-based instruments	Designed to internalize an environmental externality when individual decisions do not match social preferences because of market failure.	Interventions designed to match individual decisions with social preferences internalizing direct and learning externalities mainly in the areas of education and health.
A payment or reward in-kind is transferred	National programmes financed by public funds from taxes and donors. Private schemes might also be implemented as markets for ecosystem services. Payment should compensate for the foregone revenue and opportunity cost of adopting land management activities for the provision of ecosystem services. Cash or in-kind payments. Payment frequency regular and predictable. Designed to reduce entry barriers towards environmental protection. Payments contribute to safety nets and mitigate risk.	Financed by public funds from tax revenues and international lending from donors. Private schemes can participate from the supply side, as providers of education and health services. Payment covers the direct costs of sending children to school or to medical check-ups, and the child income lost due to school attendance rather than working. Cash or in-kind payments. Payment frequency regular and predictable. Designed to reduce entry barriers towards the use of education and health systems. Payments contribute to safety nets and mitigate risk.
Beneficiaries identified and targeted	Target to increase payment efficiency. Main criteria: environmental quality. Geographic. Target is household or plot.	Target to increase payment efficiency. Main criteria: poverty and vulnerability. Geographic. Target is household.
Beneficiaries comply with a set of conditions	Conditions designed to adjust individual behaviour to match social environmental objectives. Design includes mechanisms to verify compliance and define sanctions.	Conditions designed to adjust individual behaviour to match socio-economic objectives of the society. Design includes mechanisms to verify compliance and define sanctions.

Table 7.2 Comparison of PES and CCT schemes implemented in selected countries of Latin America, Asia and Africa. Household benefits are given in US$ and are financial and in-kind-equivalent values

	Programme Name	Household Benefit per year US$*	Targeting	Conditions	Monitoring	Sources
Latin America						
Ecuador						
CCT	Bono de Desarrollo Humano	360	Proxy means	Bimonthly health visits. 90% school attendance.	No verification of compliance.	World Bank (2009)
PES	PROFAFOR	252	Geographic	Active plantation management, including fire control and surveillance, and keeping out livestock.	At the plot level, all contracted areas are visited at least once annually. Certification companies scrutinize the annual carbon uptake. Members would legally have to reimburse the payments received if they do not fulfil the terms.	Wunder and Alban (2008)
Colombia						
CCT	Familias en Accion	552–768	Geographic Proxy means	Growth control and development checkups – 80% school attendance.	Bimonthly verification. School principal monitoring attendance.	World Bank (2009)

PES	Silvopastoral Project	607	Geographic Self-selection of individuals with minimal farm and herd size criteria	Maintain or switch to land uses that provide environmental services. Output based system. Payments are made according to the increase in an Environmental Service index.	Annual payments are made after land use changes have been monitored in the field. Switch to land uses that reduce service provision would incur payment reduction.	Rios and Pagiola (2009)
Brazil CCT	Bolsa Familia	506–703	Geographic. Means testing	Vaccine schedules, regular check-ups for children. Check-ups for pregnant and lactating women. 85% school attendance. Participation in parent–teacher meetings.	Bimonthly verification on education, using information consolidated by local municipalities. Twice a year verification on health, using information that health providers entered into a national database.	World Bank (2009)
PES	Bolsa Floresta	517	Geographic. Self-selection	Commitment to zero deforestation. Participation in dwellers' associations.	Annual monitoring of satellite images and analysed by partnering institutions. Yellow and red cards – families who have deforested a crop area up to 50% larger than	Viana (2008)

Table 7.2 (continued)

Programme Name	Household Benefit per year US$*	Targeting	Conditions	Monitoring	Sources
Latin America				the crop area in 2007 will receive a 'yellow card' and must explain to the association the reasons for having increased deforestation. Those with a yellow card who continue deforestation in the following year will receive a 'red card' and the payment will be suspended. If the new crop area is extended more than 50% (compared to the 2007 crop area), a red card will be given immediately and payments cease. The families given either two consecutive yellow cards or three in alternate years will be excluded from the programme.	

Mexico CCT	Oportunidades	371–1068	Geographic Proxy means test	Medical checkups. Attendance in health and nutrition lectures. 80% monthly and 93% annual school attendance. Completion grade 12 before age 22.	Bimonthly monitoring of compliance using data collected by local education and health service providers.	World Bank (2009)
PES	PSA-CABSA Niños Heroes	545	Geographic Self-selection	Not be receiving support from any other PES programme. Make projects comply with rules established for small-scale afforestation and reforestation projects under the Kyoto Protocol's CDM. A forest management plan, and show that PES activities were additional (i.e., the development of land-use activities for the provision of environmental services could not have been possible without payments).	A monitoring process is considered in the design but rarely done. Participants might return payments in case of no compliance. Participants were entitled to a deferral in the application of sanctions if they showed that failure to comply was due to an uncontrollable reason. No sanctions have to date been imposed.	Corbera et al. (2009)

Table 7.2 (continued)

	Programme Name	Household Benefit per year US$*	Targeting	Conditions	Monitoring	Sources
Latin America						
Nicaragua						
CCT	Programme Atencion a Crisis	325–378	Geographic Proxy means test	The Design included health conditionalities but they were not implemented. 85% school attendance. The design included a cash transfer to the teacher.	Bimonthly verification using information collected from the service providers.	World Bank (2009)
PES	Silvopastoral Project	592	Geographic Self-selection of individuals with minimal farm and herd size criteria	Maintain or switch to land uses that provide environmental services. Output based system. Payments are made according to the increase in an Environmental Service index.	Annual payments are made after land use changes have been monitored in the field. Switch to land uses that reduce service provision would incur payment reduction.	Rios and Pagiola (2009)

Asia					
Cambodia					
CCT	Cambodia Education Sector Support Project	60	Geographic. Scoring by a committee	Enrolment in school and regular attendance. Maintain a passing grade.	School monitors ongoing attendance and passing grade once a year.	World Bank (2009)
PES	Biodiversity Conservation Payments	120–160	Geographic Self selection	Stop hunting key species. Abide a land-use plan.	Monitoring by villagers. Certified by external stakeholder. Output based, no service no payment. Self enforcement within the community.	Clements et al. (2010)
Indonesia					
CCT	Programme Keluarga Harapan	65–152	Proxy means test	Medical check-ups for children age 0-6 years and pregnant and lactating women. 85% school attendance.	Sporadic monitoring of compliance but designed to be verified every 3 months.	World Bank (2009)
PES	Tlekung and Cidanau Watershed payments	42–132	Geographic based on contribution to sedimentation	Tree planting at a density of 500/ha in identified critical land under their individual ownership. Survival of seedlings and tree maintenance. Put in place and maintain high quality terracing.	A team representing the involved parties verify planting and maintenance. An external intermediary facilitates contracts, payments and compliance. Output based payments are distributed after verifying that the target was met.	Munawir and Vermeulen (2007)

Table 7.2 (continued)

	Programme Name	Household Benefit per year US$*	Targeting	Conditions	Monitoring	Sources
Asia						
Philippines						
CCT	Pantawid Pamilyang Pilipino Programme	202–342	Proxy means test	Medical check-ups for children and pregnant women. 85% school attendance.	Quarterly monitoring of compliance using monthly data collected from the service providers.	World Bank (2009)
PES	Bakun Watershed Protection	31	Geographic	Adopt land management plan and tree planting but no specific target.	Private agreement between the involved parties for monitoring purposes.	Beria et al. (2009)
Africa						
Kenya						
CCT	Cash Transfer for Orphan and Vulnerable Children	252	Geographic Community target	Vaccine schedules, regular check-ups for children with growth monitoring and vitamin A supplement. 80% school attendance. Attendance of adults in awareness sessions.	Compliance verified at least bimonthly for children's conditions using information collected from service providers and once a year for adults at awareness sessions.	Pearson and Alviar (2009)
PES	Wildlife Conservation Lease Programme	306	Geographic Self-selection	Retain ownership of their land. Leave land open, uncultivated and not	Violation of conditions involves termination of payments. Payments might be restored if	Reto-o-Reto (2006), World Bank (2008)

Mozambique					
CCT	Bolsa Escola	21	Geographic Household evaluation by the Ministry of Education	90% School attendance	subdivided. Graze livestock sustainably. Share both pasture and water among livestock and wildlife. Allow free movement of livestock and wildlife. fencing is removed. Independent monitoring and evaluation of programme implementation, lease compliance and impacts, with stakeholder participation. School attendance is registered by teachers and monitored monthly by representatives of the Ministry of Education.	Massingarela and Nhate (2006)
PES	Nhambita Community Carbon	34	Geographic after evaluation of conditions for carbon sequestration	Involvement in forest and fire management, forest rehabilitation or agroforestry. Restrict timber extraction to the amounts defined by a resource inventory.	Monitoring of compliance is carried out by community technicians with support from the carbon broker technical team. International agencies audit the carbon. Community/broker disputes are resolved by consultation, and individual/broker disputes are resolved dependent on the infraction based on guidelines defined by the contracts.	Jindal et al. (2008)

of ecosystem services (Pagiola et al. 2002; Engel et al. 2008; Muradian et al. 2010). PES are aimed at aligning private and social incentives for investment in environmental assets through direct payment or reward to individual landholders, thus facilitating the supply of improved environmental conservation outcomes.

THE TYPE OF COMPENSATION, PAYMENT OR REWARD

We argue that both payment mechanisms are broadly similar in terms of (1) the origin of the funds, (2) the estimation of the amount to be transferred, (3) the payment vehicle, and (4) the frequency of the payments.

First, there are similarities with respect to the origin of the funds. Most CCT programmes are financed by public funds from tax revenues complemented by international lending from donors. For example, the consolidation and expansion of the Mexican nationwide PROGRESA programme (currently known as Oportunidades) was partially funded by a US$1 billion loan from the Inter-American Development Bank and US$1.3 billion collateral from the Mexican Government (IADB 2002). In fact, the dependence of many CCT programmes on external financial sources is a major concern for their long-term viability, particularly as they can become increasingly an integral part of a country's social assistance strategy (Rawlings 2005).

Something similar occurs for long-term, large-scale publicly funded PES programmes such as those implemented in Costa Rica, Mexico or China, in which the state makes use of fiscal instruments, mainly taxes, and donor funds, to pay providers of ecosystem services that are strategic for those countries. For instance, the Costa Rican programme, Pago por Servicios Ambientales (PSA), was originally implemented using revenues from fuel taxes and is currently co-financed through external funds from World Bank loans and Global Environment Facility (GEF) grants which cover about 45% of the PSA total running costs (Pagiola 2008). This type of public PES scheme is seen as less efficient than when the purchaser is a private stakeholder, but public schemes are generally larger in scope and benefit from legitimacy conferred by the state, which many private schemes lack (Wunder 2005).[1]

Second, the amount of the payment is critical for the cost-effectiveness of the programme. Interestingly, despite the different programme objectives, both provide similar levels of payments (see Table 7.2). In the case of CCT, the level of payment is generally associated with the depth of poverty being addressed. An international rule of thumb is that poverty-motivated payments should represent between 20% and 40% of the minimum per capita income considered adequate (total poverty line) in order to be meaningful to the beneficiaries, but most CCT programmes transfer less than 20% (Handa and Davis 2006). For instance, with CCT programmes targeting investments in education, payments ought to cover the direct costs of sending children to schools, including fees, supplies and transport, plus the opportunity cost of their time, i.e. foregone income of children (PIDS 2007). A review of the education- and health-related CCT literature indicates that this amounts to about US$15 per month for educational needs and about US$10 for the health needs of each child enrolled in the programme. Payments may increase by up to 90% to support specific actions for disadvantaged groups, such as promoting female secondary education (Gökalp 2006; Barrera-Osorio et al. 2008). Although payments are made per child, there is normally a maximum cap on transfers to a single household. For example, in Latin American CCT programmes, the total payment per family represents 8% to 23% of the per capita income poverty line, with most programmes providing transfers of about 15%, or between 10 and 30% of a poor household's expenditure (Handa and Davis 2006).

With PES, the payment is designed to compensate the landholder for foregone revenue and the opportunity cost of adopting the land conservation activities required by the scheme. The contribution of PES to household income can be significant. For instance in Mexico's PSAH programme (Pago por Servicios Ambientales Hidrológicos), the annual payment to households is about US$27 per hectare in regions where most households earn less than US$7.50 per day. Areas with specific types of land cover could receive payments up to 30% higher. In Ecuador, the income accrued to the beneficiary households of the PROFAFOR PES programme represents over US$21 per month per household, while in the PES programme in Virilla, Costa Rica, the transferred amounts might represent about 16% of household income (Miranda et al. 2003; Munoz-Piña et al. 2008). In addition, the contribution of PES can be significant at the community level. For example, in Zimbabwe the payments from the CAMPFIRE programme for wildlife conservation may constitute up to 24% of all locally earned income, and exceed all other forms of local income and government grants (Bond 2001). A comparison of the per household annual transferred amounts of CCT and PES programmes implemented in various countries of Africa, Asia and Latin America, where both can be found, is presented in Table 7.2.

Third, the selection of the payment vehicle for CCT and PES schemes responds to the same factors, including project objectives, market assessments, cost-effectiveness and beneficiary preferences. Both social and environmental programmes use cash or in-kind transfers, e.g. food, as payment vehicles. The Mexican CCT programme Oportunidades or the Paraguayan Tekopora is an exemplar in the use of cash as the main payment mechanism.[2] By contrast, the Vulnerable Group Development project operating in rural Bangladesh, which distributes fortified food to disadvantaged women, is an example of using in-kind transfers where beneficiaries prefer food over cash and where the existing infrastructure of storage depots facilitate the logistics (Ahmed et al. 2004). Cash is the most common form of payment in PES, for example in the PROFAFOR project in Ecuador, each beneficiary landholder receives cash payments based on an estimate of the quantity of carbon sequestrated in their land (Wunder and Alban 2008). Other PES projects involve in-kind payments, such as those implemented in Santa Rosa, Bolivia, where one artificial beehive is given each year to participants in exchange for forest conservation in a buffer zone around a national park (Asquith et al. 2008).

Despite the different nature of PES and CCT programmes, cash transfers are often preferred by the implementing agencies because they give households discretion over how best to spend the money, e.g. on food, healthcare, housing or other services that are needed. In addition, cash payments can avoid the creation of secondary markets because they reduce price distortions and are more cost-effective than other methods of support. They have lower transaction costs and are flexible in that they allow policy-makers to adjust the amount over time and across populations (World Bank 2009).

Fourth, the frequency of the payments affects the final impact of the scheme. Payments in both PES and CCT schemes are regular, to make entry into the schemes easy and relatively risk-free for beneficiary households. For example, the beneficiaries of the Colombian CCT Programme Familias en Acción receive a bimonthly payment for ensuring that a child achieves at least 80% school attendance. In this case, which is typical, in order to overcome potential barriers to participation, one third of the bimonthly payment is retained in a savings account that families cannot access until a week before enrolment for the next school year (Fizbein and Schady 2009).

With PES programmes, the schedule of payments is commonly adapted to take into account the constraints on investment by landowners. For instance, an initial payment is normally made for the services already provided by current land uses. These first instalments help to remove potential entry barriers and capitalize landholders so they can finance the implementation of the desired new land management practices (e.g. Pagiola et al. 2005; Wunder and Alban 2008).

In addition, PES and CCTs play important roles as safety nets during times of economic hardship. There is some evidence about a CCT role in mitigating risks of hardship, smoothing consumption patterns and reducing the probability that the poor become poorer and increase the poverty gap (Handa and Davis 2006). PES form part of the risk management strategy of landholders because fixed, regular and predictable payments help to reduce income fluctuations when crop yields vary with weather and market conditions (Pagiola 2008).

IDENTIFYING GROUPS AS POTENTIAL RECIPIENTS OF COMPENSATION OR REWARD PAYMENTS

Although CCT and PES schemes are based on voluntary participation, they are designed and targeted to beneficiaries following a prior exercise that assesses ways of enhancing the effectiveness of the scheme in terms of performance and value for money. The choice of scheme depends on information available about the potential beneficiaries and the objective of the intervention. Both CCT and PES schemes rely on geographic as well as household (or plot) targeting approaches. Geographic targeting is relatively simple to administer because different areas are ranked by some indicative measure relevant to either CCT intervention, e.g. infant mortality rates, education levels, access to water and electricity or PES-related goals, e.g. biome distribution, biodiversity, and landscape connectivity. Resources are then allocated in proportion to their expected impacts. Hence, in terms of CCT programmes, regions with poorer human capital indicators receive higher per capita transfers. In PES programmes, larger payments are generally made for land areas having higher potential to supply ecosystem services. For example, Honduras and Peru's social programmes use geostatistical data and human development indices to identify regions where poverty interventions might best be implemented (Glewwe and Olinto 2004; Schady 2002). In an analogous way, PES programmes use georeferenced data and ecological indices to identify priority areas and to define plots for implementation of payments (Imbach 2005; Asquith et al. 2008).

At the household level, CCT programmes use different approaches to identify targets based on poverty or social category. The use of a proxy means test is the preferred option to select beneficiaries when their actual income in not known (World Bank 2009). This approach uses household data and econometric models appropriate to selected socio-economic variables, and these are used to predict household income and, in turn, to rank the eligibility of potential participants (Perez-Ribas et al. 2008). Likewise, environmental metrics are frequently used in many PES and other conservation payment programmes. Such metrics utilize site-specific biological and landscape data and models to predict ecological outcomes, e.g. enhancement of biodiversity and provision of ecosystem services, and express these as a single unit. The metric value is then used to compare and rank the expected environmental benefits of different plots and select those that provide the most benefits per dollar invested in the programme (Stoneham et al. 2003).

The Mexican CCT programme Oportunidades uses the proxy means test to classify households from previously identified geographic areas as eligible for participation ('poor') or ineligible for participation ('non-poor'). These classifications are based on information collected from socio-economic surveys (Skoufias and McClafferty 2001). An analogous example for PES is the Victorian Bushtender programme, which uses a 'habitat hectare scoring method' as the environmental metric to allocate funds for biodiversity conservation (Department of Sustainability and Environment 2008). For many PES carbon schemes like the PROFAFOR project in Ecuador, the payment amount was originally estimated using a metric based on vegetation types and soil criteria, and later complemented by field measures and model outputs (Wunder and Alban 2008).

CONDITIONALITY CRITERIA

Conditions for participation in both CCT and PES schemes, also known as 'co-responsibilities', are designed to compel individuals to conform to the programme goals. By imposing conditions, policy-makers provide incentives for households to take actions that they would not ordinarily take on their own (Das et al. 2005). However, conditions can be very expensive to implement and verify, and can also be constrained by the capacity of stakeholders to put them into practice (Schubert and Slater 2006).

In the Mexican CCT programme Oportunidades, bimonthly payments are made subject to school attendance and health controls. While these conditions can increase the demand for health and education services, they can also potentially reduce the capacity of beneficiaries to choose providers, ultimately affecting the success of the intervention. This is because the quality of services influences the decision of beneficiaries to continue with or drop out of the programme (Alvarez et al. 2008). In the case of PES schemes, the imposition of conditions can potentially increase the supply of a given ecosystem service but at the same time reduce the capacity of landholders to innovate or choose alternative management practices.

In most cases, the estimated transferred amounts per household indicate that the level of benefits from CCT is comparable with the transfers from PES programmes, suggesting the high potential of PES schemes to contribute to poverty alleviation initiatives when they are properly targeted to achieve that goal (Table 7.2). For example, in Kenya, the Cash Transfer for Orphan and Vulnerable Children provides families with about US$250 per year, with the conditions of attending regular medical check-ups and sending children to school, while a PES scheme designed to conserve wildlife has the condition that action must be taken to promote environmental protection. This scheme provides the average beneficiary family with a payment of US$306 per year, which is usually invested in medicines and school fees (World Bank 2008).

Table 7.2 also highlights the large differences in levels of benefits provided by CCT in middle income countries such as Mexico and Brazil as opposed to poorer countries like Nicaragua, Cambodia or Mozambique. Payments from PES schemes also show large differences according to the type of ecosystem services considered. Programmes such as the Ecuadorian PROFAFOR or the Mexican PSA-CABSA, established in 2004,[3] pay for global environmental benefits such as carbon sequestration and biodiversity conservation. For example, in Cambodia, biodiversity projects provide larger payments to beneficiaries than programmes providing local benefits such as watershed protection. This is because the biodiversity programmes are linked with international carbon prices or wealthy foreign tourists, while the value of the latter programme is constrained by the willingness to pay (WTP) of local communities in developing countries.

DESIGNING UNIFIED CONDITIONAL PROGRAMMES FOR ENVIRONMENTAL PROTECTION AND POVERTY ALLEVIATION

A unified CCT–PES payment scheme, called Payments for Environmental and Poverty Alleviation Services (PEPAS), is not expected to follow a simple 'one size fits all' design. Rather it needs to be adapted to reflect certain criteria such as differences in setting up both conservation and poverty alleviation goals, their relative importance, and particular environmental, socio-economic and political contexts in which the unified scheme is to be implemented. We suggest that a unified scheme should take into account the following aspects: (1) the nature and level of the payment transferred; (2) beneficiaries previously identified in a targeting exercise; and (3) whether they fulfil particular conditions designed to promote the desired objectives.

Designing the Nature of PEPAS

The level of payment

Our review suggests that the amounts transferred to the beneficiaries of CCT and PES schemes are comparable but the transfer rules differ. Payments made in CCT schemes are not at the scale required to have a significant impact on poverty (Grosh et al. 2008), and payments made in PES schemes are usually closer to the minimum willingness to accept (WTA) of environmental service providers than they are to the maximum WTP amount of the service receivers. To reach both environmental and social objectives, a unified payment scheme should define a level of payment consistent with the need to reduce the poverty level of the beneficiaries and to cover the required investments in conservation.

The level of payment in a PEPAS programme should cover the direct costs that will be incurred by landholders who participate in the programme. These will include the costs of sending children to school and the costs of implementing certain land management techniques, as well as their opportunity costs, for example, income foregone by children owing to attending school and not working, and the revenues foregone by landholders who adopt conservation activities. Due care should be taken, however, as summing these costs might not be appropriate because the direct cost of labour required for conservation activities might be the same income amount that is lost through school attendance.

The payment vehicle and frequency of transfers

The method of payment is a critical component in the design of a unified scheme. Unless rejected by the beneficiaries themselves, cash might be preferred because of the lower transaction costs and its capacity to address problems of incomplete information. For example, in a scheme aimed at reducing poverty and enhancing ecosystem services in an agricultural area, PEPAS could be transferred as investments in relevant infrastructure or technology requested by the beneficiaries. This might include improved access to markets, enhanced irrigation systems and introduction of new high-yield crops that can help farmers overcome barriers to increased revenue. However, these indirect ways of making payment might provide incentives for free-riders, resulting in the same problems that made former conservation approaches such as Integrated Conservation Development Projects unsuccessful (Wells et al. 1998). Therefore, we suggest that PEPAS should provide cash to beneficiaries who can later invest the money based on their needs and preferences, including meeting existing conditions of the scheme. Direct payments are considered the most cost-effective mechanism for conservation (Ferraro and Simpson 2002) and are also preferred in poverty alleviation initiatives (Grosh et al. 2008).

For both CCT and PES schemes, the regular and predictable frequency of payments helps beneficiary households reduce entry barriers and mitigate risk. While in general, PES schemes tend to include an up-front payment to cover, for example, the initial cost of an environmental conservation programme, CCT schemes tend to include delayed payments or accrual of savings to encourage the delivery of social objectives. We suggest that a unified scheme should be designed to include an up-front payment to cover part of the costs of the environmental component. The release of the remaining funds (remnant PES funds and the social assistance CCT component) should be designed to facilitate delivery of the objectives of the intervention. These objectives will explicitly include making resources available at the appropriate moment, e.g. funding medical check-ups when children start school, funding reforestation of a slope before the rainy season, and adopting tillage practices when harvest is finished, as well as assisting households to smooth their income and consumption cash flows. This last point requires that programme designers should have a good understanding of the poverty dynamics of the beneficiaries and of the dynamics of local environmental systems, i.e. factors such as drought frequency, seasonality, crop prices, health status and seasonal unemployment.

Targeting in PEPAS
A key aspect in the design of a unified payment scheme is to promote programme efficiency by transferring available resources to the set of beneficiaries that generate the largest societal benefits. Demonstrating efficiency gains from targeting is not a simple undertaking, considering that the poor are not a homogeneous group and that poverty is a multi-dimensional concept that includes subjective components. There are also large uncertainties about the level of supply of ecosystem services after adoption of the proposed conservation actions (Matin and Hulme 2003; Pascual et al. 2010). Errors might be expected in both environmental and socio-economic variables. In combining social and environmental goals in a unified scheme, the objective of uniform and universal provision of benefits to the whole population is weakened. This is because, although everyone might be eligible for a given level of social protection, not everyone is likely to be a provider of ecosystem services. Therefore a set of social and environmental eligibility criteria should be developed in consultation with beneficiary communities in order to define a set of rules to distinguish who is able to participate in the programme and who is eligible to receive payments. This is especially needed in regions where the incidence

140   *Values, payments and institutions for ecosystem management*

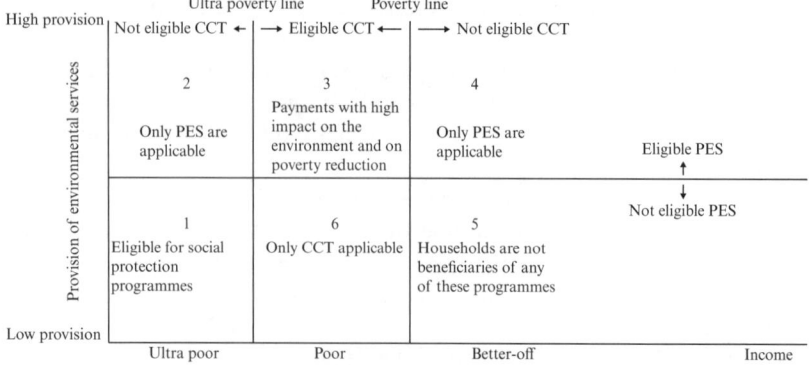

Source: Rodriguez et al. (2011).

Figure 7.1   Unified targeting for environmental and social protection payments

of poverty is high and everyone could be considered eligible for support (Semu-Banda 2009).

The possibility of a common targeting approach between PES and CCT schemes is exemplified by Figure 7.1. The vertical axis indicates the level of the provision of ecosystem services. The horizontal axis indicates the level of well-being that can be measured by different indicators. For the purpose of explanation, income is used as the main well-being variable. On the one hand, PES schemes are assessed using a metric for land suitability based on its level of provision of ecosystem services. On the other hand, CCT schemes are designed using a metric comprising a set of socio-economic criteria to target potential beneficiaries, who are usually a segment of the poor population, but not the poorest of the poor. These ultra poor are usually ineligible for CCT because their level of deprivation and vulnerability necessitates other forms of social assistance initiatives, such as non-conditioned payments, asset transfers or social mobilization initiatives (Matin and Halder 2004; Matin et al. 2008).

A horizontal threshold defined by the environmental metric demarcates land eligible for PES, based on expected provision of services. A vertical line determined by the cut-off point of the proxy means, indicates whether a household falls under the poverty threshold and hence is eligible to be a CCT beneficiary. In addition, an extra vertical line representing ultra-poverty could be drawn to determine the segment of the population that, although poor, are usually ineligible for CCT. Thus, the horizontal line of the environmental metric intersects the vertical line of the cut-off point of the proxy means as well as the ultra-poverty line, determining six regions.

Any land below the environmental metric line, by definition is not eligible for PES, since that land does not provide sufficient ecosystem services. Thus, PES should only be implemented in regions 2, 3 and 4 of Figure 7.1. However, it has been argued that payments to the poorest people (region 2) and wealthiest people (region 4) will not generate additionalities and thus will be inefficient, since these groups have either negligible environmental impact (Wunder 2007) or might use the environment for recreation, and thus are unlikely to be involved in environmental degradation activities (Kosoy et al. 2007).

The area to the right of the line of the poverty threshold (regions 4 and 5) represents situations not eligible for CCT, since these households are not poor. The situations represented by the area to the left of the ultra-poverty line (regions 1 and 2) are not eligible for CCT because the deprivation level of the households requires other types of social assistance. Thus, CCT schemes should be implemented in situations represented between these two areas, i.e. by regions 3 and 6 in Figure 7.1. While CCT schemes in these areas might be considered pure poverty alleviation programmes without an a priori environmental objective, there is evidence that in environmentally marginal areas in region 6, social payments, such as the 'Cash for Relief Programme' in Ethiopia, have had positive environmental benefits, since the beneficiaries usually invest the money in rehabilitating land, restoring its productivity and helping to regenerate livelihoods (Cuesta 2007; Standing 2007). Nevertheless, these environmental benefits are not expected from payments implemented in severely degraded areas where the rehabilitation of the land requires higher investments than those feasible with the payment under a CCT scheme. In areas like this, where environmental poverty traps are also associated with technological and infrastructure bottlenecks, larger investments and long-term political support are required to improve well-being (see Matin et al. 2008).

Since human well-being has environmental and socio-economic components and income alone is a poor indicator of welfare if not accompanied by adequate levels of provision of ecosystem services (AIHW 2007), identifying overlapping areas where CCT and PES schemes may operate jointly might also help in selection of regions for designing unified payment schemes. Communities in region 3 of Figure 7.1 are examples where investments are likely to generate the greatest social impacts for a given transfer amount. If appropriate social assistance and environmental policies are combined with adequate information systems and institutions in place to regulate PES, stakeholders can expect high levels of environmental and poverty benefits. This would increase the demand for education and health services in the target population, leading to project monitoring and coordination by environment health and education sector personnel.

The choice of targeting mechanism to select beneficiaries should take into account the transaction costs associated with designing and implementing the payment scheme. On the one hand, the cost of targeting may be high considering the requirement for both social and environmental information, but on the other hand, the public and private costs of administering the scheme should be lower because applicants are applying for a single joint programme and not for two separate PES and CCT schemes. Indirect costs of targeting arise when the selection criteria induce households to change their behaviour in an attempt to become beneficiaries of the programme, e.g. purposely changing land use or altering labour patterns to reduce their income in order to meet eligibility criteria. The resulting costs might be positive or negative depending on the relative contribution of environmental or social funds to the overall transfer.

We suggest that a cost-effective approach for unified payments should involve the coordinated use of both geographic and household (or plot) targeting. The initial step should consider a geographic target that combines georeferenced layers containing socio-economic data and environmental information. The resulting maps could prove helpful in selecting prime regions for investment, e.g. areas of high biodiversity and high incidence of poverty. Within a geographically targeted region, the selection of beneficiary households might be done using both proxy means to rank the participants for eligibility, and environmental metrics to rank the expected biodiversity benefits of different plots. From these criteria, a unified PEPAS should be implemented if there are clear correlations between the provision of ecosystem services and poverty incidence in a region. If this link does not exist, independent PES and CCT schemes should be preferred instead.[4]

Setting conditions in PEPAS
As in individual PES and CCT schemes, unified PEPAS schemes could be linked to a set of environmental and social conditions designed to adjust individual behaviour towards the delivery of conservation and poverty alleviation objectives. From a project perspective, conditional requirements are expensive to implement and monitor and, for unified payment schemes, the expected costs could be higher than for individual PES or CCT schemes. Participants may find themselves exposed to a large number of diverse conditions that are expensive and difficult to fulfil. In addition, the inclusion of both environmental and socio-economic condi-

tions can represent a challenge to implementation in areas with minimal infrastructure. For example, stakeholder skills and experience in monitoring compliance may be insufficient or housed elsewhere.

If costs are expected to be high, it may be appropriate to include an analysis of stakeholder institutions, including the ability of local stakeholders in the region to fulfil conditions required in a unified scheme. In particular, it is important to ensure that conditions can be met in practice. For example, meeting conditions of school attendance or medical check-ups will require sufficient schools and medical facilities in the region. Similarly, a condition requiring reafforestation should consider the availability of seedlings in nurseries or the capacity of landholders to produce seedlings.

A unified programme may find its costs lower than the combined costs of independent PES and CCT schemes operating in the same region. This is due to the fact that administrative costs of the joint intervention might be shared, and participating households may find it easier to demonstrate compliance within a single programme rather than across more than one. A way to reduce costs or ease monitoring may be found through, for example, agreements between different government agencies, such as ministries of health, education and agriculture. In some cases, there could be opportunities to involve private companies where they have better infrastructure and more capacity to provide services than state organizations. Such public–private alliances can help overcome the deficiencies of government agencies but should be subject to appropriate accountability and measures to minimize opportunities for political manipulation of the programme.

CONCLUSION

Donors, research and development organizations and practitioners are constantly looking to simplify the cost-effectiveness of schemes that promote environmental conservation and poverty alleviation (Muradian et al. 2010). In this context, numerous CCT and PES initiatives are being implemented at different scales with either social or environmental objectives. These have provided increasing evidence of many positive environmental side-effects of social interventions, and positive social impacts of environmental interventions (Niesten and Rice 2004; Standing 2007). Despite the similarity of some key design aspects of both types of scheme, little has been done to combine their designs into a unified payment scheme that would promote greater overall positive impacts on the environment and livelihoods.

Programmes for PES and CCT share many similarities in their concept, design and implementation. Both can be considered market-based interventions, in which a payment amount is transferred to a number of beneficiaries previously identified in a targeting exercise. The design of a unified PEPAS scheme should thus consider rewarding behaviour by targeting of payments above a fixed base level, while considering additional benefits for poorer families and households that provide enhanced environmental services. In addition, the targeting mechanism should consider the coordinated use of both geographic and household or plot targeting. There is a need for mapping of areas where poverty and high levels of provision of ecosystem services overlap and then applying proxy means tests and environmental metrics to rank households and select beneficiaries that fulfil the poverty and environmental criteria of eligibility. However, the inclusion of both environmental and socio-economic conditions might be expensive and their implementation and verification might represent a challenge in low capacity and poor infrastructure scenarios, requiring the collaboration between different government agencies and the creation of public–private alliances.

NOTES

* This article was published in *Ecological Economics* **70**(11), Rodríguez, L.C., Pascual, U., Muradian, R., Pazmino, N., Whitten S.,Towards a unified scheme for environmental and social protection: Learning from PES and CCT experiences in developing countries., 2163–2174. Copyright Elsevier BV (2011).
1. An important difference between PES and CCT is that PES schemes can also be designed and implemented as private schemes in which the buyers of the service pay the providers directly, usually on a local scale. Examples include for example those from Pimampiro, Ecuador, or Santa Rosa, Bolivia (Echavarria et al. 2002; Asquith et al. 2008).
2. Recent evaluations of these programmes suggest that the income effect of payments can only explain up to 50% of their impact. Other programme components such as talking and awareness sessions have delivered unexpectedly high impacts (Perez-Ribas et al. 2008).
3. PSA-CABSA stands for 'Programa para Desarrollar el Mercado de Servicios Ambientales por Captura de Carbono y los Derivados de la Biodiversidad y para Fomentar el Establecimiento y Mejoramiento de Sistemas Agroforestales'.
4. In practice, many PES schemes have been implemented in poor areas, and poor communities are therefore overrepresented among beneficiaries (Wunder et al. 2008). These PES schemes did not include specific mechanisms to target poverty, but the coincidence of poverty and environmentally rich areas might explain the unexpected welfare gains of these interventions. In these programmes, a second stage exercise in targeting households could increase the social benefits of the intervention beyond the unexpected poverty reduction effects of single PES schemes.

REFERENCES

Ahmed A.U., S. Rashid, M. Sharma and S. Zohir (2004). Food aid distribution in Bangladesh: leakage and operational performance. Discussion Paper 173. IFPRI, Washington DC.

Alvarez, C, F. Devoto and P. Winters (2008). Why do beneficiaries leave the safety net in Mexico? A study of the effects of conditionality on dropouts, *World Development*, **36**, 641–658.

Asquith, N. M., M.T. Vargas and Wunder, S. (2008) Selling two environmental services: In-Kind payments for bird habitat and watershed protection in Los Negros, Bolivia, *Ecological Economics* **65**, 675–684.

AIHW (2007). *Australia's Welfare 2007*. Cat. no. AUS 93. Canberra, Australia: Australian Institute of Health and Welfare.

Barrera-Osorio, F., M. Bertrand, L.L. Linden and F. Perez-Calle (2008). Conditional cash transfers in education: design features, peer and sibling effects evidence from a randomized experiment in Colombia. NBER Working Papers 13890, National Bureau of Economic Research, Inc. Cambridge, MA.

Beria, L., J. Laxman and M. van Noordwijk (2009). Can rewards for environmental services benefit the poor? Lessons from Asia, *International Journal of the Commons*, **3**(1), 82–107.

Bond, I. (2001). CAMPFIRE and the incentives for institutional change. In D. Hulme and M.W. Murphree (eds), *African Wildlife and Livelihoods. The Promise and Performance of Community Conservation*. Oxford: James Currey, pp. 227–243.

Bulte, E.H. and D. Zilberman (2008). Special Issue: Payment for ecosystem services, *Environment and Development Economics*, **13**, 245–439.

Chapman, K. (2006). Using social transfers to scale up equitable access to education and health services. DFID Background paper. Department of International Development, London.

Corbera, E., C. González Soberanis and K. Brown (2009). Institutional dimensions of Payments for Ecosystem Services: an analysis of Mexico's carbon forestry programme. *Ecological Economics*, **68**, 743–761.

Clements, T., J. Ashish, K. Nielsen, A. Dara, T. Seth and E.J. Milner-Gulland (2010). Payments for biodiversity conservation in the context of weak institutions: comparison of three programs from Cambodia, *Ecological Economics*, **69**(6), 1283–1291.

Cuesta, J. (2007). On more ambitious conditional cash transfers, social protection and permanent reduction of poverty. *Journal of International Development*, **19**, 1016–1019.

Das, J., Q.T. Do and B. Ozler (2004). Conditional cash transfers and the equity-efficiency debate. World Bank Policy Research Working Paper No. 3280. World Bank, Washington DC.

Das, J., Q.T. Do and B. Ozler (2005). Reassessing conditional cash transfer programs, *World Bank Research Observer*, **20**, 57–80.

Dasgupta, P. (2003). Population, poverty, and the natural environment. In K.-G. Mäler and J.R. Vincent (eds), *Handbook of Environmental Economics*, Vol. I. Amsterdam: Elsevier, pp. 191–247

De Janvry, A and E. Sadoulet (2004). Conditional cash transfer programs: are they really magic bullets? *Agricultural and Resource Economics Update*, **7**(6), 9–11.

Department of Sustainability and Environment (2008). *BushTender: Rethinking Investment for Native Vegetation Outcomes. The Application of Auctions for Securing Private Land Management Agreements.* East Melbourne, Australia: State of Victoria, Department of Sustainability and Environment.

Echavarria, M., J. Vogel, M. Albán and F. Meneses (2002). *Impact Assessment of Watershed Environmental Services: Emerging Lessons from Pimampiro and Cuenca in Ecuador.* Quito, Ecuador: Environmental Economics Programme, EcoDecisión.

Engel, S., S. Pagiola and S. Wunder (2008). Designing payments for environmental services in theory and practice: an overview of the issues. *Ecological Economics*, **62**, 663–674.

Farrington, J. and R. Slater (2006). Cash transfers: panacea for poverty reduction or money down the drain?, *Development Policy Review*, **24**(5), 499–511.

Ferraro, P.J. and R.D. Simpson (2002). The cost effectiveness of conservation payments. *Land Economics*, **78**, 339–353.

Fiszbein, A. and N. Schady (2009). *Conditional Cash Transfers: Reducing Present and Future Poverty.* Washington DC: World Bank.

Glewwe, P. and P. Olinto (2004). Evaluating the impact of conditional cash transfers on schooling: an experimental analysis of Honduras's PRAF Program. Final Report for USAID. International Food Policy Research Institute, Washington DC.

Gökalp, Y. (2006). Conditional cash transfers in Turkey: motivation, design, achievements, challenges, and the way forward. Presentation at the Third International Conference on Conditional Cash Transfers Istanbul, Turkey. June 2006.

Grieg-Gran, M., I.T. Porras and S. Wunder (2005). How can market mechanisms for forest environmental services help the poor? Preliminary lessons from Latin America, *World Development*, **33**, 1511–1527.

Grosh, M., C. del Ninno, E. Tesliuc and A. Ouerghi (2008). *For Protection and Promotion: The Design and Impkmentation of Effective Safety Nets.* Washington DC: World Bank.

Handa, S. and B. Davis (2006). The experience of conditional cash transfers in Latin America and the Caribbean. *Development Policy Review*, **24**, 513–536.

Inter-American Development Bank (IADB) (2002). Press release 16 January 2002. Available at http://www.iadb.org/exr/PRENSA/2002/cp1002e.htm.

Imbach, P. (2005). *Priority Areas for Payments for Environmental Services (PES) in Costa Rica.* Turrialba, Costa Rica: Centro Agronomico Tropical de Investigacion y Enseñanza (CATIE).

Jindal, R., B. Swallow and J. Kerr (2008). Forestry-based carbon sequestration projects in Africa: Potential benefits and challenges, *Natural Resources Forum*, **32**(2), 116–130.

Karsenty, A. (2004). Des rentes contre le développement? Les nouveaux instruments d'acquisition mondiale de la biodiversité et l'utilisation des terres dans les pays tropicaux, *Mondes en Développement*, **32**, 59–72.

Kosoy, N, M. Martinez-Tuna, R. Muradian and J. Martinez-Alier (2007). Payments for environmental services in watersheds: insights from a comparative study of three cases in Central America. *Ecological Economics*, **61**, 446–455.

Kosoy, N. and E. Corbera (2010). Payments for ecosystem services as commodity fetishism. *Ecological Economics*, **69**(6), 1228–1236.

Massingarela, C. and V. Nhate (2006). The Politics of What Works: A Case Study of Food Subsidies and the Bolsa-Escola in Mozambique. Background Paper for the Chronic Poverty Report 2008–09. Chronic Poverty Research Centre. Ministry of Planning, Government of Mozambique.

Matin, I. and S. Halder (2004). Combining Methodologies for Better Targeting of the extreme Poor: Lessons from BRAC's CFPR/TUP Programme. CFPR-TUP Working Paper Series No. 2. eSocialSciences. Available at: http://www.esocialsciences.org.

Matin, I. and D. Hulme (2003). Programs for the poorest: learning from the IGVD program in Bangladesh. *World Development*, **31**, 647–665.

Matin, I., M. Sulaiman and M. Rabbani (2008). Crafting a graduation pathway for the ultra poor: lessons and evidence form a BRAC program. Working Paper N 169, Chronic Poverty Research Centre (DFID), Manchester, UK.

Medeiros, M., T. Britto and F. Veras Soares (2008). Targeted cash transfer programmes in Brazil: BPC and the Bolsa Familia. Working Paper No. 46. International Poverty Centre, UNDP, Brazil.

Miranda, M., T. Porras and M. Moreno (2003). The social impacts of payments for environmental services in Costa Rica. A quantitative field survey and analysis of the Virilla watershed. International Institute for Environment and Development, London.

Munawir, S. and S. Vermeulen (2007). Fair deals for watershed services in Indonesia. Natural Resource Issues 9. International Institute for Environment and Development (IIED), London.

Muñoz-Pina, C., A. Guevara, J. Torres and J. Brana (2008). Paying for the hydrological services of Mexico's forests: Analysis, negotiations and results. *Ecological Economics*, **65**, 725–736.

Muradian, R., E. Corbera, U. Pascual, N. Kosoy and P.H. May (2010). Reconciling theory and practice: an alternative conceptual framework for understanding payments for environmental services. *Ecological Economics*, **69**, 1202–1208.

Niesten, E. and R. Rice (2004). Sustainable forest management and conservation incentive agreements, *International Forestry Review*, **6**, 56–60.

Pagiola, S. (2008). Payments for environmental services in Costa Rica, *Ecological Economics*, **65**, 712–724.

Pagiola, S., A. Arcenas and G. Platais (2005). Can payments for environmental services help reduce poverty? An exploration of the issues and the evidence to date from Latin America. *World Development*, **33**(2), 237–253.

Pagiola, S., N. Landell-Mills and J. Bishop (2002). Making market-based mechanisms work for forests and people. In S. Pagiola, J. Bishop and N. Landell-Mills (eds), *Selling Forest Environmental Services: Market-based Mechanisms for Conservation and Development*. Earthscan, London, pp. 261–290.

Pascual, U., R. Muradian, L.C. Rodriguez and A. Duraiappah (2010). Exploring the links between equity and efficiency in payments for environmental services: a conceptual approach, *Ecological Economics*, **69**(6), 1237–1244.

Pearson, R. and C. Alviar (2009). *Cash Transfers for Vulnerable Children in Kenya: From Political Choice to Scale-up*. New York: Division of Policy and Practice, United Nations Children's Fund (UNICEF).

Perez-Ribas, R., G.I. Irata and F. Vera Soares (2008). Debating targeting methods for cash transfers: a multidimensional index vs. income proxy for Paraguay's Tekopora program. International Poverty Centre. IPC evaluation note 2, Brasilia, Brazil.

PIDS (2007). Conditional cash transfers: social assistance and human development combined. Economic Issues of the day V 7, No 3. Philippine Institute for Development Studies, Makati.

Rawlings, L. (2005). A new approach to social assistance: Latin America's experience with conditional cash transfer programs. *International Social Security Review*, **58**, 133–161.

Reto-o-Reto (2006). Conservation in Kitengela: Keeping land open for people, livestock and wildlife. Policy Brief number 1. International Livestock Research Institute, Nairobi.

Rios, A.R. and S. Pagiola (2009). Poor household participation in payments for environmental services in Nicaragua and Colombia. MPRA Paper No. 13727. World Bank, Washington DC.

Rodriguez, L.C., U. Pascual, R. Muradian, N. Pazmino and S. Whitten (2011). Towards a unified scheme for environmental and social protection: learning from PES and CCT experiences in developing countries. *Ecological Economics*, **70**(11), 2163–2174.

Ruijs, A., R.B. Dellink and D.W. Bromley (2008). Economics of poverty, environment and natural-resource use. In R.B. Dellink and A. Ruijs (eds), *Economics of Poverty, Environment and Natural-Resource Use*. New York: Springer, pp. 3–15.

Schady, N.R. (2002). Picking the poor: indicators for geographic targeting in Peru. *Review of Income and Wealth*, **48**, 417–433.

Schubert, B. (2005). *Social Cash Transfers: Reaching the poorest. A contribution to the international debate based on experience in Zambia*. Eschborn, Germany: Deutsche Gesellschaft für Technische Zusammenarbeit.

Schubert, B. and R. Slater (2006). Social cash transfers in low-income African countries: conditional or unconditional?, *Development Policy Review*, **24**, 571–578.

Semu-Banda, P. (2009). Malawi: separating the ultra poor from the poor – why? Inter Press Service News Agency. Available at: http://www.ipsnews.net/news.asp?idnews=46848 (accessed 19 May 2009).

Skoufias, E. (2005). PROGRESA and Its Impacts on the Welfare of Rural Households in Mexico. Research Report No. 139. International Food Policy Research Institute, Washington DC.

Skoufias, E. and B. McClafferty (2001). Is PROGRESA working? Summary of the Results of an evaluation, The International Food Policy Research Institute, Washington DC.

Standing, G. 2007. How cash transfers boost work and economic security. Working Paper N 58. United Nations Department of Economic and Social Affairs, New York.

Stoneham, G., V. Chaudhri, A. Ha and L. Strappazzon (2003). Auctions for conservation contracts: an empirical examination of Victoria's bushtender trial. *Australian Journal of Agricultural and Resource Economics*, **47**, 477–500.

Suyanto, S., N. Khususiyah and B. Leimona (2007). Poverty and environmental services: case study in Way Besai watershed, Lampung Province, Indonesia. *Ecology and Society*, **12**, 13.

Veras Soares, F., R. Peres Ribas and G. Issamu Hirata (2008). Achievements and shortfalls of conditional cash transfers: impact evaluation of Paraguay's Tekoporã programme. IPC Evaluation note N° 3. International Poverty Centre, UNDP, Brazil.

Viana, V.M. (2008). Bolsa Floresta (Forest Conservation Allowance): an innova-

tive mechanism to promote health in traditional communities in the Amazon, *Estudos Avancados*, **22**, 64.
Wells, M., S. Guggenheim, A. Khan, W. Wardojo and P. Jepson (1998). *Investing in Biodiversity. A Review of Indonesia's Integrated Conservation and Development Projects*. Washington DC: World Bank.
Winters, P. and B. Davis (2009). Designing a programme to support smallholder agriculture in Mexico: Lessons from PROCAMPO and Oportunidades, *Development Policy Review*, **27**, 617–642.
World Bank (2008). Kenya wildlife conservation leasing demonstration project. GEF Project Brief Report 37471. World Bank, Washington DC.
World Bank (2009). Conditional cash transfers. Reducing present and future poverty. A World Bank Policy Research Report. Washington DC.
Wunder, S. (2005). Payments for Environmental Services: some nuts and bolts. CIFOR Occasional Paper 42. Center for International Forestry Research, Bogor.
Wunder, S. (2006). Are direct payments for environmental services spelling doom for sustainable forest management in the tropics?, *Ecology and Society* **11**(2) 23.
Wunder, S. (2007). The efficiency of payments for environmental services in tropical conservation. *Conservation Biology*, **21**, 48–58.
Wunder, S. (2008). Payments for environmental services and the poor: concepts and preliminary evidence. *Environment and Development Economics*, **13**, 279–297.
Wunder, S. and M. Alban (2008). Descentralized payments for environmental services: the case of Pimampiro and PROFAFOR in Ecuador. *Ecological Economics*, **65**, 685–698.
Wunder, S., S. Engel and S. Pagiola (2008). Taking stock: a comparative analysis of payments for environmental services programs in developed and developing countries. *Ecological Economics*, **65**, 834–852.
Zilberman, D., L. Lipper and N. McCarthy (2008). When could payments for environmental services benefit the poor? *Environment and Development Economics*, **13**, 255–278.

8. Exploring the potential of payments for ecosystem services for *in-situ* agrobiodiversity conservation*

Ulf Narloch, Adam G. Drucker and Unai Pascual

PAYMENT MECHANISMS FOR CONSERVING AGRO-BIODIVERSITY

Despite their crucial role in sustainable agricultural practices by contributing to agrobiodiversity conservation services, many traditional plant and animal genetic resources (PAGR) are lost at increasing rates from agricultural landscapes worldwide (Perrings et al. 2006; FAO 2007a; Hajjar et al. 2008).[1] There are many different demand and supply factors leading to a loss of agrobiodiversity on farms (Bellon 2004; Kontoleon et al. 2008). Demand for traditional PAGR decreases if inputs become available that facilitate the cultivation of commercial crops. Improved access to safety nets and non-farm incomes lowers the benefit from using traditional PAGR as a natural insurance mechanism. The supply of seeds and breeds of traditional PAGR is negatively affected by increased reliance on commercial farming systems, out-migration and erosion of traditional customs and exchange networks. We hypothesize that, in the main, farmers tend to replace a diverse pool of traditional crop varieties (also known as landraces) and livestock breeds with a few more financially profitable 'improved' (commercial) varieties and breeds, owing to the fact that incentives are often biased towards the latter, and that markets tend not to capture the public values of many agrobiodiversity conservation services (Gruère et al. 2008).

From an economics perspective, the conservation of agrobiodiversity requires that, where significant value to the public exists, this should be recognized, and mechanisms should be put in place to permit the capture of those values by the farmers who incur the conservation costs. Direct payments might prove to be most effective in conserving agrobiodiversity as they benefit wider society (Ferraro 2001; Ferraro and Simpson 2002) and are akin to the role of payments for ecosystems services (PES) in agrobiodiversity conservation (Pascual and Perrings 2007).

Although PES schemes have been hailed as a promising solution for conservation dilemmas, they have mostly been applied to conserve forests and their associated ecosystem services (Landell-Mills and Porras 2002; Engel et al. 2008; Wunder et al. 2008; Muradian et al. 2010). Nevertheless, there are examples of PES promoting ecosystem-friendly farm management practices in agricultural landscapes (Pagiola et al. 2007; Lipper et al. 2009).[2]

So-called PACS schemes (payments for agrobiodiversity conservation services) may be understood as a sub-category of agriculture-related PES that focuses on socially valuable and threatened PAGR. Interestingly, the consideration of PES for the promotion of PAGR is limited. Examples include the European Union's support for payments for threatened livestock breeds (under Regulations 1257/99 and 1750/99), as well as a Global Environment Facility-funded project in Ethiopia, which has been completed,[3] both of which are PES-like initiatives, although not described as such. The latter scheme paid farmers for conserving traditional varieties, and provided compensation based on foregone revenue owing to differences in yield between traditional and improved varieties. Wale (2008) draws on experiences from this project by analysing the financial opportunity costs of conserving traditional sorghum varieties. Similarly, Fuwa and Sajise (2009) evaluate the potential of incentive mechanisms for the conservation of rice varieties in the Philippines.

Apart from such limited examples, we are unaware of any further consideration of ad-hoc PES-like programmes in the context of PAGR or of any theoretical treatment of this important and potentially emergent topic. Acknowledging this research gap, we aim to fill it by developing a conceptual framework to shed light on the potential of PACS to result in least-cost and pro-poor agrobiodiversity conservation outcomes, especially in the context of poor rural communities in developing countries, where most threatened and valuable PAGR can be found.

In this chapter, we discuss some of the particularities of agrobiodiversity conservation, and elaborate on a potential PES-like solution to the underprovision of agrobiodiversity conservation services. We go on to consider some generic institutional constraints that would typically occur in the design of PACS schemes, discuss how to define a conservation goal, and assess PACS with regard to its potential effectiveness, efficiency and equity outcomes.

CONSERVING AGROBIODIVERSITY ON THE FARM

Agrobiodiversity conservation services have 'impure public goods' characteristics in that there is a private production component directly linked to farm-scale decision-making, and a public component (genetic information) that is not (Heisey et al. 1997; Smale et al. 2004). For farming communities, the private (i.e. rival and excludable) services associated with agrobiodiversity conservation are numerous. This is because species and varieties of crops contribute to income generation, the production of food, fodder, clothing, construction materials, medicines, fibre and fuel (Gruère et al. 2008). Additionally, risk-averse households use agrobiodiversity to spread the risk of agricultural production shocks due to weather variability or pest and disease outbreaks (Di Falco and Chavas 2009). This type of insurance service is also associated with agroecosystem resilience at a local level, and thus shares some characteristics of public goods (Heisey et al. 1997). Additional public services at a local level can be found in the socio-cultural traditions and identities associated with the use of certain traditional varieties and breeds (Nautiyal et al. 2008). Moreover, in a complex world of uncertainty and shocks due to rapid global economic and environmental changes, the intergenerational economic benefits from the conservation of genetic diversity are potentially extremely high (Bellon 2008).

Farmers contributing to conservation activities provide local and global public values as a positive externality for which they are not rewarded, while other members of society are able to benefit from these services. Consequently, there is a bias towards the provision of agrobiodiversity conservation services at a level that is less than socially optimal (Heisey et al. 1997; Pascual and Perrings 2007). This is similar to dilemmas in the provision of many other ecosystem services.

What distinguishes agrobiodiversity conservation services from many other ecosystem services is their significant direct private use value to the farmer, as PAGR are linked to the production of food and thus to income generation. It is in this context that attention has been drawn to the potential of existing agricultural market channels in promoting the use of threatened PAGR (Gruère et al. 2008). Under certain circumstances, regional and global consumers have been shown, through eco-labelling, certification, or origin schemes, to be willing to pay for agrobiodiversity-related products in order to satisfy specific tastes and preferences (Hermann and Bernet 2009; Krishna et al. 2010). However, the extent to which such marketing mechanisms are able to align the private incentives of farmers to maintain agrobiodiversity with optimal conservation levels is unclear, because public conservation values still might not be adequately incorporated in market prices.

A further common approach to tackle the loss of agrobiodiversity is found in *ex-situ* conservation strategies, for example the storage of genetic material in gene banks. Such static approaches involve the maintenance of an information stock, while *in-situ* conservation is associated with a dynamic process, i.e. not only enhancing the stock of information, but also its flow across time and space (Swanson and Göschl 1999). Although these two strategies are complementary, with neither able to ensure effective maintenance of all agrobiodiversity conservation services on its own, *in-situ* (on-farm) conservation strategies have so far gained relatively little attention (Brush 1989; Maxted et al. 2002). On-farm conservation would result in the sustained utilization of threatened crop varieties or livestock breeds within traditional farming systems, and thus the conservation focus is not on certain land uses as in most existing PES schemes, but on specific agricultural practices. Accordingly, on-farm conservation implies not only the cultivation of land and the generation and conservation of seeds, but also the maintenance of local traditions and agricultural knowledge (Brush 1995; Bellon 2004). As such, on-farm conservation of PAGR can be said to be subject to an evolutionary process in the field resulting from human selection (Perales et al. 2003). Economic interventions aiming at *in-situ* agrobiodiversity conservation should therefore be designed in such a way to directly influence farm-scale decisions to use threatened PAGR.

MATCHING SUPPLY AND DEMAND FOR AGROBIODIVERSITY THROUGH PACS

Payment for agrobiodiversity conservation services is an economic instrument that gives an incentive to farmers to use crops and livestock with lower than usual profitability but higher than usual conservation value. It tackles the market failures associated with agrobiodiversity conservation by increasing the private benefits gained from less-intensive production methods involving traditional crop varieties and livestock breeds through monetary or in-kind rewards. Given the special features of *in-situ* agrobiodiversity conservation, PACS schemes may have to incorporate the wholesale maintenance of traditional farming systems as a goal.

In highly intensive agricultural systems, the opportunity costs of managing traditional varieties and breeds can be high, so that farmers would require higher compensation payments under PACS. Therefore, PACS schemes could instead be aimed at existing areas of conservation where opportunity costs are relatively low. Such places tend to be located in comparatively remote areas of small-scale farming in developing countries. Small farms are recognized as playing key roles in the conservation of varieties and breeds with unique adaptive traits (e.g. disease resistance and drought tolerance) often bred over thousands of years of domestication across a wide range of environments (Smale 2006; Kontoleon et al. 2008).

The demand for agrobiodiversity conservation services is dispersed. Local farming communities benefit through the maintenance of traditional knowledge and culture, as well as from insurance values. Wider local benefits in terms of health and nutrition may also arise from the consumption of such crops, as local varieties are often associated with high nutritional values, while poor local consumers often tend to buy cheaper and less nutritious food products. In this context, local government may foster the use of traditional PAGR by distributing appropriate food products to schools (e.g. school-meal programmes in Tamil Nadu, India), or other public facilities.

At the global level, society as a whole is the beneficiary of conserving the intact genetic resource (option/quasi-option values) of agrobiodiversity. In addition, companies linked with agriculture may benefit through potential future product development (Di Falco et al. 2008). However, the extra revenue from agrobiodiversity conservation is normally not enough to fund larger-scale *in-situ* conservation efforts (Swanson and Göschl 2000).

As with any PES scheme, pure market transactions are unlikely to work when conservation service beneficiaries are not able or willing to pay owing to high transaction costs, difficulties in measuring service flows and incentives to free-ride (Engel et al. 2008). Therefore, governments, conservation agencies and international organizations that acknowledge the importance of agrobiodiversity conservation may have to take on the role of the service purchaser, and thus act on behalf of the service beneficiary. In contrast to conventional PES, such schemes would be government financed, as opposed to user financed (Pagiola and Platais 2007). Such funds might also be generated through businesses and agencies encouraged or required to provide funding to avoid, mitigate and offset their biodiversity impacts (Madsen et al. 2010), including those related to agrobiodiversity. Given that demand for conservation services can come from many different service beneficiaries, conservation programmes could incorporate a mixture of incentive instruments, including non-PES related marketing schemes (e.g. eco-certification schemes), in order to generate sufficient funding.

CONSTRAINTS IN THE DESIGN AND IMPLEMENTATION OF PACS SCHEMES

While PACS seems to provide a theoretically straightforward solution to the problem of agrobiodiversity loss, implementation of PACS schemes on the ground is more complex. The design of PES in general is associated with a number of generic institutional constraints (Ferraro and Simpson 2002; Wunder 2006, 2007) and PACS may share many of these. First, as with PES, PACS schemes may require the creation of new institutions in order to implement negotiation, transaction, monitoring and enforcement mechanisms. Yet any such new institutions are likely to interact with existing social systems, so that a PACS scheme would need to start with an assessment of the institutional setting in which it would operate (Narloch et al. 2012).

Second, land tenure issues need to be considered carefully. Tenure arrangements are location specific, shaped by historical and political factors, and in poor farming communities are often not solely based on private resource ownership. Many farmers whose practices have an impact on agrobiodiversity rely on public and communal lands over which they do not hold individual title (Eyzaguirre and Dennis 2007). If established in areas where land-use rights are weakly defined, PACS could potentially provoke conflict over tenure. For instance, more powerful farmers might oust smallholders from the land they use in order to take advantage of the rewards associated with the conservation programme (Landell-Mills and Porras 2002).

Third, as is the case for PES schemes, PACS schemes are expected to involve intermediaries to establish and maintain contact between different actors. They do this by providing new information, extending and linking existing networks, and supporting contract development. Although in user-financed programmes the market could play this role, in government-financed schemes, intermediaries would be able to exercise a dominant role in matching service providers and beneficiaries, and may even be able to influence prices (Kosoy and Corbera 2010). Such a situation in the context of PACS could hamper the efficient functioning of a scheme, because conservation outcomes would depend on the rules of the scheme which, in turn, might be influenced by the interests of the intermediary or by political considerations.

Fourth, with regard to monitoring and enforcement of PACS contracts, institutional arrangements need to be created that deal with baselines, verification of service delivery and sanctions in case of non-compliance. Such challenges are also faced by existing PES-like schemes, many of which are only loosely monitored, with payments not strictly conditional on service provision (Wunder 2007). Determining baselines requires the construction of easily understandable performance metrics, clearly associated with conservation services to allow evaluation of additionality over the contract period. Payments for conservation of PAGR can be linked to easily measureable and observable conservation units, such as specific animal breeds, or land area or the amount of seed associated with a certain local crop variety. However, in some cases, it may be difficult to distinguish clearly between similar breeds and varieties, particularly in the mixed cropping systems found in many indigenous communities. Furthermore, farmers often manage multiple plots distant from the village centre, in areas that are already remote, and this might add to the costs of verification.

Finally, it should be appreciated that many existing PES schemes face challenges in establishing strong scientific foundations for linking conservation activities with the provision of conservation services (Wunder et al. 2008; Kosoy and Corbera 2010). Similarly, relating the conservation of specific breeds or varieties and associated levels of genetic diversity per se is not necessarily straightforward (van de Wouw et al. 2009), and it is also unclear to what extent these can be directly linked to the wider provision of agrobiodiversity conservation services, such as the maintenance of evolutionary processes or cultural traditions. More research remains to be carried out before it is possible to define scientifically rigorous conservation goals, as will be discussed in the next section. However, it should be borne in mind that there may well be a trade-off involved in establishing such rigorous goals and the transaction costs of doing so (e.g. in terms of verification) once the practicalities of implementation in the field are taken into account (Muradian et al. 2010).

DEFINING A CONSERVATION GOAL: WHAT AND HOW MUCH?

A necessary starting point for any PACS programme is a definition of the conservation goal. This is likely to require the identification (possibly through some sort of prioritization exercise) of certain target PAGR and the definition of what might be considered to constitute a safe minimum standard (SMS) or minimum viable population. An SMS is based on the avoidance of critical thresholds at which a resource would be lost irreversibly, and aims for the maintenance of a critical level of PAGR that does not threaten the long-term *in-situ* survival of the resource. In the case of crop varieties, the estimation of an SMS is likely not only to be based on the area cultivated but also on the total of seeds available in local systems, the number of farmers of the specified resource and the degree of traditional knowledge maintained. These issues have been dealt with, at best, to a limited extent in the literature on PAGR (Bellon et al. 2003; Drucker 2006; Zander et al. 2009).

Prioritizing PAGR in PACS

Since the degree of agrobiodiversity conservation attainable in practical terms is in many cases determined by economic considerations (Kontoleon et al. 2008), there is an increasing pressure to prioritize conservation on the basis of economic values as well as on the level of threat faced, which is subject to the limited conservation funding available (Bellon et al. 2003). Given the existence of thousands of different crop species and livestock breeds, Weitzman-type decision support tools can help to target certain PAGR, so that the level of diversity conserved for a given conservation budget is maximized (Weitzman 1992, 1993, 1998).

PAGR could be prioritized for conservation based on a range of criteria which, according to Ruane (2000) may comprise the level of threat according to different risk factors, and the morphological characteristics or genetic traits of the target varieties. More specifically, it is proposed that the former should take into account such factors as the area under a specific crop variety, the number of farmers still cultivating that variety and the extent of agricultural knowledge related to its use (although the latter two factors may be closely linked in practice). A further key risk factor to consider is the amount of seed stored by farmers, because while the cultivated area may reflect the current importance of a certain variety, agricultural systems are dynamic, and vary from year to year due to rotational requirements and market conditions. Accordingly, the volume of seeds stored by households and communities can play a key role in determining the risk of extinction.

If critical values are reached in all or some of the indicated risk-categories, the species or variety may be classified as being 'under threat' or 'at risk', with priority given to those that have unique traits or are otherwise considered to embody a high degree of dissimilarity (Narloch et al. 2011). Nevertheless, there is still a high level of scientific uncertainty, especially associated with the definition of critical values and with determining the degree of dissimilarity between varieties (Ruane 2000; Bellon et al. 2003; Reist-Marti et al. 2003). Moreover, the cost needs to be taken into account of establishing the baselines necessary for carrying out the prioritization task. Given the general lack of detailed national statistics related to the status and trends of genetic resources, such activities may be both costly and time-consuming.

Furthermore, there is also a critical issue related to appropriate scales of analysis, as originally shown by Weitzman (1992, 1993). This is because a local or national-level analysis might highlight certain resources as being threatened while ignoring large, nearly identical populations across a national or international administrative border. As seeds can be exchanged between different communities and as similar varieties are likely to be managed within similar agro-ecological zones, when one crop variety or livestock breed is threatened but found in various places, the question arises of where a PACS scheme should be implemented. If PACS is understood as promoting *in-situ* conservation with a focus on maintaining local seed systems, this might imply the need to cover large geographical areas so as to conserve the threatened PAGR and traditional knowledge wherever they are under threat (Brush 1995; Stromberg et al. 2010). However, such a strategy would probably be more costly than one that aims at conservation over smaller geographical areas so as to secure *in-situ* conservation in a few selected places only. Even under these circumstances, a global conservation strategy ought to build on a decentralized system so as to integrate different conservation sites to support co-evolutionary processes in different socio-ecological systems. At the same time it should facilitate the exchange of PAGR across regions and borders, so as to maximize the safeguarded option values from agrobiodiversity conservation (Bellon 2008).

Towards a Definition of a SMS for PACS

Once PAGR have been prioritized regarding their level of threat and their uniqueness or dissimilarity, another challenge lies in defining how much of the target PAGR should be conserved. PAGR and their (uncertain) future values may be lost irreversibly if a population falls below a critical threshold. An SMS based on avoidance of this threshold can be considered a partial opportunity cost approach, which restricts the replacement of traditional varieties and breeds by commercial alternatives to an extent that does not threaten the long-term *in-situ* survival of the unique PAGR, thereby avoiding total future losses (Drucker 2006).

The application of an SMS approach is complicated by the difficulty in defining the minimum population size. In the case of domesticated animals, the Food and Agriculture Organization (FAO 1998) defines a livestock breed generally not to be at risk if there are 1000 breeding females and 20 males. As noted previously, in the case of crop genetic resources, the estimation of an SMS is likely to be based not only on the cultivated area[4] but also on the quantity of seeds available in local systems, the number of farmers choosing to grow a specific variety and the degree of traditional knowledge maintained. Additional criteria, such as geographical distribution of PAGR and associated agro-ecological factors within those locations, existing seed-exchange networks, breeding infrastructure, socio-cultural traditions and market integration could also be taken into account when establishing a workable SMS (e.g. Reist-Marti et al. 2003).

Consequently, it appears that there are many factors and underlying dynamics that would affect the definition of a SMS for PAGR, therefore PACS schemes need to draw on sound interdisciplinary research into socio-ecological dynamics in order to determine scientifically justifiable conservation goals. While it is possible that such goals might be fairly modest (e.g. conservation area goals for an individual variety might be expressed in hectares or tens of hectares rather than hundreds or thousands of hectares), to the best of our knowledge, existing research of this type is extremely limited and more work needs to be done in this area.

ASSESSING POSSIBLE PACS OUTCOMES

In addition to these design and implementation challenges facing PACS schemes, they may also prove unsuccessful if their potential outcomes are not carefully evaluated beforehand. Ideally, PACS schemes should attain their conservation goal for target PAGR (effectiveness) at least cost (efficiency) and optimum fairness (equity). As these outcomes are interlinked to a certain extent, PACS schemes share many of the challenges confronted by other PES programmes. Possible outcomes, as summarized in Table 8.1, are likely to be resource-context dependent and location specific.

Ecological Effectiveness

In general, any type of PES scheme can only be considered to have been effective in providing the necessary level of conservation services if: (1) the level of conservation services would be lower without the programme (additionality); (2) the scheme does not adversely affect other valuable non-target species, varieties or ecosystems (leakage effect); and (3) the gain is permanent (permanence or sustainability).

Under PACS, additionality will not have been achieved if farmers receive rewards for conservation efforts they would have made anyway. As with PES, PACS schemes have to focus on individual farmers or communities whose land management decisions can significantly affect the risk status (be it in the short or long term) of target PAGR (as per Wunder 2007). At first glance, PACS schemes that focus on farmers who already conserve PAGR might be considered to generate fairly low levels of additionality. However, given the existence of downward population trends for target PAGR, additionality might increase over longer time periods. Furthermore, in line with one of Weitzman's (1993) original findings, it might be expected that, given the potentially high contribution of target

160   *Values, payments and institutions for ecosystem management*

Table 8.1   A summary of criteria of potential ecological, economic and social outcomes of PACS, including site-specific considerations that need to be evaluated at the design stage of a PACS scheme

1. Ecological effectiveness	
Additionality	Scientific uncertainty in linking conservation goals with actual agrobiodiversity conservation services.
	Low in current season but potentially higher in medium to long term (due to focus on least-cost conservation).
Leakage	Agricultural expansion into non-targeted areas could lead to ecosystem impacts.
	Sustainable agricultural practices, such as rotation systems, could be undermined.
	Successful intervention might in some cases displace other valuable but non-targeted PAGR, resulting in a decline in overall diversity.
Sustainability	The scale of intervention can be designed to achieve an SMS population for threatened resources at relatively low cost.
	Depends on availability of funds, which are potentially limited if government-financed, unless an adequate endowment fund is established.
	Need to explore potential for private sector funding and niche market development.
	Interplay with other agricultural subsidies.
	Permanence of PACS undermined by inefficiencies and inequities.
2. Economic efficiency	
Payments and rewards	Households and communities involved in practical conservation are likely to be least-cost providers of agrobiodiversity.
	Competitive tender approaches may provide cost savings via opportunity-cost revealing mechanisms.
	Discriminatory pricing rules, however, might not be acceptable in many rural communities.
Implementation costs	Improving access to certain seeds and breeds or agricultural knowledge where PACS involves a change in agricultural practices.
Transaction costs	The design of a PACS programme requires a definition of the conservation goal, and may involve a trade-off between scientific precision and the costs of monitoring and verification costs.
	Programme running costs are context specific and dependent on institutional arrangements.

Table 8.1 (continued)

2. Economic efficiency	
	Potential cost savings through community-level focus.
	Trade-off in achieving socio-cultural and equity conservation goals between involvement of many small farmers and fewer but larger farmers.
3. Social equity	
Decision-making	Procedural fairness low, as most PACS would involve top-down approaches. Offer voluntary participation opportunities.
Access	Focus on marginalized groups in society, but within these groups, wealthier farmers may face fewer constraints in participating in these programmes.
Outcome	Local consumers would benefit from increasing the supply of nutritious traditional PAGR.
	Participating farmers and communities receive PACS as cash or in-kind rewards.
	Equity outcomes depend on the wealth status of the farmers and communities that are able to capture these benefits.

PAGR to overall diversity (i.e. because of their high degree of dissimilarity), it might be important to secure their continued existence prior to their actually becoming threatened. Threshold effects may also mean that such interventions will be much more cost-effective than will be the case if smaller, threatened population levels of the target PAGR are reached (Elmqvist et al. 2010). Finally, it might be considered highly unfair to offer payments to farmers who have switched to more profitable PAGR, while farmers who are still conserving the target PAGR do not obtain any reward. This might even create perverse incentives (as per Pagiola and Platais 2007), i.e. farmers might stop cultivating target PAGR in order to be considered in subsequent years for receiving rewards under PACS schemes.

Leakage under PACS might occur at different scales. Farmers that opt to cultivate target PAGR on monitored land areas under PACS might expand into previously non-agricultural areas in order to go on cultivating improved varieties or non-traditional cash crops on the unmonitored lands. This could lead to loss of other biodiversity components. Alternatively, PACS schemes might result in leakage at the farm level if farmers were to replace (with target PAGR) threatened landraces or livestock breeds not covered by the conservation programme. Such a case may be of particular concern where it is possible to develop effective reward mechanisms for some (but not all) valuable crop and livestock varieties. Furthermore, PACS, as with other interventions, might undermine existing rotation systems, when certain crop species are replaced by others. This may also create unintended effects on agrobiodiversity conservation over the longer term.

The sustainability of PACS, i.e. their permanence in providing conservation services, depends very much on the length of payment flows and thus the nature of the source of funding. Farmers may not be willing to maintain certain land uses or agricultural practices once payments dry up. This has been noted under some PES schemes (Pagiola and Platais 2007). Where these schemes are not user-financed, but rather government-financed, adequate endowment funds might need to be established in order to ensure the long-term nature of such an intervention. Identifying sources of private sector funding, including through biodiversity offset programmes and niche market development activities, could therefore be an important additional element in ensuring PACS sustainability.

A factor in favour of PACS sustainability would be where conservation goals require relatively modest areas and farmers face relatively low opportunity costs. This could translate into comparatively low levels of conservation funding being required (Drucker 2006; Zander et al. 2009), particularly when compared with other types of agricultural subsidy, hence making permanence more likely. Nonetheless, their effectiveness in reaching conservation goals over the medium and long terms is critically dependent on their efficiency and equity impacts.

Cost-inefficiencies and perceived inequalities might limit the lifespan of such interventions significantly, making effectiveness, efficiency and equity interlinked. However, there are also certain trade-offs that may need to be considered. For instance, when least-cost conservation targets and poor households do not lead to high additionality in the short-run.

Economic Efficiency and Least-cost Conservation

Given the attainment of a given conservation goal, a programme's efficiency is determined by its cost-effectiveness. Funding would be needed for three purposes: (1) payments and rewards to compensate farmers for their opportunity costs; (2) for paying for a change in agricultural practices if one-off, up-front implementation costs are required; and (3) for transaction costs associated with the functioning of the programme.

Payment levels should cover the opportunity costs of farmers arising from the forgone benefits of more financially attractive alternative agricultural practices. Such payments could be minimized when focusing on

PAGR that provide considerable private values to the farmer and high public values to wider society (Smale et al. 2004). As poor farming communities often carry out conservation in real terms, they may be expected to provide opportunities to implement relatively low-cost conservation strategies because such communities have relatively low opportunity costs.

Opportunity costs are farm specific, and shaped by location as well as individual factors, so that the efficiency challenge lies in the ability to identify least-cost service providers within selected communities. Efficiency would be maximized if farmers were not compensated in excess of their financial opportunity costs. Nonetheless, most existing PES schemes are based on fixed pricing rules, where farmers can decide how much land to conserve given a fixed price per land unit. This means that farmers might be able to receive rewards that are much higher than their opportunity costs (FAO 2007b). By contrast, substantial cost savings could be realized through competitive tenders, through which farmers are invited to bid for conservation contracts by defining a conservation area for a target PAGR and a required compensation level. Such reverse auction mechanisms involve a cost-revealing mechanism to tackle the existence of information asymmetries, as farmers have an incentive to apply for conservation contracts close to their real opportunity costs (Latacz-Lohmann and van der Hamsvoort 1997; Stoneham et al. 2003; Ferraro 2008). Yet research shows that it may be a challenge for competitive tenders to remain more efficient than fixed reward mechanisms when carried out more than once, as farmers learn from repeated auctions (Schilizzi and Lactaz-Lohmann 2007). Nevertheless, coupling conservation tenders with discriminatory pricing rules (i.e. paying farmers according to that defined in the bid as opposed to a uniform payment level) may still be associated with higher efficiency, but may raise equity concerns, as discussed below (Ferraro 2008).

While opportunity costs are permanent costs, set-up or implementation costs are often the one-off costs of specific actions required to change agricultural practices. Setting up PACS might involve costs for improving access to certain seeds or agricultural knowledge or, in the case of livestock, assistance with the rotation of male breeding animals between communities. The limited availability of seed or breeding animals (a logical consequence of focusing on threatened genetic resources) could be a constraint that will add to implementation costs and might extend time-scales for achieving an SMS. Nevertheless, such costs might be minimized where the focus is on farmers carrying out de facto conservation.

Transaction costs relate to the costs of running the scheme, including costs for programme design. Since it can be very costly to define the exact, scientifically grounded conservation goals needed to secure effectiveness, there is a trade-off between effectiveness and efficiency. Additionally, each PACS scheme will involve costs for monitoring and enforcement, although these costs may be kept to a minimum where the scheme involves participation of the farming community itself. Furthermore, where PACS schemes can be implemented at a community rather than an individual level, some cost savings might be obtainable since economies of scale tend to reduce average transaction costs related to administration. Similarly, contracting a few large rather than many small farmers could also be a strategy to reduce transaction costs (as per Pagiola et al. 2007). However, where it is considered desirable for local seed systems to be maintained, it might be convenient to maximize the number and spatial distribution of participating farmers (Stromberg et al. 2010). Furthermore, the inevitable trade-off between efficiency and equity is related to the number of participating farmers (as per Wunder 2007).

Social Equity and Pro-poor Outcomes

Many authors have highlighted the fact that PES-like interventions should have their primary emphasis on ecological outcomes to the extent that socially desirable goals may need to be traded-off, or even that existing inequities and vulnerabilities may be exacerbated (Pagiola et al. 2005; Wunder 2007; Engel et al. 2008). Nonetheless, ignoring the social dimensions of PES might prove to be destructive for poor farming communities or create a 'PES curse' by undermining the success and legitimacy of PES schemes. To a large extent, environmental and social goals are intertwined and there is a need to take equity into account in decision-making, access and outcome if schemes are to be successful in the longer term (Corbera et al. 2007; Pascual et al. 2010).

Equity in decision-making under PACS could be reached through procedural fairness criteria where different stakeholder groups have an opportunity to participate in the design of the payment schemes, or at least their interests could be taken into account. Otherwise, powerful farmers or intermediaries, who are generally more likely to intervene in decision-making processes, could quite easily favour a design that supports only their narrower interests. On the one hand, most PACS may indeed be top-down approaches, i.e. external stakeholders design the schemes and farmers accept the rules of the game if they want to participate, so that the level of fairness is generally low. On the other hand, PACS schemes might empower some farmers just by offering voluntary participation opportunities.

Equity in access could be considered in terms of the eligibility and ability of farmers to participate. The former would, of course, depend on the intentions of the scheme. As noted above, efficient PACS would aim to include farmers and communities carrying out de facto conservation and as such they may be focused on disadvantaged rural areas of developing countries. However, despite being poor from a global point of view, least-cost providers of agrobiodiversity conservation services might not always be the most disadvantaged from a regional perspective (Smale 2006). Furthermore, poorer farmers might not always be able to overcome entry barriers associated with land tenure requirements and transaction costs (Grieg-Gran et al. 2005; Wunder 2007), although this is context dependent as in some cases poor farmers are found to be able to participate in PES-like schemes notwithstanding technically difficult and costly investments (Pagiola et al. 2008).

While it is desirable for PACS schemes to support equity outcomes, it depends very much on the wealth status of benefiting farmers. As has been argued for PES (Grieg-Gran et al. 2005; Zilberman et al. 2008), not only participating farmers, but also non-participating consumers may be affected by such programmes. For instance, while many commercial varieties are cultivated for export markets, PACS that focus on often nutritious, adaptive and resistant traditional PAGR would be likely to enhance the nutrition of local consumers directly. Moreover, the poorer the farmers that receive payments from PACS, the more equitable will be the overall outcome if a premium in addition to farmers' opportunity costs is paid. The perceived fairness in the distribution of payments plays a key role in the robustness of community-based conservation programmes (Pascual et al. 2010). That said, PACS involving payment differentiation through reverse auctions may not be accepted by local farmers (Ferraro 2008). Nevertheless, even with uniform payments, there may be the risk of PACS potentially favouring relatively wealthier households within a community, as powerful resource users might be able to more easily reap the benefits of such programmes. Perceived inequalities in the distribution of benefits might even result in local conflicts. Practitioners of PACS will therefore do well to take this risk into account and seek to design PACS in a way that facilitates participation of the most disadvantaged farmers, in order not to undermine the long-term legitimacy of PACS and thus their sustainability.

CONCLUSIONS

In applying the concept of PES to *in-situ* conservation of PAGR, PACS seek to tackle the market failures associated with the public goods of agro-biodiversity by increasing the private benefits to farmers from growing or rearing threatened traditional crop varieties and livestock breeds through monetary or in-kind rewards. Payments for agrobiodiversity conservation services can provide policy-makers with an effective tool for conserving traditional PAGR while providing wider public benefits. However, this chapter draws attention to some generic constraints to be overcome in implementing such schemes. These are partly dependent on the complex institutional settings in which PACS would have to be implemented. Further difficulties are associated with the definition of a conservation goal, possibly associated with an SMS after a prioritization exercise has been carefully carried out. We argue that interdisciplinary research on socio-ecological dynamics is needed in order to complete such prioritization exercises and establish SMS for threatened and under-utilized PAGR.

As with most PES programmes, PACS schemes might to some extent need to forego scientifically rigorous conservation indicators in favour of those that are easier to implement. It is in this context that a clear trade-off becomes apparent between ecological effectiveness (i.e. scientific precision in linking agrobiodiversity conservation services with certain conservation goals) and economic efficiency (i.e. minimizing transaction costs).

For efficiency reasons, PACS schemes may focus on communities that conserve PAGR de facto, because opportunity costs remain low. Changing preferences and market conditions could, however, pose a threat to some of the last remaining areas cultivated with specific crop varieties. That said, there is a further trade-off between cost savings and the additionality of such programmes. In addition, it may be hypothesized that a PACS scheme based on competitive tenders with self-organized groups of farmers might be the most efficient approach because it reveals opportunity costs and minimizes transaction costs.

Although such approaches, coupled with discriminatory pricing rules, would maximize efficiency as well as the number of participating farmers or communities given a conservation budget, they might also lead to perceived unfairness due to payment differentiation. Payments for agrobiodiversity conservation services schemes need to consider how they might avoid exacerbating existing inequities in which better-off households face the lowest conservation costs and fewer participation constraints, and thus would be more likely to benefit from PACS. Although such concerns do not generally seem to have arisen in practice under a number of PES schemes, careful analysis might still be required to identify which kind of farmers within participating communities are able to capture the benefits of PACS schemes. Given that most threatened and underutilized PAGR are located in disadvantaged and remote rural areas in developing countries, PACS may prove to be part of rural development government aid packages. As such, PACS schemes need to be designed in a way that takes fairness considerations on board in order not to undermine the long-term legitimacy and robustness of the programmes. Much more location-specific research is needed to assess the potential trade-offs between ecological, economic and social goals.

Sustainability is another key area of concern regarding the application of PACS, particularly when compensation flows are not generated purely by market transactions. Government-financed schemes might have a limited lifespan unless adequate endowment funds can be established. That said, there is increasing promotion of market-chain development for agrobiodiversity-related products. These 'conservation-through-development approaches' can potentially be more sustainable than other theoretically more elegant approaches, as they build on existing agricultural market channels and thus could be used to generate a sustainable source of funding.

It should be noted that relying solely on market development might be a dangerous strategy for the conservation of a diverse genetic resource pool, especially as market conditions can change rapidly, and in general consumers tend to favour a narrow range of crop varieties or animal breeds. In this context, PACS schemes might be capable of providing a stronger and more flexible longer-term foundation for conservation activities than can be achieved through market development, and may be better suited to ensuring the *in-situ* conservation of safe minimum populations of important PAGR. Market-chain development approaches and PACS might thus be complementary instruments, with each enjoying a comparative advantage under different circumstances.

As found by Muradian and colleagues (2010) for PES in general, we conclude that PACS should be considered as a potentially useful tool for policy-makers. Given the many constraints, and a lack of field experience to date, further research and development remains to be carried out before PACS may become established in the policy-makers' toolbox.

NOTES

*   This article was published in *Ecological Economics*, **70**(11), Narloch, U., Drucker A. and Pascual, U., Payments for agrobiodiversity conservation services for sustained on-farm utilization of plant and animal genetic resources, 1837–1845, Copyright Elsevier BV (2011).
1. In this context, agrobiodiversity is understood as all diversity within and among plant and animal species found in domesticated systems.
2. For example: China's *Grain for Green* programme, which promotes reforestation in order to reduce soil erosion; the introduction of natural vegetation contour strips in the Philippines; integrating short-term improved fallow systems into smallholder agricultural systems in Kenya and Zambia; shade-grown coffee cultivation in Bolivia;

windbreaks in Costa Rica; and the Silvopastoral Ecosystem Management Project in Colombia, Costa Rica and Nicaragua (FAO 2007b).
3. This project was implemented through Ethiopia's Institute of Biodiversity Conservation (IBC) under the project title, 'A dynamic farmer-based approach to the conservation of African plant genetic resources' (see http://www.thegef.org/gef/project_detail?projID=351).
4. For example, under the Lazio Rural Development Programme in Italy, a crop variety may be considered highly threatened if it covers less than 1% of the regional area, has a declining trend, is grown by less than 30 farmers, is not found on the market and is not listed in commercial seed catalogues (PSR Lazio 2008).

REFERENCES

Bellon, M. (2004). Conceptualizing interventions to support on-farm genetic resource conservation. *World Development*, **32**(1), 159–172.

Bellon, M. (2008). Do we need crop landraces for the future? Realizing the global option value of *in situ* conservation. In A. Kontoleon, U. Pasqual and M. Smale (eds), *Agrobiodiversity Conservation and Economic Development*. Abingdon, UK: Routledge, pp. 56–72.

Bellon, M.R., J. Berthaud, M. Smale, J.A. Aguirre, S. Taba, F. Aragon, J. Díaz and H. Castro (2003). Participatory landrace selection for on-farm conservation: an example from the Central Valleys of Oaxaca, Mexico. *Genetic Resources and Crop Evolution*, **50**, 401–416.

Brush, S. (1989). Rethinking crop genetic resource conservation. *Conservation Biology*, **3**(1), 19–29.

Brush, S.B. (1995). *In situ* conservation of landraces in centers of crop diversity. *Crop Science*, **35**(2), 346–354.

Corbera, E., N. Kosoy and T. Martínez (2007). Equity implications of marketing ecosystem services in protected areas and rural communities: case studies from Meso-America. *Global Environmental Change*, **17**, 365–380.

Di Falco, S. and J.P. Chavas (2009). On crop biodiversity, risk exposure and food security in the highlands of Ethiopia. *American Journal of Agricultural Economics*, **91**(3), 599–611.

Di Falco, S., M. Smale and C. Perrings (2008). The role of agricultural cooperative in sustaining the wheat diversity and productivity: the case of southern Italy. *Environmental Resource Economics*, **39**, 161–174.

Drucker, A.G. (2006). An application of the use of safe minimum standards in conservation of livestock biodiversity. *Environment and Development Economics*, **11**, 77–94.

Elmqvist, T., E. Maltby, T. Barker, M. Mortimer, C. Perrings, J. Aronson, R. DeGroot, A. Fitter, G. Mace, J. Norberg, I. Sousa Pinto and I. Ring (2010). Biodiversity, ecosystems and ecosystem services. In P. Kumar (ed.), *The Economics of Ecosystems and Biodiversity: Ecological and Economic Foundations*. London and Washington DC: Earthscan.

Engel, S., S. Pagiola and S. Wunder (2008). Designing payments for environmental services in theory and practice: an overview of the issue. *Ecological Economics*, **65**(4), 663–674.

Eyzaguirre, P. and E. Dennis (2007). The impacts of collective action and property rights on plant genetic resources. *World Development*, **35**(9), 1489–1498.

FAO (1998). *Primary Guidelines for Development of National Farm Animal Genetic Resources Management Plans*. Rome: Food and Agriculture Organization of the United Nations.
FAO (2007a). *The State of the World's Animal Genetic Resources for Food and Agriculture*. Food and Agriculture Organization of the United Nations, Rome.
FAO (2007b). *State of Food and Agriculture: Paying Farmers for Environmental Services*. Rome: Food and Agriculture Organization of the United Nations.
Ferraro, P.J. (2001). Global habitat protection: limitations of development interventions and a role for conservation performance payments. *Conservation Biology*, **15**(4), 990–1000.
Ferraro, P.J. (2008). Asymmetric information and contract design for payments for environmental services. *Ecological Economics*, **65**(4), 811–822.
Ferraro, P. and R. Simpson (2002). The cost-effectiveness of conservation payments. *Land Economics*, **78**, 339–353.
Fuwa, N. and A.J.U. Sajise (2009). Exploring environmental services incentive policies for the Philippines rice sector: the case of intra-species agrobiodiversity conservation. In L. Lipper, T. Sakuyama, R. Stringer and D. Zilberman (eds), *Payment for Environmental Services in Agricultural Landscapes: Economic Policies and Poverty Reduction in Developing Countries*. Rome: Food and Agriculture Organization of the United Nations, pp. 221–238.
Grieg-Gran, M., I. Porras and S. Wunder (2005). How can market mechanisms for forest environmental services help the poor? Preliminary lessons from Latin America. *World Development*, **33**(9), 1511–1527.
Gruère, G.P., A. Giuliani and M. Smale (2008). Marketing underutilized plant species for the benefit of the poor: a conceptual framework. In A. Kontoleon, U. Pasqual and M. Smale (eds), *Agrobiodiversity Conservation and Economic Development*. Abingdon, UK: Routledge, pp. 73–87.
Hajjar, R., D.I. Jarvis and B. Gemmill-Herren (2008). The utility of crop genetic diversity in maintaining ecosystem services. *Agriculture, Ecosystems and Environment*, **123**, 261–270.
Heisey, P.W., M. Smale, D. Byerlee and E. Souza (1997). Wheat rusts and the costs of genetic diversity in the Punjab of Pakistan. *American Journal of Agricultural Economics*, **79**, 726–737.
Hermann, M. and T. Bernet (2009). The transition of maca from neglect to market prominence: Lessons for improving use strategies and market chains of minor crops. *Agricultural Biodiversity and Livelihoods Discussion Papers 1*, Bioversity International, Rome.
Kontoleon, A., U. Pasqual and M. Smale (eds) (2008). *Agrobiodiversity Conservation and Economic Development*. Abingdon, UK: Routledge.
Kosoy, N. and E. Corbera (2010). Payments for ecosystem services as commodity fetishism. *Ecological Economics*, **69**(6), 1228–1236.
Krishna, V.V., U. Pascual and D. Zilberman (2010). Assessing the potential of labelling schemes for *in situ* landrace conservation: an example from India. *Environment and Development Economics*, **15**, 127–151.
Landell-Mills, N. and I. Porras (2002). *Silver Bullet or Fools' Gold? A Global Review of Markets for Forest Environmental Services and their Impact on the Poor*. London: IIED.
Latacz-Lohmann, U. and C. van der Hamsvoort (1997). Auctions as a means of creating a market for public goods from agriculture. *Journal of Agricultural Economics*, **49**(3), 334–345.

Lipper, L., T. Sakuyama, R. Stringer and D. Zilberman (2009). *Payment for Environmental Services in Agricultural Landscapes: Economic Policies and Poverty Reduction in Developing Countries*. Rome: Food and Agriculture Organization of the United Nations.

Madsen, B., N. Carroll and K. Moore Brands (2010). State of biodiversity markets report: offset and compensation programs worldwide. Available at: http://www. ecosystemmarketplace.com/ documents/acrobat/sbdmr.pdf (accessed 3 July 2013).

Maxted, N., L. Guarini, L. Myer and E.A. Chiwona (2002). Towards a methodology for on-farm conservation of plant genetic resources. *Genetic Resources and Crop Evolution*, **49**, 1–46.

Muradian, R., E. Corbera, U. Pascual, N. Kosoy and P.H. May (2010). Reconciling theory and practice: an alternative conceptual framework for understanding payments for environmental services. *Ecological Economics*, **69**(6), 1202–1208.

Narloch, U., U. Pascual and A.G. Drucker (2011) Cost-effectiveness targeting under multiple conservation goals and equity considerations in the Andes. *Environmental Conservation*, **38**(4), 417–425.

Narloch, U., U. Pascual and A.G. Drucker (2012). Collective action dynamics under external rewards: experimental insights from farming communities in the Andes. *World Development*, **40**(10), 2096–2107.

Nautiyal, S., V. Bisht, K.S. Rao and R.K. Maikhuri (2008). The role of cultural values in agrobiodiversity conservation: a case study from Uttarakhand, Himalaya. *Journal of Human Ecology*, **23**(1), 1–6.

Pagiola, S. and G. Platais (2007). *Payments for Environmental Services: From Theory to Practice*. Washington DC: World Bank.

Pagiola, S., A. Arcenas and G. Platais (2005). Can payments for environmental services help reduce poverty? An exploration of the issues and the evidence to date from Latin America. *World Development*, **33**(2), 237–253.

Pagiola, S., E. Ramirez, J. Gobbi, C. de Haan, M. Ibrahim, E. Murgueitio and J.P. Ruiz (2007). Paying for the environmental services of silvopastoral practices in Nicaragua. *Ecological Economics*, **64**(2), 374–385.

Pagiola, S., A.R. Rios and A. Arcenas (2008). Can the poor participate in payments for environmental services? Lessons from the Silvopastoral Project in Nicaragua. *Environment and Development Economics*, **13**(32), 299–325.

Pascual, U. and C. Perrings (2007). The economics of biodiversity loss in agricultural landscapes. *Agricultural Ecosystem Environment*, **121**, 256–268.

Pascual, U., R. Muradian, L.C. Rodriguez and A. Duraiappah (2010). Exploring the links between equity and efficiency in payments for environmental services: a conceptual approach. *Ecological Economics*, **69**(6), 1237–1244.

Perales, H., S.B. Brush and C.O. Qualset (2003). Dynamic management of maize landraces in central Mexico. *Economic Botany*, **57**(1), 21–34.

Perrings, C., L. Jackson, K. Bawa, L. Brussaard, S. Brush, T. Gavin, R. Papa, U. Pascual and P. De Ruiter (2006). Biodiversity in agricultural landscapes: saving natural capital without losing interest. *Conservation Biology*, **20**(2), 263–264.

PSR Lazio (2008). Relazione di valutazione intermedia. Servizio di valutazione in itinere, comprensivo della valutazione intermedia ed ex post del programma di sviluppo rurale 2007–2013 Della Regione Lazio. Available at: http://ec.europa. eu/agriculture/rurdev/countries/it/mte-rep-it-lazio_it.pdf (accessed 4 July 2013).

Reist-Marti, S., H. Simianer, G. Gibson, O. Hanotte and J.E. Rege (2003). Weitzman's approach and breed diversity conservation: an application to African cattle breeds. *Conservation Biology*, **17**(5), 1299–1311.

Ruane, J. (2000). Framework for prioritizing domestic animal breeds for conservation purposes at the national level: a Norwegian case study. *Conservation Biology*, **14**(5), 1385–1393.

Schilizzi, S. and U. Latacz-Lohmann (2007). Assessing the performance of conservation auctions: an experimental study. *Land Economics*, **83**(4), 497–515.

Smale, M. (2006). *Valuing Crop Biodiversity: On Farm Genetic Resources and Economic Change*. Wallingford, UK: CABI Publishing.

Smale, M., M.R. Bellon, D. Jarvis and B. Sthapit (2004). Economic concepts for designing policies to conserve crop genetic resources on farms. *Genetic Resources and Crop Evolution*, **51**, 121–135.

Stoneham, G., V. Chaudhri, A. Ha and L. Strappazzon (2003). Auctions for conservation contracts: an empirical examination of Victoria's Bush Tender trials. *The Australian Journal of Agricultural and Resource Economics*, **47**(4), 477–500.

Stromberg P., U. Pascual and M. Bellon (2010). Seed systems and farmers' seed choices: the case of maize in the Peruvian Amazon. *Human Ecology*, **38**, 539–553.

Swanson, T.M. and T. Göschl (1999). Optimal genetic resource conservation: *in situ* and *ex situ*. In S.B. Brush (ed.), *Genes in the Field: On-farm Conservation of Crop Diversity*. Ann Arbor, MI: Lewis Publishing, pp. 165–191.

Swanson, T. and T. Göschl (2000). Property rights issues involving plant genetic resources: implications of ownership for economic efficiency. *Ecological Economics*, **32**(1), 75–92.

van de Wouw, M., C. Kik, T. van Hintum, R. van Treuren and B. Visser (2009). Genetic erosion in crops: concept, research results and challenges. *Plant Genetic Resources*, **8**, 1–15.

Wale, E. (2008). A study on financial opportunity costs of growing local varieties of sorghum in Ethiopia: implications for on-farm conservation policy. *Ecological Economics*, **64**(3), 603–610.

Weitzman, M.L. (1992). On diversity. *Quarterly Journal of Economics*, **107**(2), 363–405.

Weitzman, M.L. (1993). What to preserve? An application of diversity theory to crane conservation. *The Quarterly Journal of Economics*, **108**, 157–184.

Weitzman, M.L. (1998). The Noah's Ark problem. *Econometrica*, **66**(6), 1279–1298.

Wunder, S. (2006). Are direct payments for environmental services spelling doom for sustainable forest management in the tropics? *Ecology and Society*, **11**(2), 23.

Wunder, S. (2007). The efficiency of payments for environmental services in tropical conservation. *Conservation Biology*, **21**(1), 48–58.

Wunder, S., S. Engel and S. Pagiola (2008). Taking stock: a comparative analysis of payments for environmental services programs in developed and developing countries. *Ecological Economics*, **65**, 834–852.

Zander, K.K, A.G. Drucker, K. Holm-Müller and H. Simianer (2009). Choosing the 'cargo' for Noah's Ark: applying Weitzman's approach to Borana cattle in East Africa. *Ecological Economics*, **68**, 2051–2057.

Zilberman D., L. Lipper and N. McCarthy (2008). When could payments for environmental services benefit the poor? *Environment and Development Economics*, **13**, 255–278.

9. Paying for international environmental public goods

Rodrigo Arriagada and Charles Perrings

INTRODUCTION

How can we secure the provision of international environmental public goods? It is well understood that markets under-supply public goods, and there is a wealth of evidence that many environmental public goods have been systematically under-supplied over a long period of time (Millennium Ecosystem Assessment 2005). If environmental public goods occur at the scale of the nation state or below, the failure of markets to supply public goods may be offset by the actions of local or national government. There exist many national agencies with responsibilities for the provision of environmental public goods such as habitat for rare and endangered species, clean water, environmental health protection and so on. There also exist many offset or mitigation systems for securing private provision of public goods at a national level (Madsen et al. 2010). At the international level, where there is no supranational authority to take responsibility, the failure of markets to deliver environmental public goods is more difficult to offset. Depending upon the magnitude and distribution of the payoffs to be had from provision of public goods, individual countries will have a stronger or weaker incentive to commit resources to their maintenance. Doing more than that depends upon agreement between nation states (Kaul et al. 2003a; Barrett 2007).

International environmental public goods include a number that are strictly global, such as conservation of the genetic diversity on which all future evolution depends, the mitigation of climate change, the control of emerging infectious diseases, and the management of sea areas beyond national jurisdictions. They also include many that are more regional, such as the control of acid rain, the management of multi-country river basins, and the protection of international watersheds (Touza and Perrings 2011). In all cases, it is impossible for any single state to secure such goods on its own. International public good supply depends on either international coordination or international cooperation (Anand 2004).

Public goods, defined as those that exhibit both consumption indivisibilities and non-excludability, present a particularly complex category of environmental resources. Non-excludability refers to a circumstance where, once the resource is provided, none can be excluded from enjoying the benefits it confers or the harm it imposes. Indivisible consumption occurs when one country's enjoyment of the benefits does not diminish the amount available for others.

This chapter focuses on international environmental public goods where benefits extend to people in several countries, and often across multiple generations (Kaul et al. 1999). In practice, the beneficiaries of international public goods include national populations and their representatives, nation states, transnational corporations and non-governmental organizations, as well as a newly emerging set of institutions. Globalization has altered the way that members of civil society organize themselves across national boundaries. The information revolution has also stimulated new forms of social participation. New networks enable the exchange of ideas and implementation techniques. These new relationships and interactions have created a 'global environmental public' interested in asserting new rights over and new responsibilities to the resources of the planet. Its concerns span both the ethical responsibilities of individuals, organizations, countries and corporations and alternative forms of governance of the biosphere.

Following the Millennium Ecosystem Assessment (2005), we suppose that the benefits people obtain from the biosphere depend on a set of ecosystem services comprising:

- Provisioning services: products people obtain from ecosystems, such as food, fuel, fibre, fresh water and genetic resources;
- Cultural services: non-material benefits people obtain from ecosystems through spiritual enrichment, cognitive development, reflection, recreation and aesthetic experiences;
- Regulating services: benefits people obtain from the regulation of ecosystem processes, including air quality maintenance, climate regulation, erosion control, regulation of floods and droughts, regulation of human diseases and water purification;
- Supporting services: those that are necessary for the production of all other ecosystem services, such as primary production, production of oxygen, and soil formation.

These services affect human well-being in many ways: through their role in the production of consumption goods, their support of human health and security, or the satisfaction of cultural and spiritual needs. A number of these services have the characteristics of international environmental

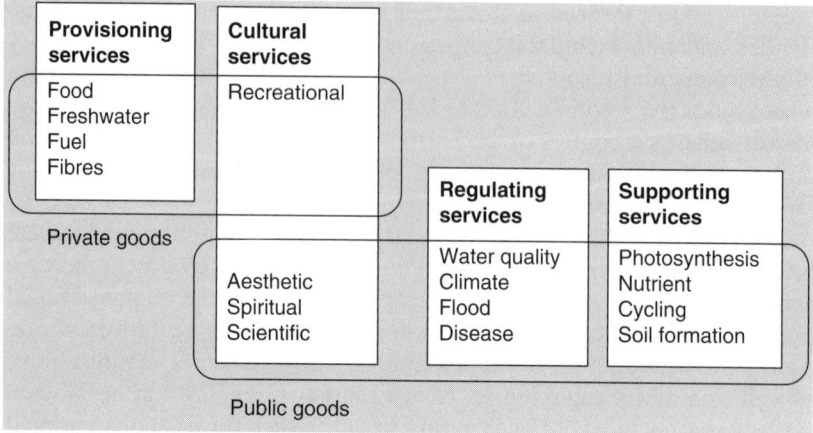

Figure 9.1 Millennium Ecosystem Services that are public goods. Most provisioning services are private goods. Most regulating and supporting services are public goods. Cultural services include both public and private goods

public goods, the most important of which involve the regulating and supporting services (see Figure 9.1).

We focus on the group of ecosystem services that are both public and international. These are services that are: 'non-rival and non-excludable ... in ... consumption, with a low potential for mediation by socioeconomic factors (i.e. a low degree of substitutability) and [their] provision cannot be restricted either to a specific population group or to a local geographical area' (Kaul et al. 1999). For global public goods, benefits must accrue to all citizens. International environmental public goods should: (1) cover more than one group of countries; (2) benefit not only a broad spectrum of countries but also a broad spectrum of the global population; and (3) meet the needs of the both present and future generations (Kaul et al. 1999; Anand 2004). International public goods generated in any one county must therefore generate spillover effects beyond a nation's boundary (Morrissey et al. 2002).

International environmental public goods can be classified further according to the degree to which they are non-excludable and non-rival, and according to their 'technology of supply' (Sandler 2004). Pure public goods are non-exclusive and non-rival. Impure public goods are either partially excludable or partially rival; the most common form of which are local public goods, particularly the local common pool resources analysed by Ostrom (1990). The technology of supply of international environmental public goods identifies the relation between the benefits they offer and the contributions of individual countries. The polar cases are 'best shot', 'weakest link' and 'simple sum'. For 'best-shot' public goods, the benefit to all countries is determined by the most effective provider. For example, the Centers of Disease Control are funded by the United States but provide information on infectious diseases to all countries. For 'weakest link' public goods, the benefits to all countries are limited to the benefits offered by the least effective provider. The best example of this is the control of infectious diseases. So for HIV and tuberculosis, the level of protection available to all countries is only as good as the control of the disease exercised in the poorest, most densely populated, and least well-coordinated country (Perrings et al. 2002). Public goods such as global warming, deforestation, or information, are said to be 'additive' (Sandler 2004; Nordhaus 2005). This implies that the benefits to all countries depend on the sum of the contributions of all countries. In the case of simple sum public goods, such as carbon sequestration, each unit of carbon sequestered has the same value no matter where it occurs. In other cases, such as habitat protection, the contribution of each hectare protected depends on its characteristics. These are said to be weighted sum public goods (Sandler 2004).

Of all the Millennium Ecosystem Assessment (MA) ecosystem services, regulating services are most often supplied as international environmental public goods. Examples include disease control, which is frequently supplied as a weakest or weaker link public good, climate regulation through, e.g. carbon sequestration, which is provided as an additive pure public good, or watershed protection, which is generally an additive but impure public good (Holzinger 2001; Dombrowsky 2007; Touza and Perrings 2011). Many international public goods are also jointly produced with local public goods. Biodiversity in tropical forests, for example, yields a set of private benefits in the form of timber and other products including medicinal plants, hunting, fishing, recreation and tourism. At the same time, tropical forests are a source of carbon sequestration, genetic information, hydrological and microclimatic regulation; commonly described as co-benefits (Perrings and Gadgil 2003).

An important feature of international environmental public goods is that their spatial extent depends partly on natural hydrological and atmospheric flows, and partly on the social linkages between countries, i.e. the flow of goods, people and information. The global reach of carbon sequestration is a property of the general circulation system, but the global reach of disease regulation is a property of the global trade and air transport systems. In fact the closer integration of the world's economic systems has rapidly increased the number of environmental public goods that are global in reach (Kaul et al. 2003b):

- New technologies increasingly enhance human mobility as well as the movement of goods, services and information around the world (e.g. the transmission of human diseases and air pollution are international environmental public bads);
- Economic and political openness have provided further impetus to cross-border and transnational activities (e.g. invasive species are international environmental public bads);
- Systematic risks have increased (e.g. the spatial correlation of agricultural risks associated with the adoption of common plant varieties and technologies is an international environmental public bad);
- International regimes are becoming more influential, often formulated by small groups of powerful nations claiming universal applicability (e.g. the General Agreement on Tariffs and Trade limits actions available to deal with international environmental public bads).

The particular concern of this chapter is that of the international environmental public goods most likely to be under-supplied relative to global demand. Its central problem is how to secure environmental public goods that (1) are provided at particular locations but offer benefits over a wider area, and (2) generate local benefits that are below the local cost of supply. We deal with the fundamental problem with international environmental public goods: the incentive that each country has to free-ride on the efforts of others. We then consider the options for addressing the problem, and review the applicability of currently popular instruments, such as payments for ecosystem services (PES), in terms of the characteristics of the public good concerned. Lastly, we draw out the implications for national and international environmental policy.

WHY ARE INTERNATIONAL ENVIRONMENTAL PUBLIC GOODS UNDER-PROVIDED?

International environmental public goods generate benefits that spill over national borders, so that the benefits (or costs) of those goods extend beyond the country of origin. If the marginal local benefits of public good provision are less than the marginal local costs, there will be no incentive to provide the public good at all. If the marginal local benefits of public good provision exceed the marginal local costs, but benefits also accrue to other countries, there will be an incentive to produce the public good, but unless the country is a 'best-shot provider' it will not be at a level

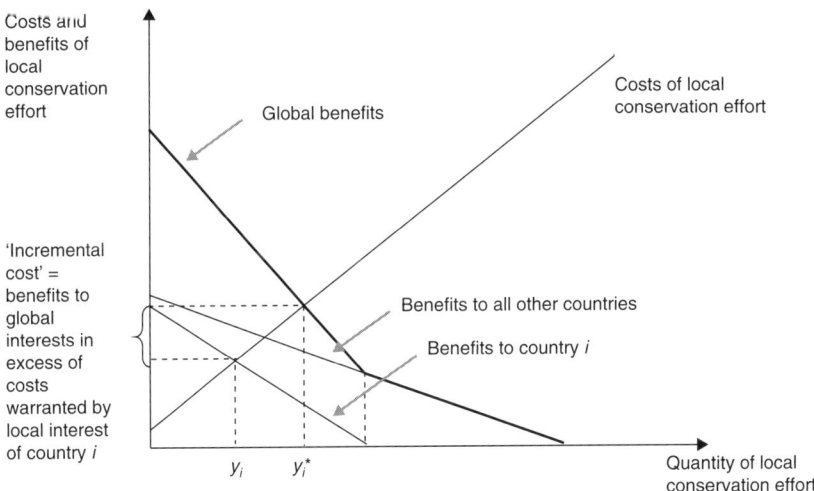

Notes: Because many of the benefits of conservation efforts represent international public goods, the market will not provide an optimal level but only y_i, the demand curve accruing only to country i. For optimal public good provision we need mechanisms to provide for non-market services, moving to y_i^*, the demand curve for all benefits to all other countries, both marketed and non-marketed. The supply curve represents the marginal cost of acquiring and managing additional units of conservation.

Figure 9.2   Efficient provision of conservation effort

that would satisfy international demand (Kanbur 2003, 2004; Ferroni and Mody 2002). We have considered elsewhere cases where the national incentive to produce international environmental public goods is sufficient to meet global demand (Touza and Perrings 2011). Here, we address the case where independent local action is not sufficient to secure efficient global supply.

Consider the conservation of endangered species. Can we rely on national action to produce the efficient amount of an international environmental public good such as the protection of iconic species? The key to understanding this lies in knowing the difference between a pure public good and a private good. For a private good, everyone pays the same price, but is free to consume as much or as little as they want. Consumers adjust the quantity they consume according to the market price. For a pure public good everyone consumes the same amount of the 'good' but is willing to pay a different price for it. Consumers adjust the amount they are willing to pay for the public good according to the quantity supplied (Batina and Ihori 2005). We illustrate the problem in Figure 9.2, in which local and global demand curves for species conservation are presented.

Global demand is represented by the vertical sum of the demand curves of the local community and the rest of the world.

The level of conservation that maximizes local net benefits is indicated by Y_i and the level of conservation that maximizes global benefits is indicated by y_i^*. y_i is given by the intersection of local supply and local demand curves, and y_i^* by the intersection of local supply and the vertical sum of local demand and demand by the rest of the world. The efficient market equilibrium for the supply of the public good requires different prices for each country. In the absence of excludability, however, neither country has an incentive to reveal the strength of their preference for conservation.

The classic public goods provision problem is that once the good is provided, it is available to everyone. Hence, users do not have an incentive to reveal their true preferences. This problem applies to international environmental public goods as well. At the first-best social optimum the public good should be supplied so that the sum of the willingness-to-pay (WTP) for those who benefit from the good is equal to the marginal production cost.

Local biodiversity conservation is an impure global public good. If there are many potential providers, each generates local benefits from its conservation effort, but also benefits from the conservation actions of others. Following Perrings and Gadgil (2003), we characterize the problem as follows. V^i denotes welfare of the ith country, assumed to depend on consumption of a vector of market goods, x^i, and biodiversity conservation, denoted Y, then the problem it faces is of the general form:

$$Max_{x^i y^i} V^i(x^i, y^i, y^1, y^2, \ldots, y^m)$$

The ith country derives direct capturable benefits from y^i, along with the global benefits offered by its contribution to the global public good. The problem to be addressed can therefore be thought of in the following terms.

$$Max_{x^i y^i} V^i(x^i, y^i, C(Y, Z))$$

where $C(Y, Z)$ denotes a conservation function that is increasing in the size of the global public good, Y, $C_Y > 0$, and the resources committed to conservation, Z, $C_Z > 0$. In the absence of cooperation, the well-being of the ith country will be maximized where:

$$\frac{V^i_{y^i}}{V^i_{x^i}} = p - \frac{V^i_c}{V^i_{x^i}} C_{y^i}$$

whereas the global optimum implies that

$$\frac{V^i_{Y^i}}{V^i_{x^i}} = p - \sum_i \frac{V^i_c}{V^i_{x^i}} C_Y$$

The extra terms in the summation term on the right hand side capture the conservation benefits that the ith country confers on others. Furthermore, if the 'cost' of conservation is denoted w, then the globally optimal level of conservation will satisfy:

$$\frac{V^i_{Y^i}}{V^i_{x^i}} = p - w \frac{C_Y}{Y_Z}$$

Notice that there are two problems to be solved. The first is the problem of international market failure. That is given by the incentive to an individual country to neglect the potential benefits it confers on other countries when it commits resources to biodiversity conservation. The second is a problem of domestic market failure. The commitment of y^i resources to local conservation requires the solution to a local public good problem. That is, $V^i = V^i(U^i_1, \ldots, U^i_m)$ and $U^j = U^j(x, y^j, y^1, y^2, \ldots, y^m)$.

The failure of markets to materialize for positive externalities and public goods has serious repercussions for welfare. In the case of international environmental public goods, the lack of payment for these public goods results in underinvestment in the protection, management and establishment of forests, for example. Apart from the loss of valuable environmental services (e.g. genetic resources, air quality maintenance, climate regulation and regulation of human diseases), resulting forest degradation frequently translates into a loss of critical timber and non-timber forest products that is essential to the livelihoods of a wide range of stakeholders (Landell-Mills and Porras 2002).

According to a study of the Office of Development Studies (ODS) of the United Nations Development Programme (UNDP), the definition of 'adequate' provision of public goods clearly varies from one global public good to another. It could, for example, correspond to the complete elimination of the global public bads that indicate under-provision (e.g. in the case of financial stability, the elimination of financial crises). Or, it could reflect what is considered possible, given the current state of technology (e.g. to control, rather than eradicate, the problem of HIV/AIDS). It could also reflect what might be a 'fair' design and shape of the good, i.e. one that would emerge if all concerned stakeholders had an effective voice in the decision-making on provision of the good (ODS-UNDP 2002).

The criterion of adequacy is not meant to indicate optimality, i.e. the balancing of marginal costs with the sum of people's marginal WTP for a particular public good (see Samuelson 1954; Cornes and Sandler 1996). Rather, it is meant to establish a relatively simple, yet reliable, yardstick for measuring the present provision of a certain good against a technical notion of adequacy.

An examination of a specific example may help clarify things. In the case of the global public good 'communicable disease control', it has been possible, given the biological characteristics of the infectious agent and available technologies, to completely eradicate certain diseases. In these cases, the criterion for adequate provision could be defined as complete eradication, or zero incidence in the 'wild'. The determination of 'adequate provision' thus is based solely on technical considerations, without reference to costs, benefits or existing preferences and WTP. Therefore, there may be cases where adequate provision may not be economically feasible within current resource frameworks. However, perceptions of 'affordability' often depend on the returns that a particular investment yields. Thus, it is important to assess the net benefits and costs of inaction against the net benefits and costs of corrective action to determine, at least approximately, whether meeting the technological requirements for adequate provision is economically desirable (UNDP 2002).

POLICY OPTIONS

In principle, the solution to international environmental public good problems of this form lies in payments designed to compensate local providers for the additional costs they incur in meeting global demand. Indeed, that is the basis on which the Global Environment Facility (GEF) was founded. The concept of incremental cost, which notionally determines the payments made by the GEF, is related to the difference between the cost a country would be prepared to bear in the provision of an environmental public good (i.e. the cost that would be warranted in terms of the national benefits generated by the public good) and the cost of meeting global demand for the same public good (Pearce 2003, 2005). This is a national payment for an environmental service that is an international environmental public good.

PES have become popular instruments for dealing with international environmental public goods, in part because they appear to satisfy the incremental cost principle (Ferraro and Simpson 2002; Goldstein et al. 2006; Ferraro and Kiss 2007; Engel et al. 2008; Wunder 2007; Pagiola

2008; Wunder et al. 2008). They are not, however, appropriate mechanisms in all cases. International PES schemes are appropriate where non-marketed ecosystem services are privately supplied in one country, but offer benefits that are public and accrue elsewhere.

To illustrate the potential pluses and minuses of PES schemes, we consider a particular problem: the impact of local deforestation on the provision of a range of international environmental public goods, including climate regulation (through carbon sequestration), protection of genetic diversity, watershed protection and timber and non-timber forest products. The current assessment of The Economics of Ecosystems and Biodiversity (www.teebweb.org) has used existing studies to estimate the mean value of both the macroclimatic regulation offered by terrestrial carbon sequestration, and the change in provisioning and cultural services offered by forest systems. Its findings are preliminary but telling. *The Economics of Ecosystems and Biodiversity* (TEEB 2009; Kumar 2010) suggests that the mean values of forest ecosystem services, in US$ per hectare per year, are dominated by regulatory functions, specifically regulation of climate (US$1965), water flows (US$1360), and soil erosion (US$694). The mean value of all provisioning services combined: timber and non-timber forest products, food, genetic information, pharmaceuticals, is US$1313. This is less than the value of water flow regulation alone. Forest conservation thus brings substantial off-site benefits that are not currently captured by forest landowners.

Governments around the world have frequently implemented forest protection policies in areas that are high in biodiversity or landscape beauty or are important for their watershed protection. However, as pressure mounts on governments to curtail spending and cut budget deficits, their ability to invest directly in the provision of public goods and services is compromised. Where public authorities have been unable to tackle the public good problem, they have searched for ways to involve non-governmental actors. Efforts to transfer responsibility for forest environmental services out of the public sector have relied on a combination of regulation and market-based approaches (Landell-Mills and Porras 2002). Experience has shown that well-designed market-based instruments can achieve environmental goals at less cost than conventional 'command and control' approaches, while creating positive incentives for continual innovation and improvement (Stavins 2003). Examples of such instruments in the forestry sector include stumpage value-based (i.e. standing-timber-based) forest revenue systems, financial and material incentives, long-term forestry concessions, trade liberalization, forest certification and the promotion of markets for non-timber forest products.

The costs and benefits associated with many human activities spill over jurisdictional boundaries, thereby generating externalities that are often reciprocal and quantitatively significant (Cornes 2008). Among payment schemes to internalize the external benefits of maintaining intact forests, Reducing Emissions from Deforestation and forest Degradation (REDD) is an effort to create a financial value for the carbon stored in forests, offering incentives for developing countries to reduce emissions from forested lands and invest in low-carbon paths to sustainable development. REDD goes beyond deforestation and forest degradation, and includes the role of conservation, sustainable management of forests and enhancement of forest carbon stocks.

Deforestation and forest degradation, through agricultural expansion, conversion to pasture, infrastructure development, destructive logging, fires, etc. account for nearly 20 per cent of global greenhouse gas (GHG) emissions. This is more than the GHG contributed by the entire global transportation sector and second only to the energy sector. There is consensus that in order to constrain the impacts of climate change within limits that society will reasonably be able to tolerate, global average temperatures must be stabilized within 2°C, although even this has been criticized for being too high (Smith et al. 2009). This will be practically impossible to achieve without reducing emissions from the forest sector, in addition to other mitigation actions. It is predicted that financial flows for GHG emission reductions from REDD could reach up to US$30 billion a year. This significant North–South flow of funds could reward real reductions in carbon emissions and also support new, pro-poor development, help conserve biodiversity and secure vital ecosystem services. Further, maintaining forest ecosystems intact can contribute to increased resilience to climate change.[1]

The REDD concept emerged out of experience with PES initiatives, which are voluntary transactions wherein buyers of environmental services compensate environmental service providers (Wunder 2007). Services include watershed protection, carbon sequestration and biodiversity conservation. True PES schemes must also have a payment system that is conditional on the actual provision of the environmental service (Wunder et al. 2008). In the case of REDD, the environmental service is the reduction of carbon emissions from forests. However, REDD differs from true PES schemes in that it will probably include official development assistance (ODA) that will not be conditional on the provision of carbon emission reductions (Dutschke and Angelsen 2008; Blom et al. 2010).

The REDD mechanism is expected to be implemented in successive phases, three of which have been proposed. Phase one will focus on strategy development and core capacity-building. Phase two is to provide support for the implementation of national policies and measures,

Paying for international environmental public goods 183

Table 9.1 *Eligibility criteria and activities of REDD implementation phases*

Phase	Eligibility criteria	Activities
1	Must be a Party to the Convention and in compliance with its commitment.	Establish policies and measures for measuring, monitoring, analysing, reporting and verifying emission reduction from the forestry sector; Develop an initial institution to address the reductions of emissions and identify necessary adjustments in forest law and governance.
2	Demonstrate commitment to implement REDD by ensuring: Transparent, rule-based forest governance; Multi-stakeholder consultations and cooperation including with indigenous people and local communities; Safeguards against the conversion of natural to plantation forests; Biological diversity protection.	Develop a comprehensive legal framework including land tenure related to collective land rights, land use planning, forest governance and law enforcement; Establish Monitoring Reporting and Verification (MRV) institutions and capacities; Develop action plans within the framework of a national low carbon development strategy.
3	Remain in compliance with the criteria of phases 1 and 2 and demonstrate that previously received compensation has been spent according to agreed guidelines.	Implement a national inventory of greenhouse gases.

Source: United Nations Framework Convention on Climate Change (2009).

together with compensation for proxy-based results for emission reductions. Phase three is a fully results-based compensation mechanism for emission reductions and removal of forested areas from commercial forestry and land-use sectors. A number of criteria need to be met and specific activities performed by participating countries in order to be eligible for the financial benefits provided for each phase (Table 9.1)

REDD is designed to value and target one important environmental service: carbon sequestration. Still, there is great potential to safeguard and bundle-in other services as well. These benefits may be, for example,

Table 9.2   Service status of selected ecosystems and their accompanying goods and services

Selected ecosystem integrity and ecosystem goods and services	Status of service
Connectivity – fragmentation of ecosystems (GBO3 2010)	–
Water quality of aquatic systems (GBO3 2010)	–
Capture fisheries (MA 2005)	–
Aquaculture (MA 2005)	+
Wild foods (MA 2005)	–
Fibre, timber (GBO3 2010)	+/–
Coverage of Protected Areas (GBO3 2010)	+
Wood fuel (MA 2005)	–
Trends in genetic diversity of domesticated animals, cultivated plants and fish species of major socio-economic importance (GBO3 2010)	–
Area of forest, agricultural and aquaculture ecosystems under sustainable management (GB03 2010)	+/–

Sources: MA = Millennium Ecosystem Assessment 2005; GBO3 = *Global Biodiversity Outlook 3.* of the Convention on Biological Diversity Secretariat 2010.

carbon-dense and biodiversity-rich natural forests, and rehabilitation of degraded forests with indigenous trees to improve watershed services. There is a need to be explicit about safeguarding these multiple benefits of REDD because they may not feature automatically in the range of strategic REDD options. The cost of these losses is felt on the ground, but might go unnoticed at the national level (TEEB 2009). Table 9.2 shows results of MA and Global Biodiversity Outlook (GBO) assessments on integrity of selected ecosystems and their accompanying ecosystem services.

Reaching an international agreement on an instrument for REDD that further emphasizes conservation, sustainable management of forests and enhancement of carbon stocks would properly reward global carbon sequestration and storage services, as well as help to maintain other valuable services provided by forests (TEEB 2009). There is growing recognition that REDD planning requires a broadened approach. A future REDD mechanism should incentivize emissions reduction in all developing countries that are forested, and should address critical non-carbon dimensions of REDD implementation, which include quality of forest governance, conservation priorities, local rights and tenure frameworks, and sub-national project potential (Phelps et al. 2010). A more comprehensive analysis to produce an optimized allocation of REDD and conservation funds within or even among tropical forest countries is technically feasible. Such analysis would allow prioritization of funds across land units. Depending on the carbon price and the baseline rate of deforestation, this would help to identify those areas naturally covered by the mechanism, those requiring additional resources if they are to benefit from the mechanism, and the 'losers', i.e. sites that are most at risk of loss or degradation as the result of pressures displaced by the mechanism. These may become new priorities for conservation and sustainable environmental management (Miles and Kapos 2008).

IMPLICATIONS FOR INTERNATIONAL ENVIRONMENTAL POLICY

Globalization is often associated with increased privateness, i.e. personal ownership of goods, in that economic liberalization is associated with the growth of the number of goods and services allocated through markets, international market integration, and enhanced private cross-border economic activity such as trade, investment, transport, travel, migration and communication. However, globalization is also about increased publicness, i.e. shared consequences, in that people's lives have become more interdependent. Events in one place of the globe often have worldwide repercussions. Moreover, a growing volume of international policy principles, treaties, norms, laws, and standards is concerned with defining common rules for an ever-wider range of activities (Kaul et al. 2003b).

Public goods are recognized as having benefits that cannot easily be confined to a single 'buyer' (or set of 'buyers'). Yet once they are provided, many can enjoy them free of charge. A clean environment is an example. Without a mechanism for collective action, these goods will generally be under-provided. In fact, many crises that dominate the international policy agenda today reflect the under-provision of global public goods. With globalization, externalities are increasingly borne by people in other countries. Indeed, issues that have traditionally been merely national are now global because of the greater interconnectedness of the planet.

Kaul et al. (2003b) suggest a reappraisal of three notions underpinning the theory of public goods. First, properties of non-rivalry in consumption and non-excludability of current benefits do not automatically determine whether a good is public or private. Some goods may be either public or private. Nevertheless, it is important to distinguish between a good having the potential of being public (that is, it having non-rival and non-excludable properties) and its being de facto public (non-exclusive and available for all to consume). Second, public goods do not necessarily have to be provided by the state. Many other actors can, and increasingly do, contribute to their provision. Third, a growing number of public goods are no longer national in scope, having assumed cross-border dimensions. Many have become global and require international cooperation to be adequately provided.

For the most part, the theoretical and empirical literature in economics has focused on two polar models related to public goods provision: the provision of pure public goods which benefit all, and the provision of local public goods which benefit only those in one community (Bloch and Zenginobuz 2007). However, there are cases where people enjoy positive spillovers from the public goods provided by others.

International payments for the local provision of public goods that offer wide public benefits can promote efficient provision of goods at a global scale. In the context of global climate regulation, the United Nations climate treaty may soon include a mechanism for compensating tropical nations that succeed in reducing carbon emissions from deforestation and forest degradation (the REDD programme). These emissions are responsible for nearly one fifth of global carbon emissions. The focus of REDD is on an ecosystem service with global benefits, and thus it could gain access to a large pool of global stakeholders willing to pay to maintain carbon in forests. Because forests offer a number of benefits aside from carbon, the scheme could potentially support the supply of ecosystem services to local stakeholders, who would otherwise be unable to afford them (Stickler et al. 2009). If well designed and implemented, REDD may offer co-benefits in the form of protection of water resources, local and regional climate amelioration, soil resources and biodiversity. While we have used REDD as an illustration of payment mechanisms, we note that the effectiveness of PES schemes depends heavily on the conditionality of payments (Arriagada and Perrings 2009) and, at least initially, this is lacking in REDD.

In summary, there are four main implications for international environmental policy:

1. Diagnosis of the public goods failure associated with particular ecosystem services is critical to the development of an appropriate international response. There are numerous cases in which the incentive structure is such that independent actions by nation states will be 'good enough' to secure the public interest (Touza and Perrings 2011). Where the technology of supply is 'best shot' or where the local benefits are high enough to lead to a level of supply that is close to the global optimum, then the independent actions of nation states will be adequate. However, where local benefits lead to a level of local supply that leaves global demand unsatisfied, then international coordination

or cooperation in the delivery of ecosystem services will be required. We note that this depends largely on the nature and strength of off-site effects. Local actions that generate significant off-site benefits or costs are most likely to require international coordination or cooperation. Off-site effects can reflect both natural (through hydrological or atmospheric flows) and social (through trade and travel) transmission of costs or benefits. Since social transmission of these effects is rapidly evolving, understanding social transmission pathways is important to the diagnosis of the public goods market failure.

2. There are two categories of international environmental public good that are most likely to be under-supplied if left to the market. The first comprises public goods that depend on many countries for their production, that are expensive to produce and that are global in reach. Examples include mitigation of climate change, and management of transboundary nutrient flows, which are currently addressed through the United Nations Framework Convention on Climate Change and the Convention on Long Range Transboundary Air Pollution. The second comprises weakest link public goods that are distributed to a large number of countries through global trade, transport and travel. Examples include the management of infectious zoonotic diseases and the control of invasive pest species, which are currently addressed through the International Health Regulations, the Sanitary and Phytosanitary Agreement, and the Convention on Biological Diversity. Both require mechanisms that change the incentives to individual countries.

3. In all cases, it is important to determine the target of international coordination or cooperation. In the case of an international environmental public good with an additive supply technology such as carbon sequestration all countries contribute to the public good, and should be targeted. On the other hand, international cooperation in the management of a weakest link international environmental public good, such as infectious disease control, should be targeted at the weakest link country. Most international environmental public goods lie somewhere in between. Particular countries are more important for the provision of some services than others (e.g. countries with high biodiversity contribute more to the global gene pool than countries having lower levels of biodiversity), so most international contributions to international environmental public goods should target some countries more than others.

4. For international environmental public goods that are supplied by specific countries, support may take the form of direct investment in supply (the GEF model) or payments for the benefits of supply

(the PES model). The fact that the GEF is under-resourced, and only weakly targeted, suggests that the second option may become the dominant mechanism for assuring local provision of international environmental public goods. We have elsewhere discussed the conditions that need to be satisfied for PES schemes to be effective (Arriagada and Perrings 2009). The most important of these is that payments for services should be conditional on the supply of those services. Where PES schemes have income transfer, poverty alleviation and public good supply objectives, conditionality may be lost altogether. It is important that the design of PES schemes fits the diagnosis of the public good problem, and the technology of public good supply.

NOTE

1. See http://www.un-redd.org/

REFERENCES

Anand, P. (2004). Financing the provision of public goods. *The World Economy*, **27**, 215–237.

Arriagada, R. and C. Perrings (2009). Making Payments for Ecosystem Services work. Ecosystem Services Economics Working Papers. United Nations Environment Programme, Nairobi.

Barrett, S. (2007). *Why Cooperate? The Incentive to Supply Global Public Goods.* Oxford, UK: Oxford University Press.

Batina, R. and T. Ihori (2005). *Public Goods: Theories and Evidence.* Berlin: Springer-Verlag.

Bloch, F. and U. Zenginobuz (2007). The effect of spillovers on the provision of local public goods. *Review of Economic Design*, **11**, 199–216.

Blom, B., T. Sunderland and D. Murdiyarso (2010). Getting REDD to work locally: lessons learned from integrated conservation and development projects. *Environmental Science and Policy*, **13**, 164–172.

Convention on Biological Diversity Secretariat (2010). *Global Biodiversity Outlook 3.* Montreal: Convention on Biological Diversity Secretariat, p. 94.

Cornes, R. (2008). Global public goods and commons: theoretical challenges for a changing world. *International Tax and Public Finance*, **15**, 353–359.

Cornes, R. and T. Sandler (1996). *The Theory of Externalities, Public Goods and Club Goods.* New York: Cambridge University Press.

Dombrowsky, I. (2007). *Conflict, Cooperation and Institutions in International Water Management: An Economic Analysis.* Cheltenham, UK and Northampton, MA, USA: Edward Elgar.

Dutschke, M. and A. Angelsen (2008). How do we ensure permanence and assign liability? In: A. Angelsen (ed.), *Moving Ahead with REDD*. Bogor, Indonesia: CIFOR.

Engel, S., S. Pagiola and S. Wunder (2008). Designing payments for environmental services in theory and practice: an overview of the issue. *Ecological Economics*, **65**(4), 663–674.

Ferraro, P. and A. Kiss (2007). Direct payments to conserve biodiversity. *Science*, **298**, 1718–1719.

Ferraro, P.J. and R.D. Simpson (2002). The cost effectiveness of conservation payments. *Land Economics*, **78**, 339–353.

Ferroni, M. and A. Mody (2002). Global incentives for international public goods: introduction and overview. In M. Ferroni and A. Mody (eds), *International Public Goods: Incentives, Measurement, and Financing*. Dordrecht, the Netherlands: Kluwer Academic Publishers.

Goldstein, J.H., G.C. Daily, J.B. Friday, P.A. Matson and R.A. Naylor (2006). Business strategies for conservation on private lands: Koa forestry as a case study. *Proceedings of the National Academy of Sciences of the United States of America*, **103**, 10140–10145.

Holzinger, K. (2001). Aggregation technology of common goods and its strategic consequences: Global warming, biodiversity, and sitting conflicts. *European Journal of Political Research*, **40**, 117–138.

Kanbur, R. (2003). IFIs and IPGs: operational implications for the World Bank. In A. Buira (ed.), *Challenges to the World Bank and IMF: Developing Country Perspectives*. London: Anthem Press, pp. 251–266.

Kanbur, R. (2004). Cross-border externalities, international public goods and their implications for aid agencies. In L. Beneria and S. Bisnath (eds), *Global Tensions: Challenges and Opportunities in the World Economy*. New York: Routledge, pp. 65–75.

Kaul, I., I. Grunberg and M. Stern (1999). Defining global public goods. In I. Kaul, I. Grunberg and M. Stern (eds), *Global Public Goods: International Cooperation in the 21st Century*. Oxford: Oxford University Press, pp. 2–19.

Kaul, I., P. Conceicao, K. Le Goulven and R. Mendoza (2003a). How to improve the provision of global public goods. In I. Kaul, P. Conceicao, K. Le Goulven and R. Mendoza (eds), *Providing Global Public Goods: Managing Globalization*. Oxford: Oxford University Press, pp. 21–58.

Kaul, I., P. Conceicao, K. Le Goulven and R. Mendoza (2003b). Why do global public goods matter today? In I. Kaul, P. Conceicao, K. Le Goulven and R. Mendoza (eds), *Providing Global Public Goods: Managing Globalization*. Oxford: Oxford University Press, pp. 2–20.

Kumar, P. (2010) *The Economics of Ecosystems and Biodiversity. Ecological and Economic Foundations*. London: Earthscan.

Landell-Mills, N. and I. Porras (2002). *Silver Bullet or Fool's Gold? A Global Review of Markets for Forest Environmental Services and Their Impact on the Poor*. London: International Institute for Environment and Development.

Madsen, B., N. Carroll and K. Moore Brands (2010). *Offset and Compensation Programs Worldwide*. Washington DC: Ecosystem Marketplace.

Miles, L. and V. Kapos (2008). Reducing greenhouse gas emissions from deforestation and forest degradation: Global land-use implications. *Science*, **320**, 1454.

Millennium Ecosystem Assessment (2005). *Ecosystems and Human Well-being: Synthesis*. Washington DC: Island Press.

Morrissey, O., D. Te Velde and A. Hewitt (2002). Denning international public goods: conceptual issues. In M. Ferroni and A. Mody (eds), *International Public Goods: Incentives, Measurement, and Financing*. Dordrecht, the Netherlands: Kluwer Academic Publishers.

Nordhaus, W. (2005). Paul Samuelson and global public goods. Working Paper. Yale University, New Haven, CT.

ODS-UNDP (2002). *Profiling the Provision Status of Global Public Goods*. New York: Office of Development Studies, United Nations Development Programme (ODS-UNDP).

Ostrom, E. (1990). *Governing the Commons: the Evolution of Institutions for Collective Action*. Cambridge: Cambridge University Press.

Pagiola, S. (2008). Payments for environmental services in Costa Rica, *Ecological Economics*, **65**, 712–724.

Pearce, D.W. (2003). The social cost of carbon and its policy implications. *Oxford Review of Economic Policy*, **19**, 362–384.

Pearce, D.W. (2005). Paradoxes in biodiversity conservation. *World Economics*, **6**, 57–69.

Perrings, C. and M. Gadgil (2003). Conserving biodiversity: reconciling local and global public benefits. In I. Kaul, P. Conceicao, K. Le Goulven and R. Mendoza (eds), *Providing Global Public Goods: Managing Globalization*. Oxford: Oxford University Press, pp. 532–555.

Perrings, C., M. Williamson, E.B. Barbier, D. Delfino, S. Dalmazzone, J. Shogren, P. Simmons and A. Watkinson (2002). Biological invasion risks and the public good: an economic perspective. *Conservation Ecology*, **6**, 1–7.

Phelps, J., M. Guerrero, D. Dalabajan, B. Young and E.L. Webb (2010). What makes a REDD country? *Global Environmental Change*, **20**, 322–332.

Samuelson, P. (1954). The pure theory of public expenditure. *Review of Economics and Statistics*, **36**, 387–389.

Sandler, T. (2004). *Global Collective Action*. Cambridge: Cambridge University Press.

Smith, J.B., S.H. Schneider, M. Oppenheimer, G.W. Yohe, W. Hare, M.D. Mastrandrea, A. Patwardhan, I. Burton, J. Corfee-Morlot, C.H. Magadza, H.-M. Füssel, A.B. Pittock, A. Rahman, A. Suarez and J.-P. van Ypersele (2009). Assessing dangerous climate change through an update of the Intergovernmental Panel on Climate Change (IPCC) 'reasons for concern'. *Proceedings of the National Academy of Sciences*, **106**(11), 4133.

Stavins, R.N. (2003). Experience with market-based environmental policy instruments. In K.-G. Mäler and J.R. Vincent (eds), *Handbook of Environmental Economics*, Vol. 1. Amsterdam: North Holland, pp. 355–435.

Stickler, C., D. Nepstad, M. Coe, D. Mcgrath, H. Rodriguez, W. Walker, B. Soares-Filho and E. Davidson (2009). The potential ecological costs and co-benefits of REDD: a critical review and case study from the Amazon region. *Global Change Biology*, **15**, 2803–2824.

TEEB (2009). *TEEB Climate Issues Update*. Nairobi: United Nations Environment Programme.

Touza, J. and C. Perrings (2011) Strategic behavior and the scope for unilateral provision of transboundary ecosystem services that are international environmental public goods. *Strategic Behavior and the Environment*, **1**(2), 89–117.

United Nations Development Programme (UNDP) (2002). *Profiling the Provision*

Status of Global Public Goods. New York: UNDP Office of Development Studies.

United Nations Framework Convention on Climate Change (2009). Revised negotiating text. Ad-hoc Working Group on Long-term Cooperative Action under the Convention, Sixth Session, Bonn, 1–12 June 2009. United Nations Framework Convention on Climate Change, Geneva.

Wunder, S. (2007). The efficiency of payments for environmental services in tropical conservation. *Conservation Biology*, **21**(1), 48–58.

Wunder, S., S. Engel and S. Pagiola (2008). Taking stock: a comparative analysis of payments for environmental services programs in developed and developing countries. *Ecological Economics*, **65**, 834–852.

10. Institution and ecosystem functions: the case of Keti Bunder, Pakistan

John M. Gowdy and Aneel Salman

INTRODUCTION: ECOSYSTEMS, ECONOMICS AND DEVELOPMENT

A consensus exists among conservation biologists that the Earth is entering a period where loss of biodiversity will be on the scale of the five major extinction episodes during the past 600 million years of complex life. It is no exaggeration to say that human activity within the past 100 years has drastically altered the course of evolution. The International Union for Conservation of Nature (IUCN) estimates that a quarter of mammal species face extinction (Gilbert 2008). At least 12 per cent of birds are threatened, together with 30 per cent of amphibians and 5 per cent of reptiles. Particularly alarming is the state of the world's oceans and coastal areas. Human-caused threats to ocean biodiversity are summarized by Jackson (2008):

> Today, the synergistic effects of human impacts are laying the groundwork for a comparatively great Anthropocene mass extinction in the oceans with unknown ecological and evolutionary consequences. Synergistic effects of habitat destruction, overfishing, introduced species, warming, acidification, toxins and mass runoff of nutrients are transforming once complex ecosystems like coral reefs into monotonous level bottoms, transforming clear and productive coastal seas into anoxic dead zones, and transforming complex food webs topped by big animals into simplified, microbially-dominated ecosystems with boom and bust cycles of toxic dinoflagellate blooms, jellyfish, and disease.

The alarm expressed by biologists about the state of the world's ecosystems is in sharp contrast to the attitude of most economists. Dasgupta (2008) puts it bluntly: 'Nature has been ill-served by 20th century economics'. Economists have frequently been guilty of 'misplaced concreteness' (Georgescu-Roegen 1971; Daly 1985) in assuming that the standard general equilibrium model of the economy can deal adequately with very long-term and dynamically complex issues like biodiversity loss and climate change. In the twenty-first century, attitudes among economists are beginning to change as (1) the magnitude of the environmental crisis becomes increasingly apparent; (2) the standard model is enriched by behavioural economics, neuro-economics and evolutionary game theory; and (3) the current financial crisis exposes the fragility of the world economic system, the growth of which has been based to a large extent on transforming natural into financial capital (Gowdy and Krall 2009; Kallis et al. 2009).

The field of ecological economics was established some 20 years ago in response to the divide between mainstream economics and those economists and biologists concerned about the degradation of the natural world. Today that divide is much smaller for two reasons. First of all, welfare economic theory is undergoing a revolution as economic models increasingly accommodate alternatives to the assumptions of 'rational economic man' and 'perfect competition'. This has broadened the concept of 'utility' beyond mere per capita income, and the concept of 'production' to include the unpaid services of nature. Second, as traditional models of economic growth are expanded to include the services of nature, the policy recommendations of mainstream and ecological economists are converging. A consensus is emerging among environmental and ecological economists around the belief that, if human economic activity is to become sustainable, damages to the Earth's supporting ecological and physical systems must be reversed.

But even if economists begin to recognize the importance of natural capital, why should the loss of biodiversity and ecosystem integrity be an economic development concern? In the past, economists have argued that only the wealthy have the luxury to be concerned about environmental quality. The received political maxim is that: the best way to protect ecosystems is to encourage economic growth so that the poor can have enough income to become 'green'. This is the so-called 'post-materialist' thesis of Inglehart (1990), Krutilla (1967) and others. Martinez-Alier (1995) points out two flaws in this thesis. First, continued material growth implies increased environmental stress. Second, compared to the wealthy, the world's poorest receive a much larger percentage of their livelihoods directly from ecosystems, and thus have a much greater incentive to preserve local ecosystems. The 'GDP of the poor' (TEEB 2010) is undervalued because so much of it depends on unpriced inputs from nature. Estimates from India suggest that ecosystem services add to only 7 per cent of measured GDP but add 57 per cent to the GDP of the country's poorest (Sukdev and Bishop 2008). Yet, in general, it is the rich who decide whether to preserve biodiversity and ecosystem functions, and they do this based on market rates of interest and investment opportunities. Although the poor have a larger stake than the rich in preserving ecosystem functions and the resulting flows of ecosystem services, a large percentage of the world's disadvantaged are in such a desperate position that they must sometimes sacrifice these services for immediate gain.

We explore the value of the services of ecosystems and biodiversity from several perspectives, including market economics, human well-being and ecosystem functioning as a biological process. A major focus is the value of regulating services; those benefits obtained from the smooth functioning of ecosystem processes, including maintaining water quality and quantity, erosion control, water purification, regulation of human disease vectors, provision of biological species, pollination and storm protection. Preliminary survey results from the village of Keti Bunder on the southeastern coast of Pakistan are used to illuminate this discussion, which deals with the functions of ecosystems and biodiversity and looks at various methods used to value them. We then discuss the potential impacts of climate change on coastal ecosystems in South Asia, before considering the relevance of some innovative ideas in development economics to ecosystem valuation. Lastly, we examine the current situation in Keti Bunder, and present some relevant preliminary results from a survey of Keti Bunder conducted by the authors.

ECOSYSTEM FUNCTIONS AND VALUES OF ECOSYSTEM SERVICES

The Millennium Ecosystem Assessment (2005) provides a useful classification of ecosystem services: supporting, provisioning, regulating and cultural. The values of ecosystems and biodiversity can be seen as layers of a hierarchy moving from market value (supporting, provisioning and regulating) to non-market value (cultural and psychological), and to the roles of biodiversity and complexity in preserving ecosystem resilience. These levels of value point to the need for a pluralistic and flexible methodology to determine appropriate policies for ecosystem use and preserving ecosystem functions (Gowdy 1997).

The Economic Value of Ecosystems and Biodiversity

Economists have used many approaches to place economic value on ecosystem services. One way is to follow the net national product (NNP) accounting framework and to calculate the 'shadow price' of a particular service. This involves using a constrained optimization model to specify the necessary conditions for a social welfare optimum including the use of environmental goods. This method has been constructively employed to estimate the benefits of environmental services but it brings with it numerous assumptions including strictly rational behaviour by all economic actors, perfect competition, near to equilibrium markets, and so on (see the discussion by Maler et al. 2008). Another approach is to make a direct estimate of the contributions of ecosystems and biodiversity to economic activity[1] including, for example, eco-tourism, recreation and the value of direct biological inputs such as fisheries and forests. Yet another way is to estimate the costs avoided by preserving an ecosystem, for example avoiding the costs of flooding consequent to the destruction of a wetland. A classic example is Chichilnisky and Heal's (1998) study of the value of the Catskill watershed in protecting the purity of New York City's drinking water. Augmenting the water purification capacity of the watershed was estimated to cost US$1.3 billion while the cost of building a new water treatment plant was estimated to be US$8 billion. Another example of ecosystem service loss is the extinction of bees in Maoxian County, China. Formerly it took two bee colonies to pollinate 100 apple trees, a task that now must done by hand by 20 to 25 people (Melillo and Sala 2008).[2]

In a simple system it may be possible to estimate the contribution of a particular ecosystem to a particular economic activity. For example, the value of mangroves to the fishing industry might be estimated by the following equation, where Q is the output (or value of the catch) of fish, S is the area of mangroves, and Xs are the other inputs (Barbier 2000):

$$Q = F(X1...X1, S)$$

Barbier (2000) gives an excellent survey of the value of mangroves as an input to the fishing industries of several countries. Complications exist with this approach owing to complex interactions and positive and negative biological and institutional feedback effects in coastal ecosystems. Perhaps the greatest drawback is the existence of large and unpredictable externally derived fluctuations both in fish stocks and market prices.

Although evidence from contingent valuation, hedonic pricing and other economic valuation tools underscores the importance of biodiversity and ecosystem functions, these methods give incomplete, lower-bound estimates of their values (Nunes and van den Bergh 2001; Maler et al. 2008). It should also be noted that environmental accounting prices do not value the environment as a whole (Dasgupta et al. 2000). Rather, these estimates place monetary values on specific economic services of specific ecosystem functions. The total value of the Earth's environment is infinite and thus is not amenable to economic analysis. If any one of the world's major biophysical systems is perturbed beyond a certain point, the existence of all human life would be threatened (Barnosky et al. 2012).

The Socio-cultural Value of Ecosystems and Biodiversity

We also know that the biological world contributes to human psychological well-being in ways that can be empirically measured (Wilson 1994; Kellert 1996) but not adequately valued in a traditional social welfare framework (Orr 1994; Norton 2005). Spiritual, cultural and aesthetic benefits of interacting with nature may be considered in a more comprehensive conception of utility such as the Bentham–Kahneman notion of utility as well-being. This goes back to Georgescu-Roegen's (1971) definition of utility as an 'immaterial flux' – the enjoyment of life.

In traditional communities, spiritual ties to the land are much stronger than in Western societies. Hunter-gatherers and traditional agriculturalists depend directly on solar energy and its immediate conversion to plants and animals for their day-to-day existence. If the direct services from nature are disturbed, the consequences are immediately apparent. Thus, all traditional societies have strict rules and regulations to protect these funds (see the examples in Gowdy 1998). Among the inhabitants of the Nicobar Islands, for example, numerous ceremonies and rituals are linked to the regulation of natural resource use (Singh 2006).

The Ecological Value of Biodiversity and Complexity to Ecosystems

Evidence suggests that biologically diverse ecosystems are more resilient to environmental shocks than less diverse ecosystems (Tilman and Downing 1994), although the relationship between resilience and biodiversity is complicated (Robinson 1992). If a system loses its resilience, it can quickly and irreversibly flip to another state (Walker et al. 2004). Furthermore, it is impossible to tell ahead of time what the loss of a species or species relationships will do to the system. In general, removing keystone species from an ecosystem will have significant (and non-marginal) effects. For example, Brock and Kelt (2004) removed kangaroo rats from a plot of land in the southwest United States of America and the result was a significant increase in plant cover, significant declines in bare ground, and declines in seed predation. The effects of removing a species or otherwise altering an ecosystem can only be known after the alteration, an effect referred to by Bromley (1998) as 'functional transparency'.

The Services of Mangrove Ecosystems

Mangrove ecosystems are one of the most important natural resources in the tropics. They provide a number of functions including protecting a pool of biodiversity and a providing a habitat for a diverse community of organisms ranging from bacteria and fungi to fish, shrimps, birds, reptiles and mammals. They are an important breeding ground and habitat for economically valuable fish and other seafood, and the mangroves themselves are a source of fuel wood and fodder for livestock. They protect the coastal zone from wind and tidal action and they reduce siltation and the strength of storm surges. Mangroves dissipate the energy and size of waves as a result of the drag forces exerted by their multiple roots and stems. Wave energy may be reduced by 75 per cent during the wave's passage through 200 m of mangrove (World Wildlife Fund-Pakistan 2005a). In October 1999, mangrove forests reduced the impact of a 'super cyclone' that struck Orissa on India's east coast, and mangroves reduced levels of destruction on 26 December 2004, when the tsunami struck Indonesia, the Andaman and Nicobar Islands, Sri Lanka and the coastlines of Kerala and Tamil Nadu. Human settlements located behind the mangrove stands suffered smaller losses compared with other areas (McLeod and Salm 2006).

Because of their important ecosystem functions, and their direct economic uses in tropical areas around the world, mangrove ecosystems are critical to the well-being of a large portion of the world's poorest people. Mangrove forests are in many ways a major focal point in the intersection between poverty, declining ecosystems and global climate change. Over the last 50 years, about one-third of the world's mangrove forests have been destroyed by human activity (Alongi 2002; McLeod and Salm 2006). Climate change threats to mangroves include decreased precipitation in some regions, increased storm damage, and rapid sea-level rise.

CLIMATE CHANGE AND ECOSYSTEMS IN SOUTH ASIA

A consensus has emerged among scientists and policy-makers that global climate change represents a major threat to the well-being of humankind and the biosphere (Intergovernmental Panel on Climate Change 2007; Stern 2007). During the past hundred years or so the average global temperature has risen by about 1°C, with much of that increase due to human activity, especially fossil fuel burning and deforestation. The rate of increase has accelerated during the past 20 years as human impacts have begun to dominate natural processes. The real concern is not what has happened so far but that global temperatures are projected to increase further, by between 1.4°C and 5–8°C by 2100 and to continue to rise long after that (Intergovernmental Panel on Climate Change 2007; Archer 2009). Scenarios of the probable consequences of such an increase differ substantially among regions, but include sea-level rise, shortages of fresh water, biodiversity loss, increased droughts and floods, more frequent and intense forest fires, more intense storms, more extreme heat episodes, agricultural disruption, the spread of infectious diseases and mass migrations. Evidence from past climate regimes indicate that even if CO_2 emissions were immediately halted, the Earth would eventually warm by 2–3°C solely because of past emissions and the inertia of the climate system (Haywood and Williams 2005; Jiang and Wang 2005). All living species are adapted to particular climate regimes and are likely to be in for a rough ride in the coming decades and centuries even if aggressive mitigation steps are taken soon (Anderson and Bows 2008; Schneider 2009).

South Asia is particularly vulnerable to the effects of climate change. About one-quarter of the world's population lives in the four countries of Pakistan, Nepal, India and Bangladesh. Hundreds of millions of people living near the coasts of South Asia will eventually be displaced by rising sea levels, increased severity of storms and shortages of fresh water. Specific physical impacts predicted from climate change models include the following:

Higher average temperatures. Analysis of climatic data shows that there has been an increase in average temperature of 0.6°C to 1.0°C in the coastal areas of South Asia since the early 1990s. During 2005, Bangladesh, India and Pakistan faced temperatures 5–6°C above the world average (United Nations Development Programme 2008). Temperature changes in the mountain ranges of the Himalayas have been even more dramatic, resulting in accelerated melting of glaciers from Nepal to the Hindu Kush (World Wildlife Fund 2008a). These higher temperatures bring a variety of risks to South Asia (Munasinghe 2008).

Changes in rainfall patterns and extreme weather events. The coastal areas of South Asia have suffered from erratic rainfall patterns in recent years. Evidence from the northern mountains is inconclusive owing to the lack of significant monitoring in that area, but the area has apparently experienced an increase in rainfall in recent years. In any case, changing rainfall patterns are expected to be an increasingly serious problem as climate change intensifies. All climate change scenarios point to increased flooding in mountainous regions from glacial melting and in coastal regions due to sea-level rise and the intensification of tropical storms. These changing physical impacts will have unknown but probably adverse consequences on agricultural output, water availability, and migration patterns.

Regional water shortages. The drinking water for much of India and Pakistan comes from the Himalayan, Karakoram and Hindu Kush glaciers, which are already beginning to melt because of warmer temperatures (Jianchu et al. 2007). Climate models indicate that this melting will accelerate in the coming years with unknown but severe consequences for drinking water, agricultural irrigation and human health. Increasing salinity in coastal regions from rising sea levels and more severe storms has already taken a toll on coastal communities (Rees and Collins 2004).

The main effects on human well-being from the physical effects of climate change are:

Loss of agricultural production. South Asian economies are heavily dependent on agriculture; the economic sector most vulnerable to climate change. Crop yields are already declining in the region, possibly due to climate change. According to Intergovernmental Panel on Climate Change (IPCC) Chairman Rajendra Pachuri: 'Wheat production in India is already in decline, for no other reason than climate change. Everyone thought we didn't have to worry about Indian agriculture for several decades. Now we know it's being affected now' (quoted in Worstall 2007). In Pakistan, agricultural yields are also declining and climate change may be the culprit. Changes in the timing of monsoons are already having an adverse effect on agriculture in Pakistan and India.

Climate-induced migration. In recent months, tens of thousands of families in India have been displaced by severe flooding. Chandrashekhar Dasgupta (2007) asks: 'If a developing country is so vulnerable even to normal seasonal variations, how will it cope with the impacts of climate change in terms of floods and droughts, sea-level rise, changes in rainfall patterns, cyclones or typhoons?'. Recent monsoons in India resulted in a major river changing course, causing the loss of thousands of human lives and tens of thousands of refugees. Water shortages in the north of Pakistan and sea-level rise along the coast will eventually result in millions of environmental refugees seeking entry to the nation's major inland cities.

Human health. South Asia's health indicators are among the worst in the world. Infant and child mortality rates are very high. According to a World Bank report (2006), environmental health risks contribute to more than 20 per cent of total disease incidence in the region. Environmental health risks will accelerate with climate change, as water shortages become more acute and as more and more environmental refugees seek comparative safety in inland urban centres. Again, it is the very poor who are most susceptible to the effects of climate change and the resulting disruption of ecosystem functions (Khan 2002).

INSTITUTIONS AND ADAPTATION: THE ECONOMICS OF SUSTAINABLE WELL-BEING

Including only measured economic output in estimates of ecosystem and biodiversity values grossly underestimates their contribution to human well-being (Banuri et al. 2002; Gowdy 2005; Maler et al. 2008). While income is a critical factor in human welfare, it is only one of several important ingredients in quality of life. A growing body of economic research uses measures of subjective well-being as indicators of social welfare (Haq 1997; Frey and Stutzer 2002; Kahneman and Sugden 2005; Agrawal 2008). These measures show that, above some minimal level, per capita income growth and well-being are frequently only weakly correlated (Frey and Stutzer 2002). Focusing only on income misses some equally important contributors to well-being and fails to capture many of the cultural and ecosystem contributions to total value.

In the 1990s, development economists began to call for approaches that go beyond increasing a nation's GDP. Sen (1999) suggested focusing on increasing the ability to live an informed and full life. Sen and Haq developed a comprehensive measure of human well-being called the Human Development Index (HDI). The HDI measures three basic dimensions of human development: health, education and income. The HDI spawned a number of related indices that go deeper in measuring the notion of 'human capabilities'. The 'capability poverty measure' (CPM) looks at three basic capabilities: nourishment and health, the capability of healthy reproduction and female illiteracy (Womenaid International 2007). The CPM measure shows that while 21 per cent of the population in developing countries is below the income poverty line, 37 per cent are below the minimum standard in terms of capability (Womenaid International 2007). Nussbaum (2000) called for a focus on 'distributive justice', that is, creating the conditions for the realization of a set of central human capabilities. Such policies promise to be more effective than simply relying on aggregate income growth alone to improve the lives of the worlds' poorest. They also offer more flexibility in adapting to environmental changes and differing cultural worldviews. With a focus on well-being, individual happiness and self-actualization, the developing world may improve its human welfare position without emulating the environmentally and psychologically destructive consumption patterns that were the drivers of past economic growth in the developed economies.

The research on subjective well-being has important implications for valuing ecosystem functions and formulating policies to protect them. Some evidence indicates that when individuals are more materially secure (not necessary financially wealthier) they are more likely to care about the environment (Rangel 2003). Development policies focusing on subjective well-being might thus pay a double dividend. People would not only be happier, they would also be more willing to support efforts to conserve ecosystem services. But even if policies are based on sustainable welfare, it may not ensure the preservation of biodiversity and ecosystems. Examples abound of societies that apparently worked well in satisfying the broad preferences of their citizens but ended in ecological collapse (McDaniel and Gowdy 1999). Humans get subjective well-being from nature but this does not ensure that individuals living today will choose to preserve those features of nature that may be essential to future generations. Sound measures of the factors contributing to human well-being are needed, but also needed are indicators of the physical and biological requirements for long-term human survival.

As mentioned earlier, the well-being of the world's poorest depends to a large extent on the direct services of ecosystems and on maintaining ecosystem functions. These functions are under threat from worldwide environmental changes as well as from inadequate institutional responses to these changes. Development policies are unlikely to be effective unless these institutional factors are understood and targeted in development initiatives.

KETI BUNDER, PAKISTAN

Keti Bunder is part of the Thatta District in Pakistan's Sindh province, located 200 km southeast of Karachi.[3] It is part of the Indus delta and its four major distributaries: the Chan, Hajamoro, Khobar and Kangri. Keti Bunder consists of 42 village clusters (called *dehs*) spread over a total area of about 60 000 hectares. The village settlements are built on mudflats between the various channels of the major Indus distributaries. The majority of the people in Keti Bunder are fishermen and belong to more than a dozen castes, most of them engaged in small-scale business and agriculture. Almost all the people except a few Pathans and Punjabis speak the Sindhi language. The Punjabis have lived in Keti Bunder for generations and are now part of the local community.

Keti Bunder's climate is typical of South Asian coastal areas. January is the coolest month, with minimum temperatures of 9–5°C while in June and July, temperatures range from 23°C to 36°C. Mild winters extend from

November to February, while the summer season extends from March to October. Average annual rainfall is 220 mm and falls mainly during the monsoon season. The northern part of Keti Bunder is a designated wildlife sanctuary harbouring a significant population of migratory and resident water birds. A recent World Wildlife Fund survey recorded five species of aquatic mammal (dolphins and porpoises), three species of large mammal, and 15 species of small mammal. A total of 69 bird species were observed including 25 resident species and 44 migratory species. Additionally, 21 species of reptile and 2 species of amphibian were recorded. Among fish, 63 species of finfish and 24 species of shellfish were observed (World Wildlife Fund – Pakistan 2008b).

Before 1950, Keti Bunder was a major port and the centre of a prosperous fishing and agricultural area. The entire area now faces a number of severe environmental problems and a resulting loss of livelihood opportunities. As a consequence of the construction of dams and other barriers upstream, which slow the downstream water flow, as well as sea-level rise, salt water intrusion from the sea has become a major problem. The area is vulnerable to cyclones and tsunamis. The intensity of cyclones has increased significantly during the last 30 years, possibly due to global warming. Thousands of Keti Bunder residents may be displaced in the next few years owing to the impacts of storms, rising sea levels, and other expected effects of climate change.

Mangroves are the key environmental resource in Keti Bunder. The Indus delta mangroves represent the largest area of arid climate mangroves in the world. An estimated 95 per cent of the mangroves in the Indus delta belong to the species *Avicennia marina*. Relatively small patches of *Ceriops roxburghiana* (from the *Rhizophora* family) and *Aegicerias corniculatum* (from the *Myrinaceal* family) are found near the mouth of the Indus at Keti Bunder. Mangroves depend on freshwater discharges from the Indus river, a small quantity of freshwater from runoff, and domestic and industrial effluents from Karachi. Mangrove ecosystems are important for maintaining the many commercial fish species along the Pakistan coast. Despite the benefits provided by mangrove ecosystems, they face continuous pressure from human activities. Water flow into the Indus delta has declined from 140 to 40 million acre-feet over the last few years and this, together with sea-level rise, has increased the salinity, which is detrimental to mangrove growth (see Figure 10.1). In 1988–89, mangroves in Pakistan covered 160 000 hectares. This cover was found to be reduced by half, to 80 000 hectares, when WWF-Pakistan surveyed the mangroves in 2002 (World Wildlife Fund – Pakistan 2005b). Overharvesting of mangroves and dumping of refuse and pollutants by the local communities also contributed to this loss. The rate of degradation of mangrove forests in

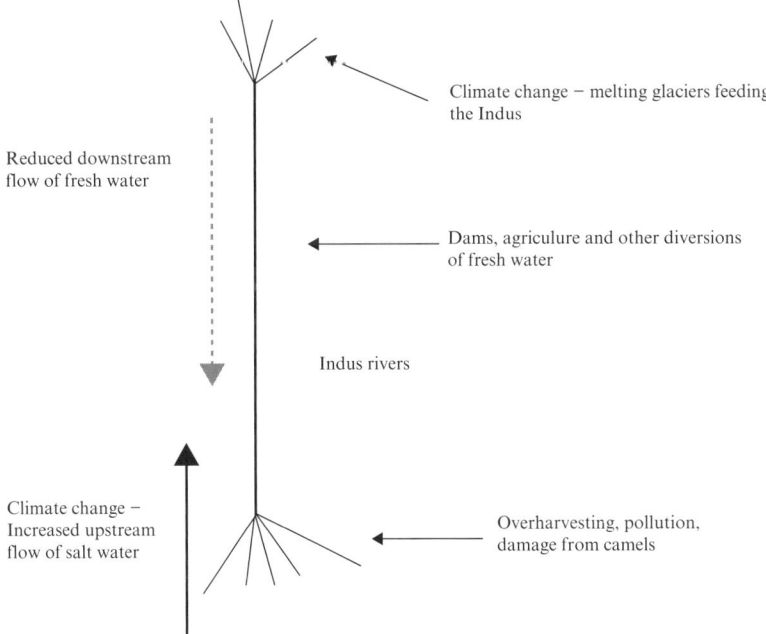

Figure 10.1   Threats to mangroves in Keti Bunder from changes in the Indus watershed

the delta was estimated to be 6 per cent per year between 1980 and 1995 (Thompson and Tirmizi 1995), and only a small percentage of mangroves are now considered to be healthy (World Wildlife Fund – Pakistan 2005b). Water flows to the delta started declining when the Punjab irrigation system was developed in the 1890s, but the construction of the Sukkur (1932), Kotri (1955) and Guddu (1962) dams on the Indus resulted in drastic reductions in water flow to the delta, representing a complete shift from agriculture to fisheries, coastal erosion and depletion of mangroves.

Fresh water for farming in Keti Bunder and other coastal regions comes almost entirely from the Indus river and its distributaries. The declining fresh water flow has negatively affected fish and shrimp breeding, and the upstream migration of the once plentiful *Palla* fish has declined dramatically. There is a water shortage in all villages in the area and water for drinking and cooking must be purchased from sellers in town. The shortage of clean drinking water is the main cause of many illnesses. A lack of health facilities also contributes to the increase in diseases. Common diseases in the community are diarrhoea or dysentery, typhoid, hepatitis B, asthma, tuberculosis, malaria, skin and eye infections and other seasonal diseases (World Wildlife Fund – Pakistan 2005b).

Economic Conditions

A few decades ago the people of Keti Bunder had multiple options for economic subsistence. But the decline in fresh water forced a major change from agriculture and livestock to fishing. A growing population has increased the pressure on natural resources, especially the mangrove ecosystems. Fishing is now the main occupation of the inhabitants of all villages and the sole source of income. The indiscriminate use of banned nets and diminishing mangrove forests has reduced the fish catch. Due to inadequate alternative employment opportunities, the pressure on fisheries resources is intense, and its demise directly affects the livelihoods of everyone in Keti Bunder. The direct economic effect of the loss of freshwater has been the complete loss of the agricultural sector. Indirect effects are the increased incidence of water-borne diseases, other effects of the lack of fresh drinking water and the disappearance of several fish species. Many of these effects are hard to quantify, and even harder to measure using market values, but if consideration is given only to market-measured income losses, the livelihoods of poor people will always be overwhelmed by the monetary gains to the wealthy.

A major internal conflict in Keti Bunder and surrounding villages involves the *Jaat* community that was previously engaged in agriculture and livestock raising but has now turned to fishing owing to land degradation. In competition with indigenous fishermen, Jaats use supposedly banned small-mesh nylon nets to get the maximum catch in the short term. These nets catch juvenile fish of many species, and drive some of them towards extinction (World Wildlife Fund – Pakistan 2008). The Jaats have also increased the camel population, and grazing by these animals damages the mangrove forests. Women have limited roles in decision-making because economic resources are owned by men. Women do have a significant role in the fishing process; usually men fish while women dry small fish for domestic use. Women are also involved in making the cotton fishing nets, but the use of fine mesh nylon nets has crowded out this domestic industry. However, the indigenous fisherwomen of the area enjoy more liberty and freedom than other women of the area. For centuries the fisherwomen have had the greater role in family matters due to the fact that the fishermen are absent for much of the time. Thus, the women of the village have to deal with all day-to-day family matters.

In April 2009, in cooperation with WWF-Pakistan, we administered a detailed questionnaire to Keti Bunder residents to help understand the perceptions of the community regarding the role of ecosystem services in community well-being, vulnerably to climate change, and possible institutional responses to environmental disruption.[4] We believe that these survey results can be generalized to many coastal village communities throughout South Asia. In consultation with WWF-Pakistan we chose two villages in creeks, Bhoori and Tippun, and one inland village, Haji Musa. Bhoori is the largest village, having 400 households. Tippun has 100 households and is significant because the 2007 cyclone levelled the entire village. Haji Musa is a small inland village, having 40 households. All the villages have common problems but are also different demographically and physically. To complement the quantitative socio-economic and ecological data collected through questionnaires, we also used in-depth interviews to collect qualitative data. Focus group discussions involving both male and female community members were used to gain deeper insights and to cross check the survey results. The interviews also provided a detailed intergenerational perspective. The survey used a random sample totalling 55 individuals in Bhoori village (39 males and 16 females), 32 individuals in Tippun village (20 males and 12 females) and 13 individuals in the inland village of Haji Musa (9 males and 4 females).

SURVEY RESULTS

The results of our survey generally confirm those of the earlier WWF survey and the impressions of the WWF field team. Based on our survey results, the main observations of Keti Bunder residents are: (1) a decline in the health of mangroves, (2) the depletion of stocks of major fish species, (3) reduced rainfall, and (4) an increase in extreme weather events. These negative trends are the result of complex interactions between exogenous physical changes (climate change and the reduced availability of fresh water) and changing patterns of resource use within the villages (overfishing and the destruction of mangroves). Another complicating socio-economic factor is the increased reliance on short-term borrowing by local fishermen, which has led to increased pressure to generate income. It also appears that the decline in ecosystem services has exacerbated ethnic tensions among the various Keti Bunder communities.

Environmental change was analysed by using a Likert scale to rank the state of mangroves, the frequency of cyclones, and rainfall currently and 20 years ago (1 poor, 5 good for the state of mangroves and rainfall; 1 less frequent, 5 more frequent for cyclones). As shown in Table 10.1, the results are quite striking. Males and females in all villages surveyed strongly agreed that the state of mangroves has deteriorated, that rainfall is less frequent, and that the frequency of cyclones has increased.

Table 10.1 Mangroves, cyclones and rainfall. Average responses of villagers to questions on perceptions of environmental change 1989–2009. Individual responses were 1 = poor or less frequent; 5 = good or more frequent

Village	Numbers and sex of participants		State of Mangroves		Frequency of cyclones		Rainfall	
			2009	20yrs	2009	20yrs	2009	20yrs
Bhoori	Male	39	2.4	4.0	5.0	2.3	1.4	4.5
	Female	16	2.6	4.1	5.0	2.6	1.3	4.2
Tippun	Male	20	2.1	4.4	4.9	2.5	1.5	4.3
	Female	12	2.0	4.5	4.7	2.0	1.5	4.3
Haji Musa	Male	9	1.7	4.3	5.0	1.9	1.6	4.3
Jaat	Female	4	1.75	4.25	5.0	2.0	1.5	4.2
		100						

In-depth interviews confirmed that the communities are aware of the role of mangroves in providing ecosystem services such as replenishing the fish stock, preventing damage from storms, preventing soil erosion, and providing wood and fodder for household consumption. The communities of Bhoori and Tippun, which are living in creeks, responded that the mangrove cover has been substantially reduced over the past 20 years. Both these communities are now replanting mangroves. One resident reported:

> When I was young, mangroves were rich and dense. The frequency of cyclones was once in a decade. People were happy and there was prosperity. There was wildlife, and birds of different kinds would migrate every year. I remember them chirping in the mornings. But things have changed in the past 40 years, there are almost no rains and the frequency of storms and cyclones has increased. Mangrove cover has been reduced and some mangrove species no longer exist. There is more poverty and the fish stock has been drastically reduced. The pressure on fish has increased as all the former farmers (Jatts) became fishermen'. (Mohammad Suleman, aged 70, Tippun)

The survey showed a near unanimous response about the lack of rainfall in recent years compared with earlier times. At present, there are almost no rains, and this is exacerbating the shortage of fresh water and reducing fish stocks and the availability of drinking water. Twenty years ago the villages were receiving an adequate amount of rain, and fresh water was abundant.

A variety of fish species inhabits the coastal waters and creeks and these are the primary source of livelihood for the local population. Fish, shrimp and crabs are key economic marine resources. Survey responses

Table 10.2 Decline in fish catch. Comments of villagers from the three villages participating in the survey

Name of fish	Status 10 years back	Present status
Suo	Boat full of the fish was caught	It has almost vanished
Dangra	Boatful was caught	Almost vanished
All	One fishing boat caught about 1000 to 2000 fish	Hardly 500 fish are caught
Paplet	Its catch was 200 to 300 kg in one trip	One boat only brings 50 fish
Seeri	Its catch was 1000 kg per fishing trip	Its catch has reduced to only 10 kg per fishing trip
Danthi	Its catch was 100 to 200 kg per fishing trip	It is only 10 to 20 kg
Goli	Its catch was up to 100 kg per fishing trip	It is only up to 5 kg
Sodi	Its catch was up to 100 kg per fishing trip	Now only up to 5 kg
Khaga	Its catch was 1000 to 2000 fish per fishing trip	Now one boat brings hardly 20 to 50 fish
Palla	Boatful was caught	Completely vanished

(Tables 10.2 and 10.3) indicate that fish and shrimp stocks have decreased drastically during the last 20 years. Reasons given by villagers for the declines include the effects of deep-sea trawlers, illegal nets and lack of freshwater from the Indus.

The availability of freshwater has reduced, but the communities themselves are involved in unsustainable fishing practices that exacerbate the problems encountered. The juvenile fish caught using banned nets are considered trash and are used for poultry feed. This activity is reducing the fish stock at an alarming rate. At times there are conflicts among the community when they fish in each other's designated zones. Sea lords and commercial trawlers forcibly occupy some of the channels and creeks, and do not allow the local fishermen to harvest in their marked territories. The coastal ecosystem is becoming fragile owing to the increase in the number of fishermen, the increase in fishing boats, and an increase in the duration of individual fishing efforts. One villager observed:

> The fish stock has declined by more than 50 per cent in recent years. Palla fish, which is very expensive, has almost vanished. Earlier, one fish might weigh 5–6 kilos but now they are around 1 kilo. The use of illegal nets has reduced the fish stock drastically. When we used Thukri (cotton) nets we would catch fewer fish

Table 10.3 Decline in shrimp catch. Comments of villagers from the three villages participating in the survey

Shrimp	Status 10 years back	Present status
Jahira	Its catch was up to 500 kg Per fishing trip	Now it is hardly 5 kg per fishing trip
Tiger	Its catch was up to 100 kg per fishing trip	Hardly one tiger shrimp is caught per fishing trip

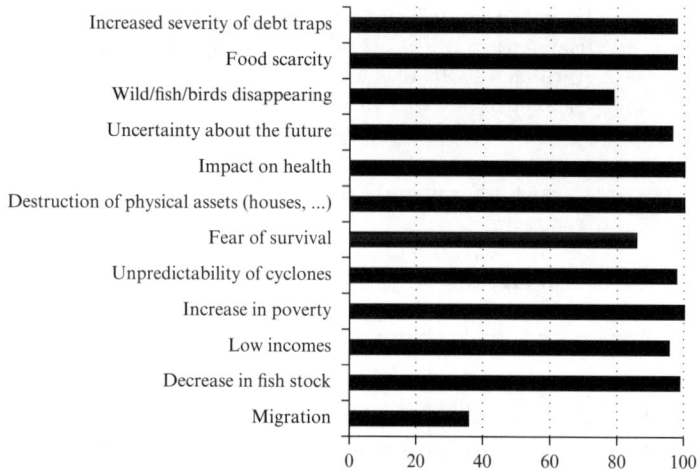

Figure 10.2 Percentage of respondents reporting negative environmental and economic changes

at a time but catch more overall. Now with the Katra nets we catch more fish at a time but less overall. The reason for the decline is that the fish are badly damaged by the nylon nets. Sea lords and big trawlers take all the fish and we are left with almost no fish. The quality of the water has changed and this has changed the taste and size of the fish.
(Yaqoob, aged 60, Haji Musa, Jaat)

There has been a direct loss of income due to declining fish stocks, as shown in Figure 10.2. One of the biggest effects of the declining catch has been the growing debt trap. This is due to increased borrowing because of the decline in, and increasing volatility of, the income stream of fishermen. Respondents to the survey indicated that in the past they used to lend money, but now they have become borrowers. Table 10.2 summarizes the negative changes reported in the survey responses.

Migration is not an option for many owing to the costs and uncertainties involved. Dollu Yar, one of the residents of Bhoori village said:

> Migration is very limited, even though extreme events have occurred, because we are poor and have no financial assets and no options. If we were financially strong we could migrate temporarily. The cost of living in urban areas is very high and we cannot afford to live there. In June 2007 we received the news that a cyclone was coming and I asked my wife if we should go out of the area. My wife replied, 'what is the assurance that our lives would be safe if we go somewhere else. If Allah wants to keep us safe we will survive here'.
> (Dollu Yar, aged 40, Bhoori)

Almost all the respondents own their houses. Since fishing is the only source of livelihood, most of the fishermen own their boats. But villagers who are poor and cannot afford a boat, or had one that was damaged in cyclones, work as helpers. Some communities still have buffalos and poultry for domestic consumption of milk and eggs. The number of livestock has been drastically reduced compared with 30–40 years ago, when freshwater and fodder was available.

Survey responses show that women are entirely dependent on the men. According to one woman:

> Due to cultural constraints our roles are restricted. We help our men-folk in domestic chores, taking care of children, embroidery and drying the fish. We do not have many employment opportunities but earlier we used to make nets, but now they are replaced by nylon nets. Financially we can contribute very little by doing embroidery but it is limited because we do not have market access. We do not have hand pumps and have to buy water from the market. I take care of water consumption and budget it. The water is not clean and it causes gastric problems. I use cotton cloth to filter it. The health facilities are poor and we get expired medicines, the quality of medicines is very bad. I am aware of the changes in the environment but weather events are unpredictable. Our traditional knowledge does not help us as earlier we could predict from the visible signs from the colour of clouds, wind direction, behaviour of birds but now we cannot. Mangroves are important in fish generation and they protect us from storms. I am replanting the mangroves as it will protect my family and hut.
> (Amnah, aged 65, Tippun)

Respondents were also aware of the connection between changing climate, the frequency of storms and the deterioration of mangrove forests. As indicated earlier, the vast majority of the villagers are illiterate. Those who manage to obtain a moderate amount of education leave the village to work in the cities. But the lack of formal education should not mask the positive effects of local education from the combination of real-world experience and science-based information campaigns by the WWF and other non-governmental organizations. The people of Keti Bunder are well aware of the importance to their future well-being of restoring damaged ecosystems.

CONCLUSION; INSTITUTIONS, POWER, AND ENVIRONMENTAL CHANGE IN KETI BUNDER

Our survey confirms these previous general observations regarding ecosystem services for the poor, and the deteriorating state of these ecosystems:

- Ecosystem functions are directly important for the well-being of the world's poorest. The residents of Keti Bunder depend entirely on the area's fisheries and mangrove ecosystems for their livelihoods. The contention that the GDP of the poor is largely composed of direct ecosystem services is confirmed in Keti Bunder.
- Coastal ecosystems are under serious threat both from environmental changes and institutional failures. To the extent that Keti Bunder is representative, both of these factors are likely to become more serious in South Asia in the future.
- Understanding local institutions and institutional failures is the key to formulating effective social and environmental policies.

More specifically, our survey and interviews confirm and extend the findings of previous research in Keti Bunder:

- A reduction in the health of Indus delta fisheries from the loss of mangroves and overharvesting due to the use of banned nets;
- The loss of agricultural land due to increasing salinity;
- A reduction in fresh water availability due to upstream diversion and salt water intrusion from rising sea levels;
- Increased vulnerability to climate events;
- Increased local conflicts over declining ecosystem services.

An ominous factor in all these observations is the increased institutional failure arising from migration, ethnic conflicts and changing power relationships. Environmental deterioration has led to a change in occupation for one group (Jaats), from agriculture to fishing, causing increased pressure on fish stocks from unsustainable fishing techniques. An apparent increase in the number of camels by the Jaats has caused conflicts over mangroves and exacerbated their destruction. This sort of resource conflict among the world's poorest is happening throughout the coastal areas of South Asia and is likely to get much worse as climate change disrupts local ecosystems. The implications are sobering not only for the future of the environment but also for social stability worldwide.

ACKNOWLEDGEMENTS

The authors would like to acknowledge the assistance of the World Wildlife Fund – Pakistan for providing logistical support in Keti Bunder and access to past surveys of the village. This work was also supported by a grant from the South Asia Network of Economic Research Institutes (SANEI). An earlier version of this chapter was presented at the International Workshop on Valuation of Regulating Services of Ecosystems, Bangalore, India, 15–16 June, 2009. The authors thank Nilanjan Ghosh, Pushpam Kumar and Mike Wood for comments on an earlier version of this chapter.

NOTES

1. It is useful to distinguish between stocks whose flows can be called forth at any desired rate (a ton of coal for example) and funds whose services can only be used at a given rate (solar energy, labour power or ecosystems) (Georgescu-Roegen 1984).
2. The macroeconomic implications of these examples are disturbing. The loss of ecosystem services can actually be an economic boon by creating jobs where none existed before. The macroeconomy does not operate by the same principles as an individual firm.
3. The following discussion containing general information about Keti Bunder is taken from the WWF-Pakistan reports: Keti Bunder Village Development (WWF-Pakistan 2005) and Fish Marketing Chain and Economic Analysis of Indebtedness of Fisher-folk of Keti Bunder (World Wildlife Fund – Pakistan 2006).
4. This research was carried out by the WWF and Emily Woodhouse of Imperial College, London.

REFERENCES

Agrawal, A. (2008). *The Role of Local Institutions in Adaptation to Climate Change*. The Social Development Department. Washington DC: World Bank.

Alongi, D. (2002). Present state and future of the worlds mangrove forests. *Environmental Conservation*, **29**(3), 331–349.

Anderson, K. and Bows A. (2008). Reframing the climate change challenge in light of post-2000 emission trends. *Philosophical Transactions of the Royal Society A*, **366**(1882), 3863–3882.

Archer, D. (2009). *The Long Thaw*. Princeton, NJ: Princeton University Press.

Banuri, T., A. Najam and N. Odeh (2002). *Civic Entrepreneurship: A Civil Society Perspective on Sustainable Development*. Vol. IV, South Asia. Islamabad: Gandhara Academy Press.

Barbier, E. (2000). Valuing the environment as input: review of applications to mangrove–fishery linkages. *Ecological Economics*, **35**, 47–61.

Barnosky, A.D., E.A. Hadly, J. Bascompte, E.L. Berlow, J.H. Brown, M. Fortelius, W.M. Getz, J. Harte, A. Hastings, P.A. Marquet, N.D. Martinez, A. Mooers, P. Roopnarine, G. Vermeij, J.W. Williams, R. Gillespie, J. Kitzes, C. Marshall, N. Matzke, D.P. Mindell, E. Revilla and A.B. Smith (2012). Approaching a state shift in Earth's biosphere. *Nature*, **486**, 52–58.

Brock, R.E. and D.A Kelt (2004). Keystone effects of the endangered Stephens' kangaroo rat. *Biological Conservation*, **116**(1), 131–139.

Bromley, D. (1998). Searching for sustainability: the poverty of spontaneous order. In Cleveland, C., R. Costanza and D. Stern (eds), *Changing the Nature of Economics*. Washington DC: Island Press, Chapter 5.

Chichilnisky, G. and G. Heal (1998). Economic returns from the biosphere. *Nature*, **391**, 629–630.

Daly, H. (1985). The circular flow of exchange value and the linear throughput of matter–energy: a case of misplaced concreteness. *Review of Social Economy*, **43**, 279–297.

Dasgupta, C. (2007). Climate-change challenge for the poor – Part I. YaleGlobal online. Available at: www.yaleglobal.yale.edu/article.print?id=9720.

Dasgupta, P. (2008). Nature in economics. *Environmental and Resource Economics*, **39**, 1–7.

Dasgupta, P., S. Levin and J. Lubchenco (2000). Economic pathways to ecological sustainability. *BioScience*, **50**(4), 339–345.

Frey, B. and A. Stutzer (2002). *Happiness and Economics: How the Economy and Institutions Affect Well-Being*. Princeton, NJ: Princeton University Press.

Gilbert, N. (2008). A quarter of mammals face extinction. *Nature*, **455**, 717.

Georgescu-Roegen, N. (1971). *The Entropy Law and the Economic Process*. Cambridge, MA: Harvard University Press.

Georgescu-Roegen, N. (1984). Feasible recipes versus viable technologies. *The Atlantic Economic Journal*, **12**, 21–30.

Gowdy, J. (1997). The value of biodiversity. *Land Economics*, **73**, 25–41.

Gowdy, J. (ed.) (1998). *Limited Wants, Unlimited Means: A Reader on Hunter-Gatherer Economics and the Environment*. Washington DC: Island Press.

Gowdy, J. (2005). Toward a new welfare foundation for sustainability. *Ecological Economics*, **53**, 211–222.

Gowdy, J. (2005). Toward a new welfare foundation for sustainability. *Ecological Economics*, **53**, 211–222.

Gowdy, J. and L. Krall (2009). The fate of Nauru and the global financial meltdown. *Conservation Biology*, **23**(2), 257–258.

Haq, M. (1997). *Human Development in South Asia*. Karachi: Oxford University Press.

Haywood, A. and M. Williams (2005). The climate of the future: clues from three million years ago. *Geology Today*, **21**, 138–143.

Inglehart, R. (1990). *Cultural Shift in Advanced Industrial Societies*. Princeton, NJ: Princeton University Press.

Intergovernmental Panel on Climate Change, (2007). Climate Change 2007: The Physical Science Basis. Contribution of working group I to the Fourth Assessment Report of the Intergovernmental Panel on Climate Change, Geneva, Switzerland. Available at: http://www.ipcc.ch.

Jackson, J. (2008). Ecological extinction and evolution in the brave new ocean. *Proceedings of the National Academy of Science*, **105**, 11458–11465.

Jianchu, X., A. Shrestha, R. Vaidya, M. Eriksson and K. Hewitt (2007). *The Melting Himalayas: Regional Challenges and Local Impacts of Climate Change on Mountain Ecosystems and Livelihoods*. Technical paper, International Centre for International Mountain Development (ICIMOD), Kathmandu, Nepal.

Jiang, D. and H. Wang (2005). Modeling the Middle Pleistocene climate with a global atmospheric general circulation model. *Journal of Geophysical Research*, **110**, D14107.

Kahneman, D. and R. Sugden (2005). Experienced utility as a standard of policy evaluation. *Environmental & Resource Economics*, **32**, 161–181.

Kallis, G., J. Martinez-Alier and R. Norgaard (2009). Paper assets, real debts: an ecological–economic exploration of the global economic crisis. *Critical Perspectives on International Business*, **5**, 14–25.

Kellert, S. (1996). *The Value of Life: Biological Diversity and Human Society*. Washington DC: Island Press/Shearwater Books.

Khan, S.R. (2002). Adaptation to climate change in the context of sustainable development and equity: the case of Pakistan. Working paper series 78, Sustainable Development Policy Institute, Islamabad, Pakistan.

Krutilla, J. (1967). Conservation reconsidered. *American Economic Review*, **57**, 777–786.

Maler, K-G., S. Aniyar and A. Jansson (2008). Accounting for ecosystem services as a way to understand the requirements for sustainable development. *Proceedings of the National Academy of Science*, **105**(28), 9501–9506.

Martinez-Alier, J. (1995). The environment as a luxury good or 'too poor to be green'. *Ecological Economics*, **13**, 1–10.

McDaniel, C. and J. Gowdy (1999). *Paradise for Sale: A Parable of Nature*. Berkeley, CA: University of California Press.

McLeod, E. and R. Salm (2006). Managing mangroves for resilience to climate change. IUCN/Nature Conservancy, Resilience Science Working Group Paper Series No. 2. Available at: http://www.iucn.org/themes/marine/pubs/pubs.htm.

Melillo, J. and O. Sala (2008). Ecosystem services. In E. Chivian and A. Bernstein (eds), *Sustaining Life: How Human Health Depends on Biodiversity*. Oxford and New York: Oxford University Press, pp. 75–115.

Millennium Ecosystem Assessment (2005). *Ecosystems and Human Well-being: Synthesis*. Washington DC: Island Press.

Munasinghe, M. (2008). Rising temperatures, rising risks. *Finance and Development*, **45**(1), 37–41.

Norton, B. (2005). *Sustainability: A Philosophy of Adaptive Ecosystem Management*. Chicago, IL: University of Chicago Press.

Nunes, P. and J. van den Bergh (2001). Economic valuation of biodiversity: sense or nonsense? *Ecological Economics*, **39**, 203–222.

Nussbaum, M. (2000). *Women and Human Development*. Chicago, IL: University of Chicago Press.

Orr, D. (1994). *Earth in Mind: On Education, Environment, and the Human Prospect*. Washington DC: Island Press.

Rangel, A. (2003). Forward and backward generational goods: why is social security good of the environment? *American Economic Review*, **93**(3), 813–834.

Rees, G. H. and D.N. Collins (2004). An assessment of the potential impacts of deglaciation on the water resources of the Himalaya. DFID KAR Project No. R7980.

Robinson, G.R., R.D. Holt, M.S. Gaines, S.P. Hamburg, M.L. Johnson, H.S. Fitch and E.A. Martinko (1992). Diverse and contrasting effects of habitat fragmentation. *Science*, **257**, 524–525.

Schneider, S. (2009). The worst case scenario. *Nature*, **458**, 1104–1105.

Sen, A. (1999). *Development as Freedom*. New York: Anchor Books.

Singh, S.J. (2006). *The Nicobar Islands*. Vienna: Czernin Verlag.

Stern, N. (2007). *The Economics of Climate Change*. The Stern Review. Cambridge: Cambridge University Press.

Sukhdev, P. and J. Bishop (2008). The economics of ecosystems and biodiversity (TEEB): a step towards biodiversity markets? Workshop on 'Capitalizing on Natural Resources: New Dynamics in Financial Markets'. Rüschlikon, Germany, September 10, 2008.

TEEB (2010). *The Economics of Ecosystems and Biodiversity: Ecological and Economic Foundations*. London and Washington: Earthscan.

Thompson, M.F. and N.M. Tirmizi (1995). Mangrove soil: its mineralogy and texture. In N. Ibrabim, M.J. Mubammed and A. Sagbir (eds), *The Arabian Sea: Living Marine Resources and the Environment*. Rotterdam, the Netherlands: A.A. Balkema, pp. 431–439.

Tilman, D. and J. Downing (1994). Biodiversity and stability in grasslands. *Nature*, **367**, 363–365.

United Nations Development Programme (2008). Human Development Report 2007/2008. Fighting climate change: human solidarity in a divided world. UNDP, New York. Available at: http://hdr.undp.org/en/reports/global/hdr2007-2008/.

Walker, B., C. Holling, S. Carpenter and A. Kinzig (2004). Resilience, adaptability and transformability in socio-ecological systems. *Ecology and Society*, **9**, 5–15.

Wilson, E.O. 1994. *Biophilia*. Cambridge, MA: Harvard University Press.

Womenaid International (2007). Capability poverty measure (CPM). Available at: http://www.womenaid.org/press/info/poverty.cpm.html (accessed 12 January 2008).

World Bank (2006). *Pakistan Strategic Country Environmental Assessment*.Vol. I. Washington DC: South Asia Environment and Social Development Unit.

World Wildlife Fund – Pakistan (2005a). Keti Bunder Village Development Plan. WWF Regional Office, Karachi.

World Wildlife Fund – Pakistan (2005b). Study on knowledge, attitudes and practices of fisherfolk communities about fisheries and mangrove resources: Keti Bunder. WWF-Pakistan.

World Wildlife Fund – Pakistan (2006). Fish marketing chain and economic analysis of indebtedness of fisher-folk of Keti Bunder. Programme Development Division, WWF-Pakistan, Karachi.

World Wildlife Fund – Pakistan (2008a). Climate change in the Northern Areas Pakistan: impact on glaciers, ecology, and livelihoods. Gilgit Conservation and Information Center, WWF-Pakistan.

World Wildlife Fund – Pakistan (2008b). Detailed ecological assessment of fauna including limnology studies at Keti Bunder. Indus for All Program, WWF-Pakistan.

Worstall T. (2007). Indian wheat yields. *The Broadsheet*, September 17, Available at: www.timworstall.com (accessed 12 December 2007).

11. How ecosystem-based restoration can yield a double dividend of adaptation to climate change and enhancement of ecosystem services

James Blignaut

INTRODUCTION

Climate change affects most if not all aspects of human society, but so does poverty and environmental degradation. An obvious question is whether climate change policy can be designed in such a way that it also addresses the ills of poverty and environmental degradation. This question has been the topic of various discussion groups, one of which led to the so-called Hartwell Paper which, among other things, states:

> The [Hartwell] Paper therefore proposes that the organising principle of our effort should be the raising up of human dignity via three overarching objectives: ensuring energy access for all; ensuring that we develop in a manner that does not undermine the essential functioning of the Earth system; ensuring that our societies are adequately equipped to withstand the risks and dangers that come from all the vagaries of climate, whatever their cause may be (LSE/ISIS 2010: 5).

This proposal for driving climate change policy forward is based on the principle of advancing human dignity. In placing human dignity and its advancement as an overriding principle, the Hartwell Paper broke with the conventional view of climate change policy that, for the most part, only considered the cost of climate change and not the opportunity it offers to address a myriad of socio-economic ills. As a way of going forward, the Hartwell Paper suggests three pathways, namely the provisioning of energy, development within sustainable ecological boundaries and climate change adaptation. It is the latter two that are of special importance here. Even if it is not explicitly stated, the concept of living within sustainable ecological boundaries presupposes the restoration of degraded or depleted natural capital (to use an economic metaphor that stands for the current stock of physical and biological natural resources). The emphasis on climate change adaptation is also notable as adaptation has not received the same level of investment or public and academic recognition as its counterpart, climate change mitigation.

Consciously or unconsciously, the Hartwell Paper ties restoration and climate change adaptation together in a framework that seeks to address the prevailing twin challenges of climate change and socio-economic disorders under the banner of 'raising-up human dignity'.

Considering climate change adaptation, restoration of degraded ecosystems, and the various socio-economic disorders conjointly is noteworthy as it signals a major leap forward in economic development thinking. Society is advancing beyond the focus on additionality, which sometimes seems obsessive and contrary to the aims of sustainability. Seeking to achieve multiple benefits on one expenditure stream is the antithesis of the additionality concept.

While we will make frequent cross-references to mitigation in this chapter, the focus is on adaptation, which is defined by the United Nations Framework Convention on Climate Change (UNFCCC) as the 'adjustment in natural or human systems in response to actual or expected climatic stimuli or their effects, which moderates harm or exploits beneficial opportunities' (http://unfccc.int/essential_background/glossary/items/3666.php).

When considering adaptation, one accepts that climate change is likely to take place, and that it will affect people's lives and the quality thereof. Solutions are therefore sought to address the impacts of climate change; solutions that could act as buffers against the most likely adverse impacts and consequences of environmental change.

This chapter will, first, provide broad parameters of climate change adaptation and the restoration of natural capital. Secondly, it will consider the value of restoring natural capital, and reflect on how it is possible to achieve both restoration and adaptation benefits.

CLIMATE CHANGE ADAPTATION AND THE RESTORATION OF NATURAL CAPITAL

The climate change adaptation literature is much less developed than that of climate change mitigation, both in terms of strategic interventions and action plans, as well as in economic terms such as benefit–cost analysis. Mitigation can, undoubtedly, contribute most towards realigning both national economies and the global economy at large onto a low carbon trajectory. Adaptation, however, potentially has a greater impact and

> **BOX 11.1 THE GLOBAL DAMAGE COST OF CARBON: HOW LONG IS A PIECE OF STRING?**
>
> What is the value of one tonne of CO_2? This question has occupied many authors (e.g. IPCC 1999, 2000a; Tol 2005; Stern 2007; Tol 2009; Kuik et al. 2009; Stage 2010; Rafey and Sovacool 2011). This is also a question that became the topic of a heated debate pertaining to the use of discount rates, or more accurately, the appropriate pure rate of time preference (PRTP), where PRTP is defined as 'the marginal rate of substitution between present and future consumption under the condition that consumption levels in both periods are equal' (Anthoff et al. 2009). The choice of PRTP is important as it drives, to a very large extent, the estimate of the likely impact of climate change on national economies (Anthoff et al. 2009; Tol and Yohe 2009; Stern 2007, 2008; Dasgupta 2007; Nordhaus 2007). It is not only the choice of discount rate that influences the estimates, but also the time period, the country focus, the income levels of countries and the distribution of income both within and among countries. It is therefore not surprising that there are very wide discrepancies in the results among many studies. This can clearly be seen from Tol (2009) who reviewed 13 studies covering the period of 1994 to 2006. While the studies diverge significantly in terms of their estimates of the anticipated impacts of climate change on national and global incomes, they do converge as to the regions likely to suffer most and those who stand to gain the most. The 13 studies estimate that, during the current century, the continent facing the most adverse impacts is Africa (to the effect of between –2 per cent and –5 per cent of gross domestic product (GDP)), with Asia and Western Europe as the continents benefitting the most from climate change (+0.5 per cent to +2 per cent of GDP).

contribution to make towards the reduction of people's vulnerability, loss of life, damage to infrastructure and nature, continued economic welfare and the improvement thereof.

While various studies consider the damage cost of climate change (see Box 11.1), few consider the potential benefits linked to adaptation. Given that a certain level of ongoing climate change is unavoidable regardless of mitigation efforts around the world, it is useful to seek measures that would quantify the relative effectiveness of such investments. More so since most countries are 'climate takers', i.e. their carbon emission load is low in comparison to the impact climate change is likely to have on them. The contributions of such countries to global mitigation options are therefore also limited. South Africa, for example, is the 13th highest CO_2 emitter among nations (according to annual emissions in 2008), but its contribution is only 1.4 per cent of combined global anthropogenic greenhouse gas emissions (UNSD 2011).

It is important to consider both the absolute and relative contributions a country makes towards climate change because a country's ability to contribute to global climate change mitigation is proportionately linked to the volume of its emissions. South Africa, for example, cannot reduce global emissions by more than 1.4 per cent even if it reduces its greenhouse gas emissions to zero. There is thus no coupling of a country's vulnerability and risk profile, on the one hand, with its emissions on the other. Countries contributing proportionately little to global CO_2 emissions can face disproportionate impacts, with little or no scope for mitigation to alleviate them. This is a dilemma beyond the reach of mitigation as a response to climate change.

To deal with the adverse impacts of climate change requires that countries focus purposefully on adaptation. It therefore goes without saying that the more a country is a 'climate taker', and can make a relatively low proportionate contribution to global mitigation efforts, the more important adaptation becomes relative to mitigation. That does not imply that a country with a proportionately low emission profile should not engage in mitigation, it merely emphasizes the importance of focusing on adaptation as well.

How much should a country invest in adaptation? A general guide is that the adaptation effort is likely to follow the anticipated relative national level of warming and precipitation change (see Figure 11.1 for one such global perspective). While there are highly localized impacts, the intensity of the change mainly relates to latitude and continentality, i.e. geographically mediated effects such as comparative size of landmass, and national exposure to specific impacts such as sea-level rise, i.e. small island states and continental states with significant coastal exposure would be more prone to these damages in relation to an international average. This raises the issue of downscaling broad climate change projections and the degree of uncertainty that it introduces. In general, the higher the degree of uncertainty, the more the precautionary principle should be applied.

The need to apply the precautionary principle is further highlighted by the fact that one of the unique characteristics of adaptation is that climate change adaptation responses have a strong local bias. The assessment of

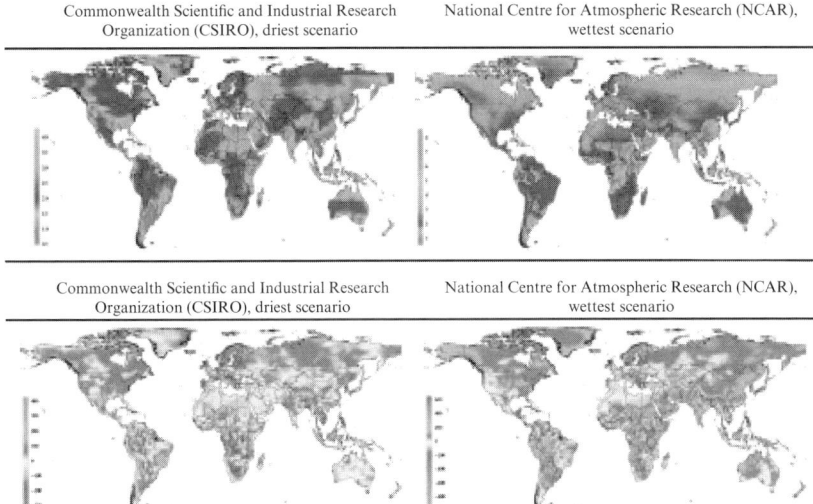

Source: World Bank (2010a); based on an International Food Policy Institute study. The map was published in World Bank (2010). The Cost to Developing Countries of Adapting to Climate Change: new methods and estimates. The Global Report on the Economics of Adaptation to Climate Change Study. Copyright World Bank Group, Washington DC.

Figure 11.1 Projected change in mean maximum temperatures and precipitation based on two climate models 2000–50

adaptation responses, and their potential benefits, should ideally flow from conducting bottom-up assessments of a suite of adaptation actions. This suite can be, and in most cases should be, very detailed and site specific. The actual benefits and costs of adaptation are therefore also highly specific both to an area and site. This contrasts with mitigation efforts, which can be driven by national and international emissions targets or actions, and can be assessed economically from a top-down perspective using sector-based analysis. The power of developing adaptation responses is that projects tend to engage with people where they live and work daily. It is, therefore, only logical to make the link between engaging in an adaptation programme that also contributes to human welfare.

On which adaptation programmes should a municipality, country or region focus? There is a range of local adaptation options possible, such as educational programmes and interventions in the health sector (e.g. the distribution of mosquito nets in areas that are malaria prone). Other interventions include investments in the built environment, such as the elevation and strengthening of bridges, roads and buildings to make provision for the possible increase in the amplitude of extreme weather events. A further, and important, adaptation effort that has to coincide with heightened awareness is the improvement in disaster management programmes, including early detection and early warning systems.

While the cost of these interventions can be estimated with a relatively high degree of accuracy, the benefits thereof, for the most part, are difficult to estimate or predict. The most common approach is to consider the avoided damage cost under conditions of 'when triggered'. For example, when there is a flood that can be withstood by an elevated and strengthened bridge, that is the time when the investment in elevating and strengthening the bridge pays off. The date (year), locality, severity and impact or damage caused by the extreme event is, however, unknown at the time the decision has to be made about whether to invest in adaptation. This is a cause for major concern within almost all adaptation benefit–cost models, given the power that is assigned to the model builder who has to anticipate the incidence frequency, severity, locality and impact of climate-change-related extreme events. In addition, the model builder has to assume a future discount rate. It goes without saying that the outcome of such a model is the result of the chosen set of assumptions rather than rigorous scientific interrogation.

The challenge faced by adaptation planning is further deepened by the fact that, in a world of increasingly scarce resources, decision-makers are hard put to invest in programmes that yield high (and short-term) returns. Investing in actions that might have a significant benefit sometime in the future (and only under specific conditions) will be considered wasteful by some, or else be given a very low priority. It is socio-politically much easier to justify investments in activities that have benefits within the term a decision-maker is in office.

One strategic policy decision is to invest in adaptation options that have benefits over and above those that only come into play under 'when triggered' scenarios. In other words, a shrewd adaptation policy is one that renders benefits irrespective of whether an extreme weather event occurs. In so-doing, the ongoing benefits could 'pay' for the adaptation policy and, in the process, also offer useful adaptation measures. One such adaptation option is the restoration of natural capital, whose benefits are self-evident from the day the investment is initiated, and is not dependent upon an adverse climate-change-induced (or any other) effect to yield benefits. Under extreme weather conditions, intact or restored natural capital can contribute much to the protection of infrastructure, and act as a safety net against poverty and even death (Dahdouh-Guebas et al. 2005; Costanza et al. 2006; EJF 2006). The potential contribution of restoration to climate change adaptation is of such significance that it warrants further assessment, to which we turn next.

THE VALUE OF RESTORING NATURAL CAPITAL

Restoring Natural Capital

Access to sufficient quantities of quality natural capital has become one of the main limiting factors of human well-being and sustainable economic development (Aronson et al. 2006). As noted above, natural capital is an economic metaphor which stands for the current stock of physical and biological natural resources. This stock consists of four different entities: (1) renewable natural capital (living species and ecosystems), (2) non-renewable natural capital (sub-soil assets, e.g. petroleum, coal, diamonds, etc.), (3) so-called replenishable natural capital (e.g. the atmosphere, potable water, fertile soils), and (4) cultivated natural capital (e.g. crops and forest plantations) (Aronson et al. 2007). The stock of natural capital delivers a suite of ecosystem goods and services upon which humankind depends for survival and which are inputs into economic processes. As a crude generalisation, the better the stock of the asset in terms of quality and quantity, the better and more services can be derived from it, and vice versa. This link between the flow of services and the stock can be seen clearly in the fact that the value of the stock is, for the most part, estimated as the discounted present value of the sum of the future income stream derived from the flows of the stock over the stock's lifetime (United Nations 1993).

De Groot et al. (2012) conducted a global assessment of approximately 225 studies regarding the range of economic values of 22 ecosystem goods and services in 10 ecosystem types. As can be seen from Figure 11.2, the annual value of ecosystem goods and services delivered by intact natural capital ranges from as little as about US$100 per hectare (open oceans) to more than US$1 million per hectare (coral reefs), with the bulk of the values between US$1000 and US$100000 per hectare per year.

The values in Figure 11.2 are indicative of the value of intact ecosystems as producers of the selected suite of 22 ecosystem goods and services, and hence their contribution to human well-being and livelihoods. It is these values that are seriously compromised under conditions of ecosystem degradation. Not only does degradation compromise the inherent integrity of an ecosystem, thereby limiting its ability to function as an integral unit (hence, becoming more vulnerable and liable to collapse over time), it also compromises human welfare through the loss of the services rendered by the ecosystems.

The only viable way to reverse the global trend of degradation (TEEB 2010), i.e. to protect and even augment the currently declining supply of natural capital, is through investing in restoration. This is true both in

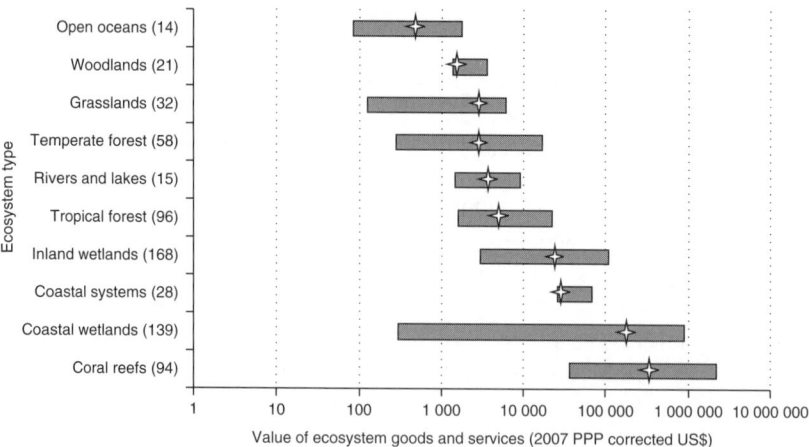

Source: De Groot et al. (2012).

Figure 11.2 Range and average of the value of ecosystem goods and services (total economic value) per ecosystem type (in US$ per hectare per year (2007/PPP-corrected))

absolute terms, and in relative terms, i.e. in relation to our rapidly growing global population.

The restoration of natural capital can be defined as any activity that integrates investment in and replenishment of natural capital stocks to improve the flows of ecosystem goods and services, while enhancing all aspects of human well-being (Aronson et al. 2007). Restoration activities may include, but are not limited to (1) the restoration and rehabilitation of impaired terrestrial, coastal, and aquatic ecosystems, (2) ecologically sound improvements to production systems on arable lands and other lands or wetlands that are managed or exploited for useful purposes, (3) improvements in the ecologically sustainable utilization of biological resources, and (4) the establishment or enhancement of socio-economic systems that incorporate knowledge and awareness of the value of natural capital into daily activities and public policy.

The restoration of natural capital has the unparalleled benefit of contributing meaningfully to climate change adaptation and human welfare through (1) ensuring or augmenting the continued delivery of ecosystem goods and services, (2) combating ecosystem fragmentation and thereby any potential future loss of ecosystem goods and services, and (3) buffering against the impacts of adverse and severe climatological events such as droughts and floods. Restoration of natural capital does these

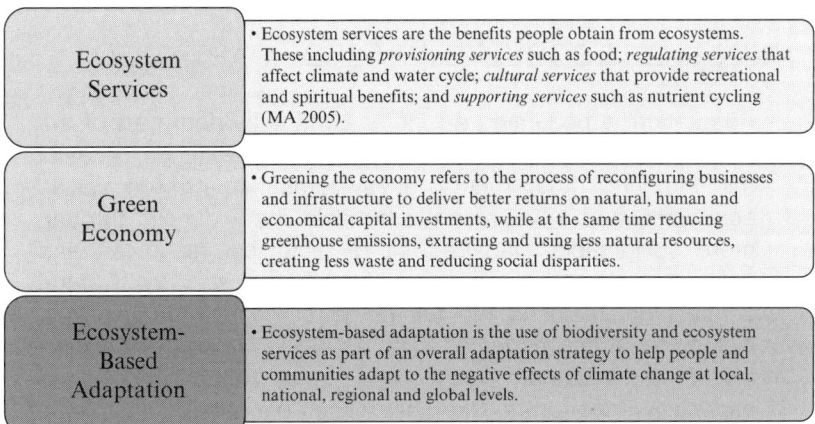

Source: UNEP (2010) Ecosystem-based adaptation program. Paris, France. This figure was published in Ecological Economics, **1**(1), De Groot, R., Brander, L., van der Ploeg, S., Costanza, R., Bernard, F., Braat, L., Christie, M., Crossman, N., Ghermandi, A., Hein, L., Hussain, S., Kumar, P., McVitte, A., Portela, R., Rodriguez, L.C., ten Brink, P. and van Beukering, P., Global estimates of monetary values of ecosystems and their services, Pages 50–61, Copyright Elsevier BV (July 2012).

Figure 11.3 Definitions and concepts of ecosystem services, green economy and ecosystem-based adaptation

things on a continuous basis from the day of engaging in the restoration exercise.

Ecosystem-based Adaptation

Based on the above exposition of the restoration of natural capital, it goes without saying that it can make a meaningful contribution to ecosystem-based adaptation. Ecosystem-based adaptation (see Figure 11.3) is the use of ecosystem goods and services within a climate change adaptation programme.

Given the overarching role ecosystems play within society as described by the Millennium Ecosystem Assessment (2005), ecosystems have a meaningful contribution to make towards climate change adaptation (see also Box 11.2 for a brief exposition). This role is eloquently described through the four main types of service ecosystems fulfil. The first of the four services is *provisioning services*, i.e. the supply of goods that people obtain from ecosystems for consumption, such as food, fuel, fibre, fresh water and genetic resources. The second is *regulating services*, which include the benefits people obtain from ecosystem

BOX 11.2 ECOSYSTEM-BASED ADAPTATION

Adaptation is becoming an increasingly important part of the development agenda. Protecting forests, wetlands, coastal habitats, and other natural ecosystems can provide social, economic, and environmental benefits, both directly through more sustainable management of biological resources and indirectly through the protection of ecosystem services. Natural ecosystems maintain the full range of goods and services, including natural resources such as water, timber, and fisheries on which human livelihoods depend. These services are especially important to the most vulnerable sectors of society. Protected areas, and the natural habitats within them, can protect watersheds, regulate the flow and quality of water, prevent soil erosion, influence rainfall regimes and local climate, conserve renewable harvestable resources and genetic reservoirs, and protect breeding stocks, natural pollinators, and seed dispersers, all of which maintain ecosystem health. Over the last decade, more and more World Bank projects have been making explicit linkages between conservation and sustainable use of natural ecosystems, carbon sequestration, and watershed values associated with erosion control, clean water supplies, and flood control. Better protection and management of key habitats and natural resources can benefit poor, marginalized, and indigenous communities by protecting ecosystem services and maintaining access to resources even during difficult times, including drought and disaster. In response to climate change, many countries are likely to invest in even more infrastructure for coastal defences and flood control to reduce the vulnerability of human settlements to climate change. As water shortages become more frequent and severe, the demand for new irrigation facilities and new reservoirs will grow. Forests, wetlands, and other natural habitats play important roles in protecting high-quality water supplies. Similarly, natural ecosystems can reduce vulnerability to natural hazards and extreme climatic events and complement, or substitute for, more expensive infrastructure investments to protect coastal and riverine settlements. Floodplain forests and coastal mangroves provide storm protection and coastal defences, and serve as safety barriers against natural hazards such as floods, hurricanes, and tsuna-

> mis, while wetlands filter pollutants and serve as water recharge areas and as nurseries for local fisheries. Traditional engineered solutions often work against nature, particularly when they aim to constrain regular ecological cycles, such as annual river flooding and coastal erosion, and could further threaten ecosystem services if the construction of dams, seawalls and flood canals leads to habitat loss. Instead, in Argentina and Ecuador, flood control projects utilize the natural storage and recharge properties of critical forests and wetlands by integrating them into 'living with floods' strategies that incorporate forest protected areas and riparian corridors. These simple and effective solutions protect both communities and natural capital (World Bank 2010b: 3–4).

processes such as air quality maintenance, protection from natural hazards (floods, storms), climate regulation, erosion control, regulation of human diseases and water purification. The third broad category of services is that of *cultural services*, which include spiritual enrichment,

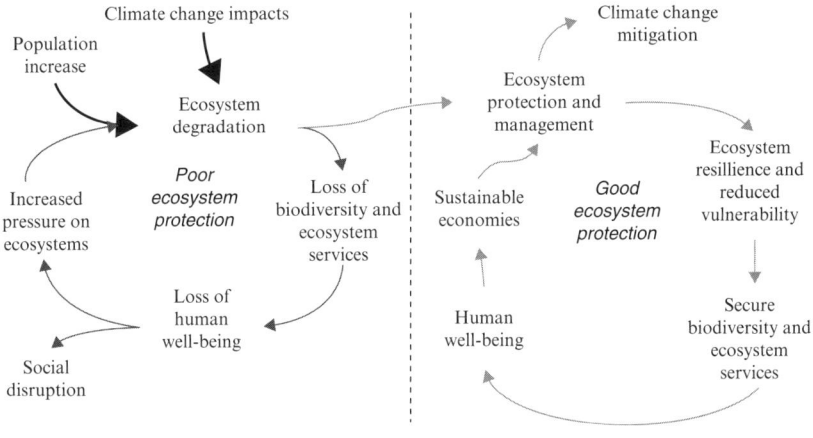

Drivers and impacts of ecosystem degradation   Ecosystem-based adaptation

Source: UNEP (2010) Ecosystem-based adaptation program. Paris, France.

Figure 11.4 Migrating from a vicious cycle of poverty, ecosystem degradation and climate change, to a virtuous one of ecosystem heath and provision of ecosystem services and human well-being

recreation, education and aesthetic experiences. The last group is *supporting services*, which are services necessary for the production of all other ecosystem services, e.g. primary production, production of oxygen, and soil formation.

Restoration has the unique feature that it can augment the world's rapidly dwindling supply of ecosystems and hence ecosystem goods and services and, in the process, develop resilient and healthy ecosystems. In turn, through an enhancement of the supply of their services, healthy ecosystems reduce vulnerabilities of people. Consequently, this benign relationship is likely to contribute to improved human well-being, which is an essential objective of achieving sustainable economies. This cycle is diagrammatically expressed in Figure 11.4. What is interesting is that the link between poor ecosystem protection, and its resultant degradation, and good ecosystem protection, is the will to change. The will to change is a function of a great number of aspects, but includes among others, the knowledge based on good information about the role ecosystems play within society, and then in particular the constructive role ecosystem restoration can play in enhancing system resilience and human livelihoods. This knowledge-information package must be supported by the institutional capacity and the funding mechanisms to affect the change. Lastly, the will to change also has to be supported by the (technocratic) ability to implement and manage a restoration programme. There is therefore a symbiotic relationship between the realms of science (that includes both information and the knowledge of how to do it) and policy (that includes the governance structures that will implement the change). Conversely, where there is a major science-policy gap, it can lead to the further deterioration of the quality of the environment. This reduces the quality of life and increases the vulnerability of people because it reduces their ability to adapt to face the impacts of climate change.

The Value of Restoration

While most benefits of ecosystem restoration (such as continued food and fodder production, and improved water quality and flow) as part of climate change adaptation, are localized, restoration also contributes towards mitigation through long-term carbon storage and sequestration. The degree to which restoration can sequester carbon is driven by several factors, namely:

- the size of the restoration project;
- the ecosystem in which the project operates;
- the climatic zone;

BOX 11.3   THE VALUE OF 1 TONNE OF CO_2

What is the value of 1 tonne of CO_2? Using a very low discount rate, Stern (2007, 2008) estimated the social cost of carbon as US$314 per tonne of carbon, or US$85 per tonne of CO_2. Many alternative views to that of Stern exist. For example, after reviewing 28 studies with respect to the damage cost of climate change under varying PRTPs, Tol (2005) found that the 'mode is US$2 per tonne of carbon (US$0.55 per tonne of CO_2), the median US$14 per tonne of carbon (US$3.82 per tonne of CO_2), the mean US$93 per tonne of carbon (US$25.4 per tonne of CO_2), and the 95th percentile US$350 per tonne of carbon (US$95.5 per tonne of CO_2)'. He concluded that 'the marginal damage costs of carbon dioxide emissions are unlikely to exceed US$50 per tonne of carbon (US$13.6 per tonne of CO_2), and probably much smaller' (Tol 2005). He also stated that '[i]f we use a pure rate of time preference of 3%, corresponding to a social rate of discount of 4–5%, close to what most western governments use for most long-term investments, the combined mean estimate is US$16 per tonne of carbon (US$4.4 per tonne of CO_2), not exceeding US$62 per tonne of carbon (US$16.90 per tonne of CO_2) with a probability of 95 per cent' (Tol 2005). In 2009, Tol conducted another review, this time of more than 200 studies, in which he concluded that 'for a standard discount rate, the expected value is US$50 per tonne of carbon (US$13.6 per tonne of CO_2), which is much lower than the price of carbon in the European Union but much higher than the price of carbon elsewhere' (Tol 2009). It should be noted that these values are for 1995. In a paper drawing the debate to a close, Anthoff et al. (2009) noted that the most likely social cost of carbon is approximately US$41 per tonne of carbon (US$11.2 per tonne of CO_2), if one ignores uncertainty and equity. If uncertainty and global income differentials are taken into consideration, the value lies somewhere between US$61 per tonne of carbon (US$16.6 per tonne of CO_2) and US$206 per tonne of carbon (US$56.2 per tonne of CO_2). Based on the above, while outliers and exceptions are obviously possible, a realistic range of between US$15 and US$50 per tonne of CO_2 can be accepted.

- the type of restoration work conducted;
- the degree of degradation at the time restoration commenced (the so-called baseline); and
- the total biomass carbon in undisturbed or intact conditions (the so-called reference or total carbon value).

Based on the above, the rate of carbon sequestration (under conditions of restoration) typically follows an S-shaped curve, which starts to rise slowly, before accelerating and then flattening off as the system becomes restored and reaches a steady-state of some sort. While there is debate as to the amount of carbon that is sequestered in any ecosystem, most studies use the very conservative (by own admission) IPCC guidelines to estimate the most plausible amount of carbon that a restoration project can sequester (IPCC 2000b).

Not only is the amount of carbon that an ecosystem can sequester, following restoration, a subject of much debate, so is the value of a tonne of carbon (see Box 11.3). Currently, a range between US$15 and US$50 per tonne of CO_2 seems realistic. A survey designed to estimate the carbon mitigation value of 55 restoration projects in developing countries found that the combined restoration area of the projects covers more than 15 million hectares, with tropical forests accounting for 55 per cent of the area, followed by grasslands in arid or semi-arid zones (26 per cent).

A survey of carbon sequestration and restoration projects

This survey was conducted to develop a database of restoration case studies in Africa and other developing countries. The studies were identified following a detailed literature review of restoration and rehabilitation projects found on *ScienceDirect*, *Google Scholar*, *Google* and the *Global Restoration Network* websites (see also Aronson et al. 2010). The total identified hits were reduced by searching for those that had 'restoration' and 'rehabilitation' in the title, abstract or keywords. The list was further reduced to include only Africa or developing countries; literature on projects in Europe and America were not considered. All the remaining documents were screened to ensure that they referred to actual restoration projects. Articles and case studies that only described experimental plot designs or reviewed policy implications were not considered for the database. Following this elimination process, a set of 83 case studies were identified for the final list, a summary of which is provided in Table 11.1. They were subsequently categorized in Excel under the following headings:

- Country;
- Area under restoration;

Table 11.1 Summary of project database on restoration in developing countries

Region	Arid grasslands		Temperate wet grasslands		Tropical dry grasslands		Inland wetlands and riparian		Temperate wet forests		Tropical forests		Other areas		Total	
	# of case studies	ha	# of case studies	ha	# of case studies	ha	# of case studies	ha	# of case studies	ha	# of case studies	ha	# of case studies	ha	# of case studies	ha
Africa	2	4 000 000			1	160 000					27	8 161 808	2	633	32	12 322 441
Asia											2	60 030			2	60 030
India/China			1	2670							1	11 000	2	20 020	4	33 690
South America	2	500					2	528 050	2	16	12	71 071	4	2 000 036	22	2 599 673
East Europe			1	3											1	3
Other							2	2 005 670	1	36 450			1	6 000	4	2 048 120
Total	4	4 000 500	2	2673	1	160 000	4	2 533 720	3	36 466	42	8 303 909	9	2 026 689	65	17 063 957

Note: ha = hectares of restoration.

Table 11.2 Summary: carbon sequestrated, or its potential, from 55 ongoing restoration projects in developing countries

Carbon sequestration rates by ecosystem according to IPCC guidelines		Number of cases	Restoration area	CO_2 sequestrated	Value (@ $15/t)	Value (@ $50/t)
Ecosystem	t C/ha/a	#	ha	t	$ (mil)	$ (mil)
Arid/semiarid grasslands	0.3	4	4 000 500	2 640 570	39.6	132.0
Temperate wet grasslands	2	2	2 673	11 762	0.2	0.6
Tropical dry grasslands	1.5	1	160 000	528 048	7.9	26.5
Inland wetlands and riparian	1	4	2 533 720	5 574 691	83.6	278.9
Temperate wet forests	3	3	36 466	240 697	3.6	12.0
Tropical forests	4.6	41	8 303 909	84 043 197	1260.6	4202.0
Total		65	15 037 268	93 038 966	1395.6	4652

Notes: t C/ha/a = tonnes of carbon sequestered per hectare per year; ha = hectares of restoration; t = tonnes; $ = US dollars; (mil) = million.

- Ecosystem type;
- Cost of restoration;
- Who executed the restoration;
- Who financed the restoration;
- Type of restoration;
- Purpose of restoration.

One of the key outcomes in most restoration projects is that they contribute meaningfully to carbon sequestration. Below, a summary is provided of a range of carbon sequestration rates from around the world and from various sources. Using the IPCC guidelines (see Table 11.3 below), it is possible to calculate the most plausible amount of carbon that 55 of the 83 restoration projects (that provided detailed data with respect to area, as listed in the database), have sequestered or potentially can sequester. This excludes the ten studies in undefined ecosystems, the carbon sequestration capacity of which can therefore not be determined. The IPCC states that the values they list are very low; much lower than in most other studies. In determining the total amount of carbon that therefore can be or has been sequestered, this survey uses the maximum values according to the IPCC. The results are summarized in Table 11.2. Since it is unlikely that

Double dividends from ecosystem-based restoration 231

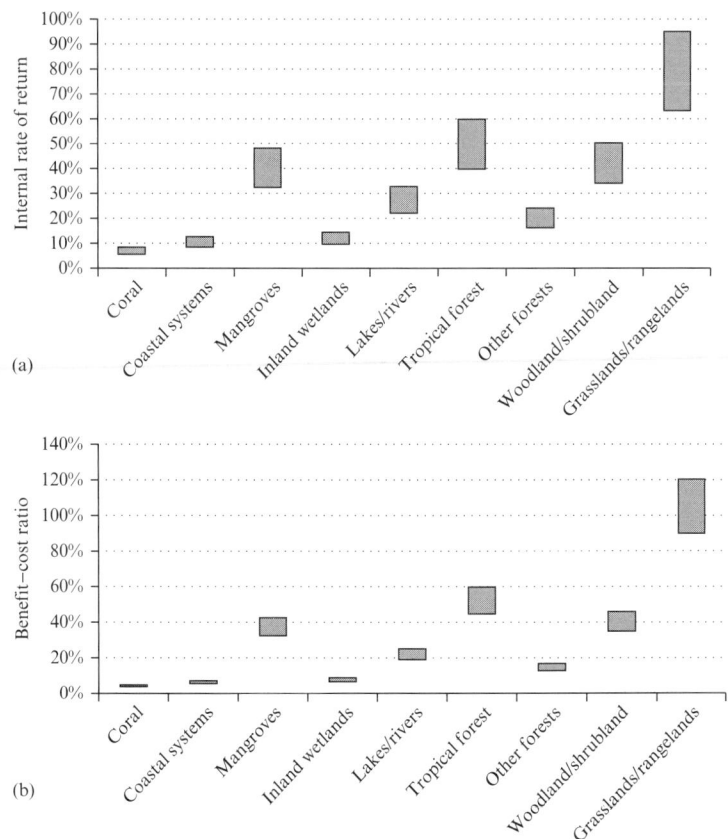

Note: In (a) the most plausible range of the internal rate of return of restoration across nine biomes is provided. The bottom of the bars indicates an 80 per cent value of the values presented in TEEB (2009) with the top of the bars representing 120 per cent of the values. (b) presents the benefit–cost ratios using the same dataset, with the worst-case scenario (bottom of the bar) being 80 per cent of the value and the best-case scenario 120 per cent of the value.

Source: TEEB (2009).

Figure 11.5 Internal rate of return (IRR) and benefit–cost ratios of restoration across nine major ecosystem types, based on approximately 325 case studies (values in 2007 US$ per hectare)

Table 11.3 Mean values for total carbon stocks of different forest types in tropical Africa (in tonnes of carbon per hectare)

Forest type	Mean	Std Dev	Std Error
Tropical moist intact forest	240	87.8	20.70
Tropical moist degraded forest	113	7.1	7.09
Tropical dry intact forest	251	43.2	17.65
Tropical dry plantation forest	158	24.1	12.03
Temperate moist intact forest	147	80.96	21.64
Temperate moist degraded forest	87	16.77	9.68
Temperate dry plantation forest	120	31.98	12.09

Source: Rayden et al. (2010).

Table 11.4 Mean values for total carbon stocks of different agricultural land uses in tropical Africa (in tonnes of carbon per hectare)

Land use	Mean	Std Dev	Std Error
Grassland/Savannah	64	19.9	6.29
Annual agriculture	69	15.5	6.31
Perennial agriculture (not Africa)	76	11.0	6.50
Agroforestry or mixed systems	95	29.4	8.88

Source: Rayden et al. (2010).

restoration projects will sequester the volume of carbon that intact ecosystems would, we assume that the carbon sequestered is equal to 60 per cent of the IPCC guideline. The carbon dioxide potential of these restoration projects is therefore calculated as the IPCC carbon sequestration rate by ecosystem × 3.667 × 0.6 (where 3.667 converts C to CO_2). This totals to more than 93 million tonnes of CO_2 sequestered, of which 90 per cent stems from tropical forests. At a carbon value of US$15 per tonne of CO_2, this equates to a total value of US$1 395.6 million per year.

Given the significance of the numbers depicted in Table 11.2, it comes as no surprise that TEEB (2009) found positive internal rates of return for restoration in seven of the nine ecosystems considered (see Figure 11.5a). Ecosystems with the lowest internal rates of return in this study were coral reefs and coastal systems, mainly because of the high restoration costs. Ecosystems that showed the best internal rates of return following restoration were woodlands, tropical forests and grasslands; mainly because of the rate and the level of ecosystem goods and services produced, and grasslands because of their relatively low cost of restoration. Restoration

Table 11.5 A schematic of aboveground biomass and carbon sequestration in different forests in various countries

	Tropical rain forest		Dry evergreen forest		Mixed deciduous forest	
	AGBM (t ha^{-1})	C-stock (t C ha^{-1})	AGBM (t ha^{-1})	C-stock (t C ha^{-1})	AGBM (t ha^{-1})	C-stock (t C ha^{-1})
Thailand	275.46	137.76	140.58	70.29	96.28	48.14
Thailand	358.00	179.00	126.00	60.30	311.00	155.50
Thailand	–	–	252.00	126.00	–	–
Thailand	–	–	–	–	31.95–175.50	15.97–87.75
Malaysia	225–446	112.5–223	–	–	–	–
Cameroon	238–341	119–170.5	–	–	–	–
Sri Lanka	153–221	76.5–110.50	–	–	–	–

Notes: t C ha^{-1} = tonnes of carbon per hectare; t ha^{-1} = tonnes per hectare; AGBM = above ground wet biomass.

Source: Terakunpisut et al. (2007).

can yield benefit–cost ratios ranging from 3 to 90 (see Figure 11.5b), which makes it a good investment when resources are scarce.

Rates of carbon sequestration vary greatly, leading to a range of carbon budgets according to soil and substrate type, ecosystem type and land use. Understanding of these can be an important tool in planning restoration programmes. Various published sources provide useful data on carbon budgets. A summary of these is given in Tables 11.3–11.10.

The rates of carbon gain in Table 11.6 refer to the average accumulation rate from the time a practice is started until carbon storage again reaches a new equilibrium. Almost invariably, the rates in Table 11.6 are lower than those from published studies, which are usually measured for time intervals shorter than that needed to reach saturation. As shown, some of the rate values have greater uncertainty than others, largely because of a paucity of studies in many regions. The net effect of the practices on climate forcing is also affected by impacts on other greenhouse gases (GHGs) (Table 11.6). For example, where the proposed practice increases N_2O or CH_4 emissions, the rate of carbon sequestration alone overestimates the net benefit of the practice to the atmosphere.

Natural capital, and the restoration of it, has a meaningful role to play not only in adaptation and mitigation, but also in sustained human development (Blignaut and Aronson 2008; Blignaut 2009). Given this sig-

Table 11.6 *Summary of potential rates of carbon gain and associated impacts for various activities*

Activity	Eco-zone[a]	Key practices[b]	Rate[c] (t C ha^{-1} a^{-1})	Confidence[d]	Duration[e] (years)	Other GHGs[f]	Associated impacts
Cropland management	Boreal	Ley/perennial forage crops, organic amendments	0.3–0.6 (0.4)	M	40	+N$_2$O	Increase food production, improved soil quality
	Temperate – dry	Reduced tillage, reduced bare fallow, irrigation	0.1–0.3 (0.2)	H	30	+N$_2$O	Increased food production, improved soil quality, reduced erosion, possibly higher pesticide use
	Temperate – wet	Reduced tillage, fertilization, cover crops	0.2–0.6 (0.4)	H	25	+N$_2$O	Increased food production, improved soil quality, reduced erosion, possibly higher pesticide use
	Tropical – dry	Reduced tillage, residue retention	0.1–0.3 (0.2)	L	20	+N$_2$O	Increase food production, improved soil quality, reduced erosion, possibly higher pesticide use
	Tropical – wet	Reduced tillage, improved fallow management, fertilization	0.2–0.8 (0.5)	M	15	+N$_2$O	Increased food production, improved soil quality, reduce erosion, fertilizers often unavailable, possibly higher pesticide use

	Climate	Practices	Range (mean)	Confidence	%	Gases	Co-benefits/Effects
	Tropical – wet (rice)	Residue management, fertilization, drainage management	0.2–0.8 (0.5)	L	25	++CH_4, +N_2O	Increased food production
	Tropical	Improved management	0.5–1.8 (1.0)	M	25	+N_2O	
Agroforest management	Temperate – dry	Grazing management, fertilization, irrigation	0–0.3 (0.1)	M	50	?CH_4 +N_2O	Increased energy use, salinity, higher productivity
Grassland management	Temperate – wet	Grazing management, species introduction, fertilization	0.4–2.0 (1.0)	M	50	?CH_4 ++N_2O	Higher productivity, acidification, erosion, reduced biodiversity
	Tropical – dry	Grazing management, species introduction, fire management	0.1–1.5 (0.9)	L	40	−CH_4 ++N_2O	Reduced soil degradation, higher productivity, woody encroachment (reduced productivity)
	Tropical – wet	Species introduction, fertilization, grazing management	0.2–3.9 (1.2)	L	40	−CH_4 ++N_2O	Increased productivity, reduced biodiversity, acidification

Table 11.6 (continued)

Activity	Eco-zone[a]	Key practices[b]	Rate[c] (t C ha^{-1} a^{-1})	Confidence[d]	Duration[e] (years)	Other GHGs[f]	Associated impacts
Forestland management	Boreal and temperate – dry	Forest regeneration, fertilization, plant density, improved species, increased rotation length	0.1–0.8 (0.4)	L	80	+N$_2$O, +NO$_X$	Leakage (rotation length), high cost efficiency
	Temperate – wet	Forest regeneration, fertilization, species change	0.1–3.0 (1.0)	L	50	+N$_2$O, +NO$_X$	Leakage (rotation length), reduce biodiversity
	Tropical – dry	Forest conservation, reduced degradation	(1.75)	L	40		Ecological improvement, high cost efficiency
	Tropical – wet	Reduced degradation	3.1–4.6 (3.4)	L	40		Environmental improvement

Wetland management	All	Restoration	0.1–1 (0.5)	L	100	++CH_4 ?N_2O	Increase in water quality, decrease in flooding, increased biodiversity
Restoration of degraded land	All	Restoration of eroded lands, saline soil reclamation	0.1–7 (0.25)	M	30	+N_2O	Increased productivity, may be expensive
Urban land management	All	Tree planting	(0.3)	M	50		Increased biodiversity
Conversion to agroforestry	Tropics	Conversion from cropland to grassland at forest margins	1–5 (3)	L	25		Improved biodiversity, CH_4 sinks, poverty alleviation, food security
Conversion (cropland to grassland)	Temperate – dry	Marginal cropland reseeded to grassland	0.3–0.8 (0.5)	H	50	–N_2O, –CH_4	Enhanced biodiversity, reduced erosion
	Temperate – wet	Surplus cropland seeded to grassland	0.5–1.0 (0.8)	M	50	–N_2O, –CH_4	Enhanced biodiversity, reduced erosion

Table 11.6 (continued)

Notes:
t C ha^{-1} a^{-1} = tonnes of carbon per hectare per year; GHGs = greenhouse gases; N$_2$O = nitrous oxide; NO$_x$ = other oxides of nitrogen; CH$_4$ = methane; ? = possibly includes.

a 'Wet' vs. 'dry' based on potential evapotranspiration:replication ratio. 'Wet' <1 and 'dry' >1; Tropical = latitude <30.
b List of practices that may yield largest gains in carbon stocks, roughly in order of importance.
c Range of carbon increase rates that might reasonably be expected to occur in response to adaptation of best-possible complement of key practices. Actual rate will depend on previous management practices (e.g. rates of gain may be higher in a carbon-depleted system), climate, ecosystem properties (e.g. soil carbon gain may be favoured by higher clay content), and many other factors. Value in parentheses is a default estimate. Rates for tropical forest management to recover carbon stocks on degraded forest apply only to the present area of degraded forest (as of 1990) as reported by FAO (1996). That is, closed canopy forest having full biomass stocks are excluded.
d Relative reliability of rate estimates. Generally, confidence increases with the number of studies conducted in the activity-ecozone grouping. L = Low, M = Medium, and H = High.
e An estimate of the time required for the system to approach a new steady state after the adoption of the new practices.
f Relative magnitude of potential effects on emission of N$_2$O, CH$_4$, and other GHGs. '+' denotes increased emission; '–' denotes reduced emission; number of '+' or '–' denotes relative magnitude of possible effects.

Source: IPCC (2000b).

Table 11.7 Rates of potential carbon gain under selected practices for cropland (including riceland) in various regions of the world

Practice	Country/Region	Rate of carbon gain (t C ha^{-1} a^{-1})	Time* (a)	Other GHGs and impacts	Notes
Improved crop production and erosion control	Global	0.05–0.76	25	+N$_2$O	a
Partial elimination of bare fallow	Canada	0.17–0.76	15–25		b
	USA	0.25–0.37	8	?N$_2$O	q
Irrigation water management	USA	0.1–0.3			c
Fertilization, crop rotation, organic amendments	USA	0.1–0.3		+N$_2$O	c
Yield enhancement, reduced bare fallow	Tropical and subtropical China	0.02	10		e
Amendments (biosolids, manure, or straw)	Europe	0.2–1.0	50–100		f
Forages in rotation	Norway	0.3	37		d
Ley-arable farming	Europe	0.54	100		m
Conservation tillage	Global	0.1–1.3	25	?N$_2$O	a
	UK	0.15	5–10		g
	Australia	0.3	10–13		h
	USA	0.3	6–20		i
		0.24–0.4			c
	Canada	0.2	8–12		r
	USA and Canada	0.2–0.4	20		j
	Europe	0.34	50–100		k
		0.5	10		l
	Southern USA	0.2	10–15		s

239

Table 11.7 (continued)

Practice	Country/Region	Rate of carbon gain (t C ha^{-1} a^{-1})	Time* (a)	Other GHGs and impacts	Notes
Rice and management					
Organic amendments (straw, manure)		0.25–0.5		++CH_4	n
Chemical amendments				–CH_4, +N_2O	o
Irrigation-based strategies				–CH_4	

Notes:
* Time interval to which estimated rate applies. This interval may or may not be time required for the ecosystem to reach a new equilibrium.
a Lal and Bruce (1999). Estimates of carbon gain shown represent range of values presented by the authors for various regions throughout the world.
b Dumanski et al. (1998). Estimates presented are for the 0–30 cm layer. Estimated carbon gain for the 0–100 cm layer are twice those for the 0–30 cm layer. Estimated rates of carbon gain are higher for conversion of fallow to forages (0.48–0.76 Mg C ha^{-1} a^{-1}) than for conversion to cereal crops (0.17–0.52 Mg C ha^{-1} a^{-1}).
c Lal et al. (1999).
d Singh et al. (1994). Reported rate is from one long-term study.
e Li and Zhao (1998). Rate of carbon gain is based on total carbon gain (0.7 Tg C a^{-1} for 10 years) and total cropland area in the region (~40 Mha) reported by the authors.
f Smith et al. [1998, including data from Smith et al. (1997a, b) and Powlson et al. (1998)]. Carbon gain from manure, sewage sludge, and straw incorporation assumes that carbon in these materials would otherwise all be lost as CO_2. Rates reported here were calculated from annual mitigation potential (Tg C a^{-1}) and area values in source reference.
g Mean carbon accumulation rate in four sites sampled after 5–10 years; from literature data compiled and cited by Paustian et al. (1997).
h Mean carbon accumulation rate in two sites sampled after 10 or 13 years; from literature data compiled and cited by Paustian et al. (1997).
i Mean carbon accumulation rate in 22 sites sampled after 6–20 years. Includes one site from Canada.
j Bruce et al. (1999). Rates of carbon accumulation assume 'best management practices', including no-till.

k   Smith et al. (1998). Based on data from 14 sites in UK and Germany, ranging in duration from 2 to 23 years. Rates reported here were calculated from annual mitigation potential (46.6 Tg C a^{-1}) and area of arable land (135 × 10^6 ha).
l   Franzluebbers et al. (1998). Increase in soil carbon in soybean/wheat double crop vs. soybean (average across tillage treatments).
m   Smith et al. (1997b). Rates reported here were calculated from annual mitigation potential (73 Tg C a^{-1}) and area of arable land (135 x 10^6 ha).
n   Net local increase in carbon stored in organic matter; likely small net carbon gain regionally, depending on fate of organic amendments if not applied as fertilizer.
o   Addition of sulfate, nitrate, or iron decreases activity of methanogens by providing alternative electron acceptors and restricting availability of substrates in submerged soils (e.g. Hori et al. 1990). Amendments tend to reduce CH$_4$ emissions by 0–77 per cent in different experiments (Schutz et al. 1989; Lindau et al. 1993; Wassmann et al. 1993; Denier van der Gon and Neue 1994). Amendments will probably result in net loss of organic carbon, though estimates were not reported.
p   Drainage of field during cropping season. Oxygen availability stimulates CH$_4$ oxidation and reduces CH$_4$ emission (Yagi et al. 1997). Reduced CH$_4$ emission may be offset by increased CO$_2$ emission (soil carbon loss).
q   Peterson et al. (1998); values for increased carbon levels with continuous crop rotations vs. wheat-summer fallow rotations for four experiments in Montana and Colorado.
r   Janzen et al. (1998); mean for six experiments with no-till vs. tilled treatments and continuous crop rotations. No apparent increases in soil carbon with no-till were found for wheat-summer fallow rotations.
s   Potter et al. (1998); mean for no-till vs. conventional tillage treatments at three sites (11 crop rotations) in Texas.

Source: IPCC (2000b).

Table 11.8 Rates of potential carbon gain under selected practices for grasslands in various regions of the world

Practice	Country/Region	Rate of carbon gain (t C ha^{-1} a^{-1})	Time* (a)	Other GHGs and impacts	Notes
Reduce degradation	Global	0.5	20	Increase sustainability.	a
	Global	0.024–0.24	50	Generally likely to reduce methane emissions via reductions in animal numbers and increases in diet quality. Possible improvement in biodiversity, particularly in set-aside lands.	b
	Global	0.41	110		c
-Improve grazing management	Global	0.22	40		d
	Global	0.7	50+		e
	Australia	0.24	30		f
-Protected lands and set-asides	USA CRP	0.52	50		g
	China	1.3			h
Increase grassland productivity	Global	0.51		+N$_2$O. Reduced erosion if grazing management appropriate.	d
Fertilization	Global average	0.23		++N$_2$O, off-site nutrient impacts, acidification.	d
	N. Australia	0.50	40+		i
Irrigation	Global average	0.16	10	Associated fossil fuel emissions, salinization risks.	d
Improved species and legumes				+N$_2$O, ?CH$_4$. Risk of introduced species becoming weeds in adjacent areas. Biodiversity loss from native pastures.	
-Legumes	Global	1.09			d
-Grasses	Global	3.34			d
-Conversion from native pasture	Global	0.36			d
	South America	2.8–14.4			j

Fire management	Orinoco Plains, South America	1.4		k
	NE Australia	0.56	50	$-CH_4$, $-N_2O$. Reduced agricultural production, ? biodiversity depending on site	l

Notes:

t C ha^{-1} a^{-1} = tonnes of carbon per hectare per year; N_2O = nitrous oxide; CH_4 = methane; Mha = million hectares; Gha = billion hectares.

* Time interval to which the estimated rate applies. This interval may or may not be the time required for the ecosystem to reach a new equilibrium.

a Glenn et al. (1993), 5172 Mha of drylands, halophyte storage over 5 years.
b Paustian et al. (1998). Assumes potential carbon sequestration of 1–2 kg C m^{-2} on an arbitrary 10–50 per cent of moderately to highly degraded land (1.2 Gha globally; Oldeman et al. 1992).
c Keller and Goldstein (1998). Revegetation of agricultural and pastoral drylands.
d Conant et al. (2001). Based on literature review.
e Ojima et al. (1993). Regressive 50 per cent consumption vs. sustainable 30 per cent land management for grassland and rangelands.
f Ash et al. (1996). Average estimate 0–10 cm soil carbon only for transfer from deteriorated condition sustainable across northern Australia.
g McConnell and Quinn (1988); Gebhart et al. (1994); Barker et al. (1995); Burke et al. (1995).
h Li and Zhao (1998). Change in top 20cm of soil.
i Dalal and Carter (1999). Phosphorus and sulfur fertilization over 56Mha in northern Australia.
j Fisher et al. (1994, 1995). Introduced grass/legume pastures with deep, dense root systems.
k San Jose et al. (1998). Extrapolated from one site to 28Mha region.
l Burrows et al. (1998); Burrows et al. (1999). Aboveground and below-ground biomass extrapolated from 47 sites to 60 Mha region.

Source: IPCC (2000b).

Table 11.9 *Carbon uptake rates and time-averaged system carbon stocks and differences in carbon stock from land transformation at margins of humid tropics. Summary of 116 sites with different land uses before and after slash-and-burn located in Pedro Peixoto (Acre) and Theobroma (Rondnia), Brazil; Ebolowa, M'Balmayo, and Yaounde, Cameroon; Jambi and Lampung, Sumatra, Indonesia; and Yurimaguas and Pucallpa, Peru*

Land-use practice	Carbon uptake rates (t C ha^{-1} a^{-1})			Duration (years)	Carbon stocks (time-averaged) (t C ha^{-1})			Differences in Modal carbon stocks (time-averaged) (t C ha^{-1})	
	Low	Modal	High		Low	Modal	High	Forest	Pasture/Grasslands
Primary and logged forest	n/a[b]	n/a[b]	n/a[b]	?	192	230	276	–	−201
Cropping after slash-and-burn	−76	−92	−112	2	39	46	52	−184	+17
Crops/Bush fallow	2	3	4	4	32	34	36	−196	+5
Tall secondary forest fallow	5	7	9	23	95	112	142	−118	+83
Complex agroforest	2	3	4	25–40	65	85	118	−145	+56
Simple agroforest	5	7	9	15	65	74	92	−156	+61
Pasture, *Imperata* grassland	−0.2	−0.2	−0.6	4–12	27	29	31	−201	–

Notes:
t C ha^{-1} a^{-1} = tonnes of carbon per hectare per year; t C ha^{-1} = tonnes of carbon per hectare; Mg = million grammes (one tonne).
[a] Sanchez (2000). Calculated from data of Woomer et al. (1999) and Palm et al. (2000), assuming the following time-averaged soil carbon stocks (in Mg C ha^{-1}): 40 for primary/logged forest and crops after slash-and-burn, 35 for tall secondary forest fallow and complex agroforest, 30 for bush fallow and simple agroforest, and 25 for pasture and *imperata* grassland.
[b] Not available; probably close to zero.

Source: IPCC (2000b).

Table 11.10 *Measured carbon sequestration (accumulation) rates of temperate grasslands*

Location	Grassland system	Management	Technique	Climate	CO_2 sequestration (t C ha^{-1} a^{-1})
Switzerland	Semi-natural calcareous grassland	2 cuts a^{-1} (low nutrient)	C labelling in OTC	600 ppm CO_2	<0.8
Switzerland	Sown *Trifolium repens*	4 cuts a^{-1}	C labelling in FACE	600 ppm CO_2	4.6±2.2
	Sown *Lolium perenne*	4 cuts a^{-1}		600 ppm CO_2	6.3±3.6
	Sown *T. repens*	4 cuts a^{-1}	C-enriched soil (C_4) in FACE	Ambient CO_2	1.9
				600 ppm CO_2	3.1
Netherlands	Sown *L. perenne*	10 cuts a^{-1}	C balance	Ambient CO_2	2.29
		High N (800 kg ha^{-1} a^{-1})		2 × ambient CO_2	3.90
France	Sown *L. perenne*	5 cuts a^{-1}	Soil C analysis	Ambient CO_2	0.6 (LN), 0.8 (HN)
		Low N (160 kg ha^{-1} a^{-1}) (LN) and high N (530 kg ha^{-1} a^{-1}) (HN)		2 × ambient CO_2	1.4 (LN), 1.9 (HN)
UK	Species-rich grassland on limestone soil in solar dome	4–5 cuts a^{-1}	C balance	Ambient CO_2	1.2±0.5
				Ambient CO_2 + 250 ppm CO_2	2.6±0.3
	Species poor community on peaty gley in solar dome	No cutting/grazing	C balance	Ambient CO_2	6.4±0.6
				Ambient CO_2 + 250 ppm CO_2	8.7±0.2

Table 11.10 (continued)

Location	Grassland system	Management	Technique	Climate	CO_2 sequestration ($t\ C\ ha^{-1}\ a^{-1}$)
USA California	Annual grassland on serpentine	No cutting/grazing	C labelling in OTC	$2 \times$ ambient CO_2	2.66 ± 0.6
	Annual grassland on sandstone	No cutting/grazing	C labelling in OTC	$2 \times$ ambient CO_2	3.03 ± 0.4
USA Central Texas	Perennial grassland	Converted from arable to grassland for 6–60 a	Soil C analysis	Ambient	0.45

Notes: $t\ C\ ha^{-1}\ a^{-1}$ = tonnes of carbon per hectare per year; OTC = open-top chambers; FACE = free-air CO_2 enrichment; LN = low nitrogen; HN = high nitrogen.

Source: Jones and Donnelly (2004).

Table 11.11 *Representative Carbon Sequestration rates and saturation periods for key agricultural and forestry practices*

Activity	Representative carbon sequestration rate in US (t C acre^{-1} a^{-1})	Time over which sequestration may occur before saturating (assuming no disturbance, harvest or interruption of practice)
Afforestation[a]	0.6–2.6[b]	90–120+ years
Reforestation[c]	0.3–2.1[d]	90–120+ years
Changes in forest management	0.6–0.8[e] 0.2[f]	If wood products included in accounting, saturation does not necessarily occur if C continuously flows into products
Conservation or riparian buffers	0.1–0.3[g]	Not calculated
Conservation from conventional to reduced tillage	0.2–0.3[h] 0.2[i]	15–20 years 25–50 years
Changes in grazing land management	0.02–0.5[j]	25–50 years
Biofuel substitutes for fossil fuels	1.3–1.5[k]	Saturation does not occur if fossil fuel emissions are continuously offset

Notes:

t C acre^{-1} a^{-1} = tonnes of C per acre per year; there are roughly 2.47 acres to the hectare. Any associated changes in emissions of methane (CH_4) nitrous oxide (N_2O) or fossil CO_2 not included.

a   Values are for average management of forest after being established on previous croplands or pasture.
b   Values calculated over 120-year period. Low value is for spruce-fir forest type in Lake States; High value for Douglas Fir in Pacific Coast. Soil carbon accumulation included in estimate.
c   Values are for average management of forest established after clear-cut harvest.
d   Values calculated over 120-year period. Low value is for Douglas Fir in Rocky Mountains; high value for Douglas Fir in Pacific Coast. No accumulation in soil carbon is assumed.
e   Select examples, calculated over 100 years. Low value represents change from 25-year to 50-year rotation for loblolly pines in Southeast; high value is change in management regime for Douglas Fir in Pacific Northwest.
f   Forest management here encompasses regeneration, fertilization, choice of species and reduced forest degradation. Average estimate here is not specific to US, but average over developed countries.
g   Assumed that carbon sequestration rates are same as average rates for lands under USDA Conservation Reserve Program.

Table 11.11 (continued)

h   Estimates include only conversion from conventional tillage to no-till for all cropping systems except for wheat-fallow systems, which may not produce net carbon gains. Estimates of changes in other greenhouse gases not included.
i   Assumed that average carbon sequestration rates are same for conversion from conventional till to no-till, mulch till or ridge till. Estimates of changes in other greenhouse gases not included.
j   Low end is improvement of rangeland management; high end is changes in grazing management on pasture, where soil organic carbon is enhanced through manure additions. Estimates of flux changes in other greenhouse gases not included.
k   Assumes growth of short-rotation woody crops and herbaceous energy crops, and that burning this biomass offsets 65–75 per cent of fossil fuel in CO_2 emissions. Estimates of changes in other greenhouse gases not included.

Source: http://www.epa.gov/sequestration/rates.html.

nificance, how should one consider climate change adaptation? We turn to this next.

ADAPTATION BENEFITS OF RESTORATION

As mentioned earlier, the cost of climate change adaptation should, preferably, follow a bottom-up approach, since adaptation responses are often locally relevant and site specific. To provide a regional or global assessment of such costs would imply a significant increase in research and also policy focus among all climate change stakeholders and the adaptation community at large. Ideally, to assess the relative costs and benefits of adaptation it is necessary to project four costs, namely (1) non-climate-change cost increases in the infrastructure or process considered (i.e. costs that would have occurred in the absence of climate change); (2) the predicted damage costs of climate change without adaptation measures; (3) damage costs after adaptation investments have been made; and (4) the cost of adaptation investments (Stern et al. 2006). Another option is to consider the economic optimality argument (Hughes et al. 2010). The economic optimality argument assumes that the adaptation costs would rise to equal the impact costs; a fair assumption for a resource that is limited in supply. Still, this would, in most cases, require detailed adaptation plans, be it on the municipal level or at a larger spatial scale. However, even this level of resolution would not remove the uncertainty embedded within the analysis.

At a larger scale, several attempts have been made to provide a top-down valuation assessment of the costs and benefits of adaptation (World

> ## BOX 11.4 WEAKNESS OF THE MCKINSEY ADAPTATION COST CURVES
>
> Inadequate incorporation of inherent uncertainty. It is not possible to predict climate impacts with certainty and to deliver the definitive prevention measures required to mitigate or adapt to these impacts.
>
> The complete absence of restoration from the analysis and hence also the insufficient acknowledgement of the value of ecosystem goods and services in both reducing climate impacts and in appropriate adaptation responses.
>
> Subjective use of discount rates and the assumption that a single discount rate can be used across all parties affected.
>
> Poor alignment between financial value and local priorities and the associated failure to address equity issues or to identify the most vulnerable communities. The approach adopted by McKinsey Company (2009) led to the prioritization of high value real estate and the discounting of the needs and assets of the poor where these were not reflected in market value.
>
> Ignored institutional responses (described by the World Bank (2010b) as 'soft options ... early warning systems, watershed management, community preparedness, zoning legislation'), and treating each adaptation option as discrete.
>
> Failure to link climate adaptation with local development priorities, and in so-doing misses the potential value of ancillary benefits from climate adaptation. This is in stark contrast to the position taken by the Hartwell Paper (LSE/ISIS 2010) that actively sought such co-benefits.
>
> Results difficult (as well as expensive) to apply at the local level, thereby making it difficult to apply with the requisite urgency (Stafford et al. 2011).
>
> *Source:* Adapted from Cartwright et al. (2011).

Bank 2006, 2010a; UNDP 2007; Parry et al. 2009). Global aggregates, however, do not assist local decision-makers in identifying and prioritizing adaptation options. To meet this need, McKinsey attempted to rank climate change adaptation options according to cost-benefit ratios (McKinsey Company 2009). While the effort should be applauded, this study also has some weaknesses (see Box 11.4).

BOX 11.5   DISTRIBUTION OF RESTORATION CASE STUDIES

In a study reported in Aronson et al. (2010) it is noted that most (72 per cent) academic papers from the period January 2000 to September 2008 that covered restoration case studies did so for high-income countries. It is important, however, to note that restoration sought to improve the cultural (28.5 per cent), provisioning (26 per cent), regulating (29 per cent) and supporting (16.5 per cent) services provided by natural and semi-natural ecosystems (see Table 11.12). Activities to restore biodiversity, or plant or animal populations, were classified as restoring cultural services, because threatened, endangered, or otherwise 'special' species, are valued as part of human natural heritage. There was little evidence that such restoration would increase material, supporting or regulating services.

The human well-being objective of most restoration was aimed towards improved social relations (30 per cent) and security (30 per cent). Fewer studies envisaged restoration outcomes that resulted in better health (15 per cent) or a greater supply of goods (25 per cent) (see Table 11.13).

Table 11.12   *Distribution of restoration case studies reported in 13 peer-reviewed journals over the period January 2000 to September 2008: ecosystem goods and services*

	Low income	Low middle income	Upper middle income	High income	Not specified or others	Total
Supporting	16.3%	12.8%	12.4%	16.4%	20.2%	16.3%
Regulating	23.8%	35.2%	30.1%	29.3%	26.0%	29.2%
Provisioning	41.3%	31.3%	29.5%	25.0%	25.1%	26.2%
Cultural	18.8%	20.7%	28.0%	29.4%	28.7%	28.3%
	100%	100%	100%	100%	100%	100%
	n = 80	n = 179	n = 193	n = 1902	n = 331	n = 2685

Source: Adapted from Aronson et al. (2010).

Table 11.13 Distribution of restoration case studies reported in 13 journals over the period January 2000 to September 2008: human well-being

	Low income	Low middle income	Upper middle income	High income	Not specified or others	Total
Material	38%	26%	26%	23%	28%	25%
Health	10%	14%	15%	15%	18%	15%
Security	31%	38%	30%	31%	25%	31%
Social relations	21%	23%	28%	31%	29%	30%
	100%	100%	100%	100%	100%	100%
	n = 77	n = 181	n = 184	n = 1841	n = 325	n = 2608

Source: Adapted from Aronson et al. (2010).

The proportion of papers focused on provisioning services was significantly higher (Chi square = 13.01, $p < 0.01$) in low-income countries (41.3 per cent) as opposed to high-income countries (25 per cent). The converse was true for research dealing with restoration to provide cultural services (18.8 per cent for low-income countries compared to 29.4 per cent for high-income countries). Only 37 per cent of the research papers from high-income countries reported on restoration to improve agricultural productivity compared with 53 per cent of upper middle-income, 43 per cent for low middle-income and 56 per cent of low-income countries, with the differences being statistically significant (Chi square = 16.06, $p < 0.01$).

Therefore, it is self-evident that, from a technical perspective, the application of a cost–benefit analysis to climate change adaptation is problematic owing to the difficulty of establishing an accurate 'damage function'. It is not possible to say precisely 'what' climate change impacts will be experienced and 'when', given the uncertainty around future impacts and the complex ways in which these impacts are manifested over time and space. In this light, it is very difficult to put a monetary value on the damage caused by climate change or the benefit of adaptation options that avoid (usually only partially) this damage, without making far-reaching and difficult-to-substantiate assumptions.

These difficulties can be avoided by embarking on an adaptation strategy that yields continuous benefits, such as restoration. Restoration activities aimed at restoring provision (e.g. food, fodder, water and medicinal material) and regulating (e.g. soil stabilization, soil productivity, waste assimilation, water flow, climate regulation) services will, in all likelihood, contribute more to climate change adaptation than cultural and habitat services. This, incidentally, is also the focus of restoration activities within developing countries (see Box 11.5).

From the above it is evident that there is a natural fit between adaptation and the restoration of natural capital. There are three reasons for this:

- Restored areas (and for that matter protected areas or land under wildlife management) act as buffers against losses of ecosystem goods and services, and by maintaining them they reduce vulnerability;
- Restored areas are a source of ecosystem goods and services that act as a highly valuable insurance policy against future disasters;
- Restored areas act as a bank of genetic material and seed for future use and restorative action.

CONCLUSION

Within the climate change debate, mitigation tends to attract the most attention. It is, however, adaptation that potentially has the greatest to offer in terms of socio-economic development.

The following recommendations can therefore be made:

- climate change adaptation measures are often good socio-economic development interventions. By considering adaptation and socio-economic development as potential joint outcomes of a single investment stream, policies and interventions are not hamstrung by the additionality condition that has hampered development of the mitigation market. This has the expected benefit of addressing several society-wide ills in concert;
- an adaptation strategy needs to be sought that will render benefits irrespective of whether there is an extreme event. This is an imperative consideration given scarce resources. Policy-makers are hard-pressed to consider interventions that offer immediate benefits. Investing in interventions that are either risky or whose benefits are

uncertain is unlikely to gain much traction. However, investments that render both immediate and ongoing adaptation benefits and that contribute to socio-economic development have to be considered extremely favourably;
- the restoration of degraded natural capital is the only augmentation option to the world's declining stock of natural capital and hence, ecosystem goods and services. It has also been shown that restoration offers excellent returns on investment with benefit–cost ratios ranging from 2 to 27. Investments in natural capital are of paramount importance. As an adaptation strategy, restoration reduces people's vulnerability as it acts as a buffer against losses of ecosystem goods and services. It is a highly valuable insurance policy against future disasters, and it safeguards the genetic bank of biological material and seed for future use. In addition, it sequesters carbon and therefore contributes meaningfully to climate change mitigation.

Therefore, it can be concluded that while restoration is surely not the only adaptation or socio-economic development intervention option, it can and should play its role on the international stage. By and large, restoration is a bottom-up strategy, and it is at the bottom where people live and where they are vulnerable.

ACKNOWLEDGMENTS

Leandri van der Elst, Anton Cartwright, Guy Midgley, James Aronson and Pushpam Kumar are hereby acknowledged and thanked for their comments on earlier drafts.

REFERENCES

Anthoff, D., R. Tol and G. Yohe (2009). Risk aversion, time preference, and the social cost of carbon. *Environmental Research Letters*, **4**, 024002.

Aronson, J., J. Blignaut, J. Milton and A. Clewell (2006). Natural capital: the limiting factor. *Ecological Engineering*, **28**, 1–5.

Aronson J., S. Milton and J.N. Blignaut (eds) (2007). *Restoring Natural Capital: Science, Business and Practice*. Washington DC: Island Press, p. 384.

Aronson J., J. Blignaut, S. Milton, D. Le Maitre, K. Esler, A. Limouzin, C. Fontaine, M. De Wit, W. Mugido, P. Prinsloo, L. Van der Elst and N. Lederer (2010). Why restore? A meta-analysis of recent papers (2000–2008) in *Restoration Ecology* and 12 other scientific journals. *Restoration Ecology*, **18**(2), 143–154.

Ash, A.J., S.M. Howden, J.G. McIvor and N.E. West (1996). Improved rangeland management and its implications for carbon sequestration. In N.E. West (ed.) *Proceedings of the Fifth International Rangeland Congress*, Salt Lake City, UT, USA, 23–28 July 1995, pp. 19–20.

Barker, J.R., G.A. Baumgardner, D.P. Turner and J. Lee (1995). Potential carbon benefits of the Conservation Reserve Program in the United States. *Journal of Biogeography*, **22**, 743–751.

Blignaut, J.N. (2009). Fixing both the symptoms and causes of degradation: the need for an integrated approach to economic development and restoration. *Journal of Arid Environments*, **73**, 696–698.

Blignaut, J.N. and J. Aronson (2008). Getting serious about maintaining biodiversity. *Conservation Letters*, **1**(1), 12–17.

Bruce, J.P., M. Frome, E. Haites, H. Janzen, R. Lal and K. Paustian (1999). Carbon sequestration in soils. *Journal of Soil and Water Conservation*, **54**, 381–389.

Burke, I.C., W.K. Lauenroth and D.P. Coffin (1995). Soil organic matter recovery in semiarid grasslands: implications for the Conservation Reserve Program. *Ecological Applications*, **5**(3), 793–801.

Burrows, W.H., J.F. Compton and M.B. Hoffmann (1998). Vegetation thickening and carbon sinks in the grazed woodlands of north-east Australia. *Proceedings of the Australian Forest Growers Conference*. Lismore, New South Wales, Australia, pp. 305–316.

Burrows, W.H., M.B. Hoffmann, J.F. Compton, P.V. Back and L.J. Tait (1999). Allometric relationships and community biomass estimates for some dominant eucalypts and associated species in Central Queensland woodlands. *Australian Journal of Botany*, **48**(6), 707–714.

Cartwright, A., M. Mander, M. De Wit and J. Blignaut (2011). Cost–benefit analysis of municipal adaptation plans for eThekwini municipality: inception report. Inception report submitted to the eThekwini Municipality under Contract: 1N- 5931. eThekwini Municipality, Durban.

Conant, R.T., K. Paustian and E.T. Elliot (2001). Grassland management and conversion into grassland: Effects on soil carbon. *Ecological Applications*, **11**, 343–355.

Costanza, R., W.J. Mitsch and J.W. Day Jr. (2006). A new vision for New Orleans and the Mississippi delta: applying ecological economics and ecological engineering. *Frontiers in Ecology and Environment*, **4**(9), 465–472.

Dahdouh-Guebas, F., L.P. Jayatissa, D. Di Nitto, J.O. Bosire, D. Lo Seen and N. Koedam (2005). How effective were mangroves as a defence against the recent tsunami? *Current Biology*, **15**(2), R443–R447.

Dalal, R.C. and J.O. Carter (2000). Soil organic matter dynamics and carbon sequestration in Australian tropical soils. In R. Lal, J.M. Kimble and B.A. Stewart (eds), *Global Climate Change and Tropical Ecosystems*. Boca Raton, FL: Advances in Soil Science. CRC Press, pp. 283–314.

Dasgupta, P. (2007). Commentary: the Stern Review's economics of climate change. *National Institute Economic Review*, **199**, 4–7.

De Groot, R., L. Brander, S. van der Ploeg, R. Costanza, F. Bernard, L. Braat, M. Christie, N. Crossman, A. Ghermandi, L. Hein, S. Hussain, P. Kumar, A. McVittie, R. Portela, L.C. Rodriguez, P. ten Brink and P. Van Beukering (2012). Global estimates of the value of ecosystems and their services in monetary units. *Ecosystem Services*, **1**(1), 50–61.

Denier van der Gon, H.A.C. and H.U. Neue (1994). Impact of gypsum application on methane emission from wetland rice field. *Global Biogeochemical Cycles*, **8**, 127–134.

Dumanski, J., R.L. Desjardins, C. Tarnocai, C. Monreal, E.G. Gregorich, V. Kirkwood and C.A. Campbell (1998). Possibilities for future carbon sequestration in Canadian agriculture in relation to land use changes. *Climate Change*, **40**, 81–103.

EJF (2006). Mangroves: nature's defence against tsunamis. A report on the impact of mangrove loss and shrimp farm development on coastal defences. Environmental Justice Foundation, London. Available at: http://www.ejfoundation.org/pdf/tsunami_report.pdf (accessed 8 August 2011).

FAO (1996). Forest resources assessment 1990: survey of tropical cover and study of change processes. FAO Forestry Paper 130, Rome.

Fisher, M.J., I.M. Rao, M.A. Ayarza, C.E. Lascano, J.I. Sanz, R.J. Thomas and R.R. Vera (1994). Carbon storage by introduced deep-rooted grasses in the South American savannas. *Nature*, **371**(15), 236–238.

Fisher, M.J., I.M. Rao, C.E. Lascano, M.A. Ayarza, J.I.Sanz, R.J. Thomas and R.R. Vera (1995). Pasture soils as carbon sink. *Nature*, **376**(6540), 473.

Franzluebbers, A.J., F.M. Hons and D.A. Zuberer (1998). *In situ* and potential CO_2 evolution from a Fluventic Ustochrept in southcentral Texas as affected by tillage and cropping intensity. *Soil and Tillage Research*, **47**, 303–308.

Gebhart, D.L., H.B. Johnson, H.S. Mayeux and H.W. Polley (1994). The CRP increases soil organic carbon. *Journal of Soil and Water Conservation*, **49**(5), 488–492.

Glenn, E., V. Squires, M. Olsen and R. Frye (1993). Potential for carbon sequestration in the drylands. *Water, Air and Soil Pollution*, **70**, 341–355.

Hori, K., K. Inubushi, S. Matsumoto and H. Wada (1990). Competition for acetic acid between methane formation and sulfate reduction in the paddy soil. *Japanese Journal of Soil Science and Plant Nutrition*, **61**, 572–578. (In Japanese with English summary.)

Hughes, G., P. Chinowsky and K. Strzepek (2010). The costs of adaptation to climate change for water infrastructure in OECD countries. *Utilities Policy*, **18**, 142–153.

IPCC (1999). *Economic Impact of Mitigation Measures*. Geneva: IPCC.

IPCC (2000a). *Sectoral Economic Costs and Benefits of GHG Mitigation*. Geneva: IPCC.

IPCC (2000b). Additional human-induced activities: Article 3.4. In R.T. Watson, I.R. Noble, B. Bolin, N.H. Ravindranath, D.J. Verardo and D.J. Dokken (eds), *IPCC Special Report on Land Use, Land-Use Change And Forestry*. Cambridge: Cambridge University Press, Chapter 4. Available at: http://www.grida.no/publications/other/ipcc_sr/?src=/climate/ipcc (accessed 20 June 2011).

Janzen, H.H., C.A. Campbell, R.C. Izaurralde, B.H. Ellert, N. Juma, W.B. McGill and R.P. Zentner (1998). Management effects on soil C storage on the Canadian prairies. *Soil Tillage Research*, **47**, 181–195.

Jones, M.B. and A. Donnelly (2004). Carbon sequestration in temperate grassland ecosystems and the influence of management, climate and elevated CO_2. *New Phytologist*, **164**, 423–439.

Keller, A.A. and R.A. Goldstein (1998). Impact of carbon storage through restoration of drylands on the global carbon cycle. *Environmental Management*, **22**(5), 757.

Kuik, O., L. Brander and R. Tol (2009). Marginal abatement costs of greenhouse gas emissions: a meta-analysis. *Energy Policy*, **37**, 1395–1403.

Lal, R. and J.P. Bruce (1999). The potential of world cropland soils to sequester C and mitigate the greenhouse effect. *Environmental Science and Policy*, **2**, 177–185.

Lal, R., J.M. Kimble and R.F. Follett (1999). Agricultural practices and policies for carbon sequestration in soil. *Recommendation and Conclusions of the International Symposium*, 19–23 July 1999, Columbus, OH.

Li, Z. and Q. Zhao (1998). Carbon dioxide fluxes and potential mitigation in agriculture and forestry of tropical and subtropical China. *Climatic Change*, **40**, 119–133.

Lindau, C.W., P.K. Bollich, R.D. DeLaune, A.R. Mosier and K.F. Bronson (1993). Methane mitigation in flooded Louisiana rice fields. *Biology and Fertility of Soils*, **15**, 174–178.

LSE/ISIS (2010). A new direction for climate policy after the crash of 2009. The Hartwell Paper. London School of Economics/Institute for Science, Innovation and Society, University of Oxford. Oxford and London. Available at: http://eprints.lse.ac.uk/27939/1/HartwellPaper_English_version.pdf (accessed 25 July 2011).

McConnell, S.G. and M.L. Quinn (1988). Soil productivity of four land use systems in southeastern Montana. *Soil Science Society of America Journal*, **52**, 500–506.

McKinsey Company (2009). Shaping climate resilient development: a framework for decision making. A Report of the Economics of Climate Adaptation. ECA Working Group, New York.

Millennium Ecosystem Assessment (2005). *Ecosystems and Human Well-being: Synthesis*. Washington DC: Island Press.

Nordhaus, W.D. (2007). A Review of the Stern Review on the economics of climate change. *Journal of Economic Literature*, **45**(3), 686–702.

Ojima, D.S., B.O.M. Dirks, E.P. Glenn, C.E. Owensby and J.O. Scurlock (1993). Assessment of C budget for grasslands and drylands of the world. *Water, Air and Soil Pollution*, **70**, 95–109.

Oldeman, L.R., V.W.P. van Engelen and J.H.M. Pulles (1992). The extent of human-induced soil degradation. In L.R. Oldeman, R.T.A. Hakkeling and W.G. Sombroek (eds), *World Map of the Status of Human-Induced Soil Degradation: An Explanatory Note*. Wageningen, the Netherlands: International Soil Reference and Information Centre.

Palm, C.A., R.J.K. Myers and S.M. Nandwa (2000). Combined use of organic and inorganic nutrient sources for soil fertility maintenance and replenishment. In R.J. Buresh, P.A. Sanchez and F. Calhoun (eds), Replenishing Soil Fertility in Africa. SSSA Special Publication 51, Soil Science Society of America, Madison, WI, pp. 193–218.

Parry, M., N. Arnell, P. Berry, D. Dodman, S. Fankhauser, C. Hope, S. Kovats, R. Nicholls, D. Satterthwaite, R. Tiffin and T. Wheeler (2009). *Assessing the Costs of Adaptation to Climate Change: A Review of the UNFCCC and Other Recent Estimates*. London: International Institute for Environment and Development and Grantham Institute for Climate Change.

Paustian, K., O. Andren, H.H. Janzen, R. Lal, P. Smith, G. Tian, H. Tiessen, M. Van Noordwijk and P.L. Woomer (1997). Agricultural soils as a sink to mitigate carbon dioxide emissions. *Soil Use and Management*, **13**(4), 230–244.

Paustian, K., C.V. Cole, D. Sauerbeck and N. Sampson (1998). CO_2 mitigation by agriculture: an overview. *Climatic Change*, **40**, 135–162.

Peterson, G.A., A.D. Halvorson, J.L. Havlin, O.R. Jones, D.J. Lyon and D.L. Tanaka (1998). Reduced tillage and increasing cropping intensity in the Great Plains conserves soil C. *Soil and Tillage Research*, **47**, 207–218.

Potter, K.N., H.A. Torbert, O.R. Jones, J.E. Matocha, J.E. Morrison, Jr. and P.W. Unger (1998). Distribution and amount of soil organic C in long-term management systems in Texas. *Soil and Tillage Research*, **47**, 309–321.

Powlson, D.S., P. Smith, K. Coleman, J.U. Smith, M.J. Glendining, M. Korschens and U. Franko (1998). A European network on long-term sites for studies on soil organic matter. *Soil and Tillage Research*, **47**, 263–274.

Rafey, W. and B.K. Sovacool (2011). Competing discourses of energy development: The implications of the Medupi coal-fired power plant in South Africa. *Global Environmental Change*, **21**(3), 1141–1151.

Rayden, T., P. Ramani, T. Lander, J. Ebeling and R. Nussbaum (2010). *Terrestrial Carbon: Emissions, Sequestration and Storage in Tropical Africa: FPAN African Tropical Forests Review of the Scientific Literature and Existing Carbon Projects.* Oxford: Forests Philanthropic Action Network.

Sanchez, P.A. (2000). Linking climate change research with food security and poverty reduction in the tropics. *Agriculture, Ecosystems and Environment*, **82**(1–3), 371–383.

San Jose, J.J., R.A. Montes and M.R. Farinas (1998). Carbon stocks and fluxes in a temporal scaling from a savanna to a semi-deciduous forest. *Forest Ecology and Management*, **105**, 251–262.

Schutz, H., A. Holzapfel-Pschorn, R. Conrad, H. Rennenberg and W. Seiler (1989). A 3-year continuous record on the influence of day-time, season, and fertilizer treatment on methane emission rates from an Italian rice paddy. *Journal of Geophysical Research*, **94**, 16405–16416.

Singh G., N.T. Singh and I.P. Abrol (1994). Agroforestry techniques for the rehabilitation of degraded salt-affected land. *Land Degradation Development*, **5**, 223–242.

Smith, P., D.S. Powlson, M.J. Glendining and J.U. Smith (1997a). Potential for carbon sequestration in European soils: preliminary estimates for five scenarios using results from long-term experiments. *Global Change Biology*, **3**, 67–79.

Smith, P., D.S. Powlson, M.J. Glendining and J.U. Smith (1997b). Opportunities and limitations for C sequestration in European agricultural soils through changes in management. In R. Lal, J.M. Kimble, R.F. Follett and B.A. Stewart (eds), *Management of Carbon Sequestration in Soil*. Boca Raton, FL: CRC Press, pp. 143–152.

Smith, P., D.S. Powlson, M.J. Glendining and J.U. Smith (1998). Preliminary estimates of the potential for carbon mitigation in European soils through no-till farming. *Global Change Biology*, **4**, 679–685.

Stafford, M., L. Horrocks, A. Harvey and C. Hamilton (2011). Rethinking Adaptation in a 4°C World. *Philosophical Transactions of the Royal Society*, **369**, 196–216.

Stage, R. (2010). Economic valuation of climate change adaptation in developing countries. *Annals of the New York Academy of Sciences*, **1185**, 150–163.

Stern, N. (2007). *The Economics of Climate Change: The Stern Review*. Cambridge: Cambridge University Press.

Stern, N. (2008). The economics of climate change. *American Economic Review*, **98**, 1–37.

Stern, N., S. Peters, V. Bakhshi, A. Bowen, C.S. Cameron, C.D. Catovsky, S. Cruickshank, S. Dietz, N. Edmondson, S. Garbett, L. Hamid, G. Hoffman, D. Ingram, B. Jones, N. Patmore, H. Radcliffe, R. Sathiyarajah, M.C. Stock, V.T. Taylor, H. Wanjie and D. Zenghelis (2006). *Stern Review on the Economics of Climate Change.* Cambridge: Cambridge University Press.

TEEB (2009). *TEEB Climate Issues Update.* Nairobi: United Nations Environment Programme.

TEEB. (2010). *The Economics of Ecosystems and Biodiversity: Ecological and Economic Foundations.* London and Washington: Earthscan.

Terakunpisut, J., N. Gajaseni and N. Ruankawe (2007). Carbon sequestration potential in aboveground biomass of thong Pha Phum National Forest, Thailand. *Applied Ecology and Environmental Research,* 5(2), 93–102.

Tol, R. (2005). The marginal damage costs of carbon dioxide emissions: an assessment of the uncertainties. *Energy Policy,* 33(16), 2064–2074.

Tol, R. (2009). The economic effects of climate change. *Journal of Economic Perspectives,* 23(2), 29–51.

Tol, R. and G. Yohe (2009). The Stern Review: a deconstruction. *Energy Policy,* 37, 1032–1040.

UNDP (2007). *Human Development Report 2007/08.* New York: Palgrave McMillan.

UNEP (2010). *Ecosystem-Based Adaptation Programme.* Paris: UNEP.

UNSD (2011). *Millennium Development Goals Indicators: Carbon Dioxide Emissions (CO_2).* New York: United Nations Statistics Division. Available at: http://mdgs.un.org/unsd/mdg/SeriesDetail.aspx?srid=749&crid (accessed 8 August 2011).

United Nations (1993). *A System of National Accounts.* New York: United Nations.

Wassman, R., H. Pappen and H. Rennenberg (1993). Methane emission from rice paddies and possible mitigation strategies. *Chemosphere,* 26, 201–217.

Watson, R.T., I.R. Noble, B. Bolin, N.H. Ravindranath, D.J. Verardo and D.J. Dokken (eds), *IPCC Special Report on Land Use, Land-Use Change And Forestry.* Cambridge: Cambridge University Press. Available at: http://www.grida.no/publications/other/ipcc_sr/?src=/climate/ipcc.

Woomer, P.L., C.A. Palm, J. Alegre, C. Castilla, D.G. Cordeiro, K. Hairiah, J. Kotto-Same, A. Moukam, A. Ricse, V. Rodrigues and M. van Noordwijk (2000). Slash-and-burn effects on carbon stocks in the humid tropics. In R. Lal, J.M. Kimble and B.A. Stewart (eds), *Global Climate Change and Tropical Ecosystems.* Boca Raton, FL: CRC Press, pp. 99–115.

World Bank (2006). *Investment Framework for Clean Energy and Development.* Washington DC: World Bank.

World Bank (2010a). The cost to developing countries of adapting to climate change: new methods and estimates. *The Global Report on the Economics of Adaptation to Climate Change Study.* Consultation Draft. Washington DC: World Bank Group.

World Bank (2010)b. *Convenient Solutions to an Inconvenient Truth: Ecosystem-Based Approaches to Climate Change.* Washington DC: World Bank Group.

Yagi, K., H. Tsuruta and K. Minami (1997). Possible options for mitigating methane emission from rice cultivation, nutrient cycling. *Agroeconomics,* 49, 213–220.

12. The ethical foundations of cultural diversity in ecosystems and their role in economic valuation

Ian Parker

INTRODUCTION

The problem of 'economic valuation', in which monetary value is assigned to environmental factors, attempts to connect the domain of the economic, i.e. the accumulation and management of capital, seen as a priority under capitalism, with the perceptions and preferences of individual subjects. When this problem is taken for granted, the main concern in research-oriented to policy is how cultural diversity should be included alongside specific calculable biological resources (Nunes and van den Bergh 2001).

Economic valuation usually entails abstraction of people from their social context, i.e. people are treated as individual subjects rather than social actors, in order that their own specific assessments of value might be aggregated. This aggregation of viewpoints is assumed necessary for economic purposes, and for policy initiatives that attempt to recruit and mobilize individuals and populations to work together in ways that enhance rather than diminish cultural diversity. Protection of the environment has been recognized in Article 12.2.b of the International Covenant on Economic, Social and Cultural Rights (United Nations 1966) and in Article 37 of the Charter of Fundamental Rights of the European Union (European Union 2000) as a human rights issue. In this context, public policy is an institutional response to address relevant social issues, but it has also been noted by feminist theorists that policy operates as a set of strategies to maintain a system, in other words, strategies of power (Puleo 2005). This contradictory disposition to affect behaviour of the population is related to political language that infers conclusive action (Butler 1997).

The underlying problem elaborated in this chapter is that the domain of the 'economic' is itself often experienced by social actors in capitalist economies as an abstract and mysterious domain from which they are separated. As such it becomes viewed as beyond the scope of the individual to comprehend or influence. This lack of interest and ability is also gender-oriented when economic expertise pertaining to the market is confined to stereotypically masculine concerns, while femininity is constituted in such a way as to be defined and experienced within more immediate social relationships (Haug 1992). This division into domains of technical control of large systems independent of social actors on the one hand and of ostensibly intuitive affective networks of care-giving on the other does not necessarily correspond to the sexes, but to constructions of gender which tend to distribute people into different categories (Butler 1990). This distribution itself, and the sense that is made of gender within those categories, also varies from culture to culture (Tazi 2004a).

The ethical consequences of the separation of the individual, who is required to place value on discrete components of an ecosystem, from the domain of the economic are immense. Culture comprises all the different individual contradictory positions within it, and an understanding of particular cultural practices and local conditions evidently has to include attention to the particular impact that degradation of cultural diversity has on women of different classes (Pini and Leech 2011).

In order to approach this topic, it is also useful to take account of the importance of diversity in the perspectives of different social actors in cultures, which can be seen as equivalent to 'ecosystems'. In other words, human culture can be seen as equivalent to an ecosystem, i.e. a multiplicity of actors and interactions comprising the whole, and biodiversity can be seen as equivalent to cultural diversity. Rather like the way an organism can viably only sit within its niche, it is the problem of the separation of the individual from the physical economy that makes economic valuation so difficult. For example, gender vulnerability is related to social vulnerability; women are more likely than men to be affected by the conditions of social and environmental deterioration, and they have less ability to recover or adapt to it. The particular position of women is associated with the fragility of communities, social groups and the environment when exposed to hazardous events, economic and environmental instability, and lack of resilience (Oswald Spring 2009).

The guiding principles for such questioning of economic valuation are provided by an emerging current of work inside the discipline of psychology that itself draws on cultural concerns with ethics and social contradictions (Parker 2011a). That current of 'critical psychology' is here turned to the task of assessing how conceptions of nature, and the mobilization of individuals and populations around those conceptions, inform policy.

UNITY AND DIVERSITY WITHIN CULTURAL AND ECONOMIC SYSTEMS

The concept of an ecosystem operates as a conceptual category alongside that of biodiversity in contemporary understanding of global economics (Kumar 2010). The notion of a cultural system as a representation of a social whole that includes some subjects and marginalizes others is in tension with the motif of cultural diversity applied to relations between human beings in their manifold relations with nature.

Consensual Conceptions of Societal Systems

The tension between the assumption that there is a totality of experience of those participating in a social group and the assumption that there should be diversity of subject positions has consequences for how agreement over the valuation of resources may be conceptualized.

One social anthropological tradition that was dominant in the West when the discipline of anthropology was founded, conceived of the task of understanding what a culture was and what it wanted to be through a compilation of accounts (Nadar 1997). Understanding in this tradition was tied to management in an evaluative process which also thereby romanticized the unity of the group in question (even as it also informed and reflected colonial strategy). Descriptions of human beings within particular ecosystems, e.g. as gatherer-hunters, navigates that fraught history.

Contradictory Conceptions of Societal Diversity

Against this orientation, which emphasized consensus, were dissenting voices from the 'structural anthropology' tradition, which provided a different way of respecting diversity, if not also conflict.

Studies from within this second tradition showed that disagreement in a community may often be structured not only in such a way as to reflect the power relations between different sub-groups, but also in setting the very terms over which the disagreement was negotiated. Competing views of what the community was, and how it was structured, meant that there was no common ground on which to compare different viewpoints (Lévi-Strauss 1958 [1972]).

The collection of data on the valuation of biological resources by members of a community therefore has to take into account the range of competing conceptions of cultural diversity. A lesson of this alternative strand of social anthropology is that the stance of dominant groups is that their own view is the one that should set the terms by which the arguments of others should be evaluated.

There are thus implications for methodological approaches in social research that assume consensus of meanings of terms between participants, and between participants and researchers (Kagan et al. 2011). Critiques of the ideological role that such consensus plays have drawn on understandings of gender and colonialism that give privilege to certain standpoints e.g. those of the oppressed, over the taken-for-granted worldviews of people in positions of power (Tuhiwai Smith 1999).

There are also profound consequences for how procedures of valuation, explicitly the requirement for certain social actors to produce an economic valuation of an ecosystem service, intersect with cultural and political processes.

Competing Conceptions and Social Functions of 'Nature'

Naturalistic metaphors have often been employed as social constructs as part of corporatist (and fascist) rhetoric to delegitimize the argument that there are different competing economic interests and corresponding perceptions of what really is of value in a society (Harvey 1996). The social body is then often portrayed by its own rhetoric in a gendered way, i.e. as a masculine entity whose balanced internally coherent self-concept is disturbed by an intrusion of dissent from what is conceived of as 'other' (Harcourt 2009). This history now gives a social dimension to debates over ecosystem assessment (Yuval-Davis 1997). Contrasting cultural conceptions of nature indicate that ideological and political histories shape how an ecosystem and biodiversity might be understood today by people of different cultures.

Representations of nature in Africa, if it is possible to make a generalization across a continent, have according to some accounts, been developed from within a colonial relationship where, for the colonialists, nature was both an obstacle and a resource. The reclaiming of nature by African writers as 'something that was once in balance with the human beings who lived in harmony with it' also comes from within the colonial relationship while also defying such a relationship (Ouédraogo 2005). In India, claim some writers, there was and could once again be continuity between nature and spirituality, such that they are both experienced as a totality of relationships that was broken and might now be repaired in conditions of post-colonialism (Srivastava 2005). It is said that in the Arab world, the microcosmic and macrocosmic aspects of reality are mirrored across each other such that nature becomes an expression at each level of a 'universal cosmic soul' (El-Bizri 2005).

The contradictory aspects of nature are emphasized in reflections on nature from various cultures. In China, according to some accounts, there are conceptions of nature that play a role in different cosmological systems, and in different conceptions of 'balance' that are inherent in each of them, but there has until recently been no abstracted notion of 'nature' as such (Shao-Ming 2005). The history of 'nature' in the briefer history of the United States of America has been traced through Puritanism, reflecting the religious standpoint of early colonial settlers, the Western Enlightenment, reflecting enduring cultural ties with Europe, and then Romanticism as a local reaction to the ideology of the nation as on a frontier into which it was expanding (Marx 2005). Europe, in contrast, has been characterized as traversing and transcending 'fables of nature' which comprise 'positivist anti-naturalism' in which knowledge is thought to require a separation from pre-scientific ideas about nature as such; 'technophobic anti-naturalism' in which direct experience is predisposed against any external force, including a notion of nature that might intrude on that experience; and 'technophilic and artificialist naturalism' in which the pervasiveness of nature calls for the elaboration of a multiplicity of perspectives that channel it (Zaoui 2005).

Different or contradictory conceptions of nature, as elaborated from within different cultures, are testimony to the importance of an individual's view of the intimate relationship between humans and nature, but also of the disagreement inherent in global systemic responses to the question of biodiversity.

The always-present, if often obscured, role of conflict in social systems is conceptualized in contemporary social theory as a basic underlying dynamic that structures political debate in positive ways (Butler et al. 2000). When we are concerned with ecosystems and the value of biodiversity, attention needs to be given to the influence of conflict and the different ways and degrees to which human subjects as members of communities are separated from the false consensus that frames the terms of debate; full agreement is fictional. That is, social context operates as symbolism materially embedded in systems of production and consumption that is not completely accessible to any particular individual inhabiting it. This brings us also to the different relations that individuals as social actors have in relation to the 'economic' and to their own economic activity.

SEPARATION IN AND FROM THE SOCIAL REALM

The process of valuation of ecosystem services for individual, community and global well-being, and the rendering of these values into economic valuation, rests upon what economists must presume to be an engaged interested evaluation by the individual who lives with the environment under scrutiny. However, there are good conceptual grounds for suspecting that such engaged and interested evaluation is made impossible by the norms of already-existing relations between individuals and the contemporary economic forms of life they inhabit.

It is possible to identify four different aspects of separation that render economic valuation of environmental resources by individuals into something that must, of its very nature, be partial and distorted (Kovel 2007). This account draws on and elaborates early analyses of different forms of alienation in capitalist society (Ollman 1976).

Those analyses are all the more relevant today in a world that is organized as a global economic system by capitalist forms of production and consumption (Went 2000). They have particular pertinence to critical assessment of economic valuation as a research and policy tool that conceives of all forms of culture as amenable to a perspective derived from Western political economy (Slovic 1999).

What should be noted is that those early analyses presuppose, as do those of conflict provided by structural anthropology, that there are underlying constitutive features of the human subject's relation to the world that should be taken seriously in any attempt to produce a representation of what they believe they need (and what they believe they need of others).

Separation of Individuals from Creative Labour

A first separation, which we derive from analyses of the division that opens up between an employee and their own creative labour, concerns the attempt and failure of members of communities who must produce from their environment goods for sale in a market-place that is increasingly embedded in the international market. The employer–employee relationship, which would once be thought to be patently anachronistic (applicable only to late nineteenth century and early twentieth century capitalism) is now actually a dominant economic model in which individuals are now positioned as 'employees' or 'employers' in familial, community and informal work and trade networks. The separation of 'work', as an activity necessary to trade, from 'leisure' itself constitutes each microeconomical relationship as part of macro-economical global capitalism (Kothari 2005). Flows of capital sedimented in the products of labour (commodities) separates the labourer from the fruits of their labour so that the commodity comes to stand in a relationship of 'otherness' to them that is exacerbated by the globalization of development (Katz 2004).

The moment of production of nature as economic activity therefore introduces into the equation an element that is overlooked in economic valuation. If it is taken into account, it is done so as if from the position of someone who purchases and circulates nature as a product, as opposed to someone who has been a co-producer with the environment of what is then taken to be an ecological resource.

Attempts to conceptualize natural capital have attempted to overcome this gap by including quantitative and qualitative studies of the ecological, economic and socio-cultural dimensions of health and well-being (Chiesura and de Groot 2003). This is then supplemented in models of ecological resilience and ecosystem rehabilitation, which emphasize sustainability and aim to bring ecosystems back to a balanced state, but which do not adequately include the diverse meanings given to those ecosystems by cultures (Brand 2009).

Separation between Social Actors

A second separation is a function of competition between social actors, such that their position is constituted and understood as individuals or as members of competing familial units. Only then are they able to relate to the world as a system of economic resources, which is then given value as such. This separation makes it difficult for a common viewpoint to be elaborated even as a unified sub-set of the kind of community that is always characterized by social divisions.

Studies of social dimensions of ecosystem management have drawn attention to the problem of incorporating dimensions of understanding by social actors in such a way as to avoid the idea that there is consensus in the community or that each individual exemplifies every aspect of community life, which is a form of reductionism. Those using the notion of 'social capital' have attempted to plug the gap between the individual and the economic, but there are suspicions that this is a terminological nicety that draws attention away from real underlying economic inequalities, and may be no more than a passing fad (Ostrom 2000). The modelling of 'mind maps' focuses on ecocentric, anthropocentric, interdisciplinary or complex systems, for example, but in each case the diversity of interpretations made is not adequately conceptualized, and it has been noted that there is a lack of consensus about what the 'social dimension' actually is (Glaser 2006).

Separation from the Subject's Own Nature

A third separation concerns the division that opens up between a subject constituted as a separate individual and their own nature, and an index of this separation is that the labouring body (human capital) is often not included as one of the ecological resources in an ecosystem.

Conditions of production and consumption that require that the subject adheres to linear, goal-directed and managed (sometimes self-managed) activity tend to turn the body into an apparatus experienced separately from the subject. Failure of the body in conditions of bad health or as a result of injury is then viewed as a liability not only for those who buy labour power but also for those who sell it. This then sets the scene for failures in health provision to regard those who suffer as victims unable to change their circumstances (Crawford 1977). Public health debate then has to disentangle social-systemic problems from the actual needs of individuals (Robertson 1998).

Studies of payments for ecosystem services (PES) have drawn attention to the problem of 'commodity fetishism' in conditions where ecological functions are subject to trade, in which there is then necessarily the expectation that there is a standard unit of exchange, and where relationships are mediated by supply and demand (Kosoy and Corbera 2010). Again, social capital has been deployed as a concept to avoid these problems, but there is then a danger that social relationships and social networks themselves become the focus of attention as tangible units, and thus oversimplified and abstracted from their material contexts of generation, application and accountability (Woolcock and Narayan 2000).

Separation from Nature as Such

The fourth separation comes to provide an epistemological and ontological frame for the individual as a particular kind of social actor to make sense of their position in relation to the first three, and has particularly important consequences for the economic valuation of ecosystems and biodiversity. This fourth separation considers reflection as a distortion of reality that warrants specific kinds of economic relationships, and naturalises them so that they are then taken for granted. Here, the human subject is alienated from nature.

The enclosure of cultivable land such that a difference opens up between the value given to it based on the worst product it may realise for the owner and the actual surplus produced, as 'differential rent', is a source not only of intensification of production at the expense of the environment, but also of a utilitarian conception of nature itself (Tanuro 2010).

This ideological reflection of the fourth form of separation leads us to the domain of ethics. Many analyses of the impact of economic liberalism in Europe in the nineteenth century, including those outside the Marxist tradition, have made the point that commodification of natural resources led to profound distortions of the way that human capacities and potentialities were then understood (Polanyi 1944 [2002]).

The separation of human beings from their own nature has been the particular concern of 'ecofeminists' who link the motifs of equality, harmony and environment with the position of women and the historical construction of cultural values associated with femininity (Bloodhart and Swim 2010). This brings into the equation the impacts that market systems have on agriculture, on the position of women, and thus on the reification of nature as such (Prügl 2008).

The various types of separation that Marx terms 'alienation' create positions in the social order for particular kinds of stakeholder who then judge how the world should be understood by economists and what choices might be made about the world as a constellation of valued resources to be exploited.

ETHICS AND ECONOMICS

There is an inescapable ethical consequence of economic valuation, specifically with the disproportionate weight often accorded to technological procedures over the environment (Spash 1997).

Economic valuation is embedded in a specific ethical system, utilitarianism, that is but one of the dominant moral frameworks used by Western subjects to evaluate their own behaviour and the behaviour of others. It coexists with at least two other systems of ethics that have different consequences for how the subject who carries out an evaluation of the resources at their disposal should be understood. Needless to say, these are not the only three systems of ethics relevant to an understanding of the relationship of people to ecosystems. These are dominant ethical standpoints, intimately related to the forms of alienation that have come to characterize the Western world. They conceal antagonistic interests that are embedded in hostile social relations (Badiou 2001). Alternative systems of ethics outside these frameworks call upon quite different conceptions of truth (Tazi 2004b).

Underlying presuppositions of conceptual paradigms that we use to grasp the concepts of ecosystems, biodiversity and nature, need to be reflected upon if we are to understand how they function and how we might wish them to function (Rajchman 1991).

The Good as Ideal Standard

The first paradigm is one that presumes that we can all agree on what constitutes a 'good', and what falls below it as the ideal standard. This is a notion of ethics to be found in the Aristotelian tradition, which has its roots in early Greek philosophy and which is still often viewed as at the root of West European, and so also a 'Western', moral worldview (Aristotle 2004). That the 'origins' of West European civilization should be traced back to ancient Greece is itself a contested and, some would say, ideological claim. It means that the character of the human being as 'subject', i.e. as a locus of sentience and agency, is thereby treated as always necessarily located in the individual, and is then universalized, with other cultural traditions seen as less advanced than those of the West (Seshadri-Crooks 2000).

A number of theoretical frameworks in the discipline of mainstream psychology are compatible with this notion of the good and of ethics, and this is not surprising given that the discipline was founded upon the abstraction and examination of the individual separated from others (Parker 2007). Specifications of essential underlying human nature, perhaps deriving from the secular humanist tradition, for example, would presume that we all know what it is to do right and wrong, and that knowledge should therefore be grounded in a shared sense of what is good and what is bad.

The problem is that the good as an ideal measure is not something that already exists somewhere to be intuited and aimed at, but it is something that is constructed by us. Or, it is constructed for those in a culture who are then subject to dominant notions of what is good. Here, the good functions as an ideal which we set up ourselves based on our own idiosyncratic experience of what is delightful and alluring to us: an image of what we would like to be or a point from which we would like to seen as likeable. In addition, each notion of the good that we too easily conflate with what is beautiful to us, is suffused with powerful ideas unavailable to an immediate understanding of what is desirable. What may be good to us as an ideal that is suffused with unconscious fantasy, then, may turn out to be quite horrific to other people.

The measurement of well-being and sustainability by governments has attempted to go beyond standard economic models to bring in indicators based on subjective indicators, but definitions of 'life satisfaction', 'quality of life' or 'well-being' are themselves notoriously subjective, reflecting the assumptions built into the measurement instruments, usually questionnaires, and the demand characteristics of those who complete them as subjects (Thomas and Evans 2010). Attempts to impose 'measurement' without thinking through the 'concepts' that guide it, can lead to an impoverished view of how individuals and communities actually relate to the environment (Schaffrin 2011).

Duty of the Subject to do Good

The second paradigm appeals to an imperative to follow the right course of action, which we assume to be potentially if not actually present in each individual human being. This is the way of Kant, who is at the heart of the Western Enlightenment tradition (and again, by default, central to conceptions of morality in the West). It is expressed in his formulation of the 'categorical imperative', in which we are asked to assess our actions according to the maxim that we should imagine it to be applicable to and carried out by all other human beings. This maxim is designed to bring some measure of universality directly into the moral decision-making of any particular individual (Kant 1785 [2009]).

Again, this is an ethical paradigm that dominant traditions in Western developmental psychology, that focus on the individual, have adopted and then struggled with (Piaget 1932 [1977]). The image of the person as containing within themselves a conscience by virtue of which they are able to participate in a society as a civilized enlightened human being then also sets the stage for the inclusion of those who agree with prevailing taken-for-granted views of what is important, and marginalizes those who do not.

The problem with this paradigm, which operates according to a notion of the existence of conscience under law, is that some people who carry out the most horrific actions feel themselves to be following some version of a moral injunction and to be in conformity with what human nature is like (Ahmed 2004).

Such debates over the nature of morality can be found reflected in contemporary discussions of ecological diversity. Some of the most radical interventions into debates over the nature of biodiversity, conservation and political ecology have drawn on ethnography as an approach that respects the autonomy of local ecological and cultural movements. These interventions have typically operated in a discursive mode that connects environmental degradation with violence, appealing to the moral sentiments of those involved to believe that they do good (Parenti 2011).

Distribution of Goods

Now we come to the third paradigm, which is concerned with the calculation of goods, and of costs and benefits of actions for different individuals. This is the universe of Jeremy Bentham, which presumes that it is possible to determine what will be good for people (and bad for them) and to arrange roles and responsibilities so that the greatest possible good is distributed among them (Bentham 1823 [2009]).

The discipline of psychology has constructed a number of influential models of human behaviour to warrant this version of ethics. Here, some of the strands of work in the behaviourist tradition, which seem to refuse to adopt a specific moral standpoint, do actually rest on notions of what are healthy and unhealthy patterns of behaviour, and how 'contingencies of reinforcement' (patterns of reward) might be set up to bring benefit to people.

The problem with this paradigm is that it rather conveniently overlooks, as does behaviourism itself, what the stakes are for the individuals or groups that arrange the distribution of goods. Some neutral position outside the system is presupposed such that decisions can be made which are not themselves affected by certain benefits. A critical assessment of utilitarianism would ask what the individual who arranges things gets out of it. The problem rests not so much at the level of motivations or particular decisions as at a deeper systemic level, of structurally distributed powers, that enter into the definition of the very position accorded to those who will decide.

Critical interventions into debates over the valuation of ecosystem services that have sought to bring in a 'psycho-cultural' perspective have noted the inadequacy of market-based methods, and have stressed that the 'ecological identity' of individuals can only be approached by also questioning dominant psychological models of the individual (Kumar and Kumar 2008)

What these three ethical systems have in common is the assumption in the West, that a 'subject', as a locus of sentience and agency, should be conceptualized as an individual. The category of 'subject' in history in different cultures actually takes different forms, with human beings becoming agents in relation to others in communities, families, networks and organizations (Badiou 2001). In some cultures, the reduction to the level of the individual, i.e. a subject bounded by their own skin and made to perceive, cognize and act separately from others, is thereby felt to be a violation, when it is enforced as a model to which every individual has to conform, of what it is to be genuinely human (Kamin 1974). The separation enforced by Western psychologists, for example, sets terms on which those other cultures are described which, in turn, stand in contrast to the terms on which those cultures would understand themselves.

Utilitarian ethics as the historical culmination of philosophical debate proceeding from the Aristotelian Greek tradition through to the Kantian Enlightenment tradition thus underpins the attempt to configure subjects as individuals. The third Benthamite tradition appeared at exactly the moment that capitalism assumed dominance in England, not merely as an economic system but also as a culturally hegemonic series of assumptions about the nature of the human subject; as individuals contractually obligated to sell their own labour power to others, or as individuals motivated to buy that labour power and harness it for profit. Utilitarian ethics is thus a reflection and warrant for the assumption that each individual should give economic value to the world, including to what they conceive of as nature (Kovel 2007).

CONSTRUCTING FOUNDATIONS FOR ALTERNATIVES TO ECONOMIC VALUATION

A consequence of an adequate conception of the social structure in which social actors are separated in different ways from nature is that policy-relevant 'foundations' for comprehending social dimensions of economic valuation of ecosystems and diversity must also be constructed in such a way that they can be contested.

Constructs and Foundations

One way of stepping back and taking some conceptual distance from foundations that can be contested, rather than assumed to be consensual and then taken for granted (and so then functioning ideologically), is to place emphasis on constructs as such.

Alternative approaches to ethics that attend to the 'discursive' nature of ecosystem valuation and environmental policy have turned to traditions that deconstruct the relationship been facts and values, and the forms of rationality and decision-making that reinforce the hierarchies that privilege a certain kind of expert knowledge and certain kinds of disciplinary assumptions and conceptual norms (O'Hara 1996).

Different constructions of what counts as 'natural' behaviour of human beings in ecosystems, such that the forms of valuation move beyond mere economic valuation to a more pluralist debate within conditions of cultural diversity, are not necessarily grounded in one single agreed idea of what nature is, but in a variety of ways of viewing it.

Second Nature

The problem of 'second nature' from within feminist and social theory provides a different perspective on how the elements of an ecosystem may be abstracted, as part of economic valuation.

Thinking and acting are embodied, and as dependent on biology as breathing and digestion. This biological nature is transformed when we struggle to understand ourselves and others. The historical social relationships through which we carry out that struggle produce a 'second nature', which comprises needs, demands and desires that are peculiarly human, and which become human at the moment we articulate them to others and ourselves. Critical studies of primatology, for example, have shown how masculine fantasies of control over women's 'nature' were reproduced in the 'motherless monkey studies' to make it seem as if the researchers had discovered this 'nature' when they were actually constructing it (Haraway 1989). The studies make clear that the claims of the experimenters to be 'discovering' how monkeys behave were false since they were actually creating conditions in which the monkeys could not behave otherwise. This is gendered 'second nature', which functions ideologically to oppress women, and makes them seem more primitive, and closer to nature than men. Such studies thus demonstrate the importance of differentiating between underlying biological, natural processes (a 'first nature' that cannot ever be captured and shown exactly as it is, precisely because a description of it has to be in a particular language infused with cultural and ideological preconceptions) and our 'second nature'.

Such research is conducted from what has been termed a 'standpoint', here as 'feminist standpoint' research. This works on the assumption that social actors in oppressed or marginalized positions in a social system have, by virtue of their position, access to a view of the operations of the system that is obscure to those who benefit from existing arrangements because those arrangements are taken for granted and unproblematic to them (Harding 2003). This tradition makes the important point that no single individual, by virtue of their own specific and contradictory position, can grasp the totality of social relations; every view of the social field is a view from somewhere.

This conception of 'second nature' connects with the argument that those who refuse the dominant ideological ethical paradigms in capitalist society, and who are often caricatured as wanting to reject any biologically-grounded human needs, must have some notion of human nature if they are to argue against injustice and unequal distribution of resources under capitalism (Geras 1983).

Human Nature

Such an appeal to human nature, however, does not necessarily mean that one subscribes to the view that each separate individual is driven by their nature to compete for resources with other individuals. The key characteristic of human nature, according to critical traditions of work in anthropology, psychology and social theory is, in contrast, defined by the social context in which a human being comes to reflect upon themselves and act alongside (even if sometimes against) other subjects. Again, this allows for a conception of the subject as not only being formed within the boundaries of a separate body, as it is in cultures that privilege the individual in economic and social policy programmes, but as being formed in collectives that transcend the individual (Badiou 2001).

This view of human nature, and nature in general, relies on the notion of a mediated relation of the subject to others, to an ecosystem and, indeed, to themselves, which is why psychoanalytic models of subjectivity have often been favoured in such critical traditions (Parker 2011b). A mediated conception of the human subject is one in which there is acknowledgement that there is something beyond the self-conscious reasoning mind, that which psychoanalysts call the unconscious, but which can also be conceptualized as the set of social–economic relations which condition how the subject understands themselves (Jameson 1981). Mediation is provided by a variety of different kinds of representation of society to its members, and this then means that an ecosystem, as with any social system, needs to be configured as something diverse, as a contradictory rather than consensually understood form of cultural diversity.

CONCLUSIONS

The attempt to recruit subjects as individuals who are invited to give an economic valuation of an ecosystem thus also recruits those subjects into a particular partial representation of the ecosystem. The abstraction of elements of the ecosystem into different components that can thereby be given economic value draws them into a certain worldview underpinned by a particular conception of ethics. This is a worldview in which economic value is assigned, assessed and balanced between social actors according to utilitarian criteria. The point here is not so much that such a construction of reality is false, though from some political standpoints it would be considered to be so, but that it is ideological (Harvey 1996). The point is that at a deeper epistemological and ontological level, such a construction of reality sets in place certain foundations for policy that obscure the competing viewpoints that make up an ecosystem.

Cultural diversity, on the other hand, requires acknowledgement of the different competing terms of debate which construct foundations for policy that are conceptualized as contradictory rather than consensual. Nature itself is thus conceptualized as mediated by competing cultural representations of it on a global scale, and is contested within each particular culture from the standpoint of different social actors, who are subject to policy initiatives, perhaps including those who ostensibly benefit from policy. Forms of alienation in which there is a variety of intersecting separations from nature thus provide the conditions of possibility for knowledge about economic value, and what we may conceptualize as forms of 'counter-knowledge' (Foucault 1977). There are implications here for the way that research is conducted by those seeking to develop policy, and for the way that such research invites participants into one supposedly consensual worldview in which they each, one-by-one, give economic value to their world.

There are also implications for the reflexive positions of those who research, develop and implement policy (and those who write review and position papers about such questions). An academic or administrative position is itself partial, and embedded in a set of practices in which, in the best of cases, some conception of what is good and bad guides the work. Policy operates within ethical paradigms that also insist upon a certain view of the world, one that confirms and sustains the work of the policy-maker. Dominant available systems of ethics, whether they be Aristotelian, Kantian or Benthamite, limit the policy-maker to a particular sub-cultural universe, and a dominant system also invites treatment of this universe as the only one, or, at least, as the best one. This much has been noted by Latin American theorists of ethics and subjectivity, who have then attempted to transcend these conceptions (Burton and Flores Osorio 2011).

A culture treated as part of an ecosystem thus risks overlooking the construction of different forms of second nature, the always mediated representational aspect of nature. It also risks reinforcing the worldview of the policy-maker understood by them to be consensually validated each time they succeed in gathering data about economic valuation of resources from participants in policy-relevant research projects. The evidence from a number of different studies of cultural diversity which attend to contradictions between and within communities, taken alongside the conceptual problems with notions of consensus underpinned by Western ethical paradigms reviewed in this chapter, is that a debate-mediated conception of second nature that respects and extends biodiversity would be preferable. Stepping back from the partial, limited and reductive notion of economic valuation would not mean that one single alternative paradigm should be implemented in its place. Rather, it means that policy itself would need to be elaborated and contested by social actors who are able to construct different foundations for democratic inclusive cultural diversity, albeit one that is itself without foundations.

ACKNOWLEDGEMENTS

I would like to acknowledge, with many thanks, constructive feedback on earlier versions of this work from Brendon Barnes (University of the Witwatersrand, South Africa), Erica Burman and Mark Burton (Manchester Metropolitan University, UK), Gregorio Iglesias Sahagún (Universidad Autónoma de Querétaro, Mexico) and Rosario González Arias (Universidad de Oviedo, Spain).

REFERENCES

Ahmed, Sara (2004). *The Cultural Politics of Emotion*. Edinburgh: Edinburgh University Press.

Aristotle (2004). *The Nicomachean Ethics* (originally approximately 350 CE). Harmondsworth, UK: Penguin.

Badiou, Alain (2001). *Ethics: An Essay on the Understanding of Evil*. London: Verso.

Bentham, Jeremy (2009). *An Introduction to the Principles of Morals and Legislation* (originally published 1823). Dover, New York.

Bloodhart, Brittany and Janet Swim (2010). Equality, harmony, and the environment: an ecofeminist approach to understanding the role of cultural values on the treatment of women and nature, *Ecopsychology*, **2**(3), 187–194.

Brand, Fridolin (2009). Critical natural capital revisited: ecological resilience and sustainable development, *Ecological Economics*, **68**, 605–612.

Burton, Mark and Osorio Flores and Mario Jorge (2011). Introducing Dussel: the philosophy of liberation and a really social psychology, *Psychology in Society*, **41**, 20–39.

Butler, Judith (1990). *Gender Trouble: Feminism and the Subversion of Identity*. London and New York: Routledge.

Butler, Judith (1997). *Excitable Speech: A Politics of the Performative*. London and New York: Routledge.

Butler, Judith, Ernesto Laclau and Slavoj Žižek (2000). *Contingency, Hegemony, Universality: Contemporary Dialogues on the Left*. London: Verso.

Chiesura, Anna and Rudolf de Groot (2003). Critical natural capital: a sociocultural perspective. *Ecological Economics*, **44**(2–3), 219–231.

Crawford, R. (1977). You are dangerous to your health: the ideology and politics of victim blaming, *International Journal of Health Services*, **7**(4), 663–680.

El-Bizri, Nader (2005). The conception of nature in Arabic thought. In Nadia Tazi (ed.), *Keywords: Nature: For a Different Kind of Globalization*. New York: Other Press, pp. 63–92.

European Union (2000). Charter of Fundamental Rights of the European Union. Available at: http://www.europarl.europa.eu/charter/pdf/text_en.pdf (accessed 2 November 2011).

Foucault, Michel (1977). *Language, Counter–Memory, Practice: Selected Essays and Interviews*. Oxford: Blackwell.

Geras, Norman (1983). *Marx and Human Nature*. London: Verso.
Glaser, Marion (2006). The social dimension in ecosystem management: Strengths and weaknesses of human-nature mind maps, *Human Ecology Review*, **13**(2), 122–142.
Harcourt, Wendy (2009). *Body Politics in Development: Critical Debates in Gender and Development*. London: Zed Books.
Haraway, Donna (1989). *Primate Visions: Gender, Race, and Nature in the World of Modern Science*. London and New York: Routledge.
Harding, Sandra (ed.) (2003). *The Feminist Standpoint Theory Reader: Intellectual and Political Controversies*. London: Routledge.
Harvey, David (1996). *Justice, Nature and the Geography of Difference*. Oxford: Blackwell.
Haug, Frigga (1992). The Hoechst Chemical Company and boredom with the economy. In Frigga Haug (ed.) *Beyond Female Masochism: Memory-Work and Politics*. London: Verso, pp. 13–30.
Jameson, Fredric (1981). *The Political Unconscious*. London: Methuen.
Kagan, Carolyn, Mark Burton, Paul Duckett, Rebecca Lawthom and Asiya Siddiquee (2011). *Critical Community Psychology: Critical Action and Social Change*. Cambridge: Blackwell.
Kamin, Leon (1974). *The Science and Politics of IQ*. Harmondsworth, UK: Penguin.
Kant, Immanuel (2009). *Groundwork of the Metaphysic of Morals* (originally published 1785). New York: Harper.
Katz, Cindi (2004). *Growing Up Global: Economic Restructuring and Children's Everyday Lives*. Minneapolis, MN: University of Minnesota Press.
Kosoy, Nicolás and Esteve Corbera (2010). Payments for ecosystem services as commodity fetishism, *Ecological Economics*, **69**(6), 1228–1236.
Kothari, Uma (ed.) (2005). *A Radical History of Development Studies: Individuals, Institutions and Ideologies*. London: Zed Books.
Kovel, Joel (2007). *The Enemy of Nature: The End of Capitalism or the End of the World?* 2nd edn. London: Zed Books.
Kumar, Pushpam (ed.) (2010). *The Economics of Ecosystems and Biodiversity: Ecological and Economic Foundations*. London: Earthscan.
Kumar, Manasi and Pushpam Kumar (2008). Valuation of the ecosystem services: a psycho-cultural perspective. *Ecological Economics*, **64**(4), 808–819.
Lévi-Strauss, Claude (1972). *Structural Anthropology* (originally published 1958). Penguin, Harmondsworth.
Marx, Leo (2005). The idea of nature in America. In Nadia Tazi (ed.), *Keywords: Nature: For a Different Kind of Globalization*. New York: Other Press, pp. 37–62.
Nadar, Laura (1997). The phantom factor: impact of the Cold War on anthropology. In Noam Chomsky (ed.), *The Cold War and the University: Toward an Intellectual History of the Postwar Years*. New York: The New Press, pp. 107–146.
Nunes, Paulo and Jeroen van den Bergh (2001). Economic valuation of biodiversity: sense or nonsense? *Ecological Economics*, **39**, 203–222.
O'Hara, Sabine (1996). Discursive ethics in ecosystems valuation and environmental policy, *Ecological Economics*, **16**(2), 95–107.
Ollman, Bertell (1976). *Alienation: Marx's Conception of Man in Capitalist Society*, 2nd edn. Cambridge: Cambridge University Press.

Ostrom, Elinor (2000). Social capital: a fad or a fundamental concept? In Partha Dasgupta and Ismail Serageldin (eds), *Social Capital: A Multi-Faceted Perspective*. Washington DC: World Bank, pp. 172–214.

Oswald Spring, Úrsula (2009). A HUGE gender security approach: towards human, gender, and environmental security. In Hans Günter Brauch, Patricia Kameri-Mbote, Úrsula Oswald Spring, Navnita Chadha Behera, John Grin, Bécher Chourou, Czeslaw Mesiasz and Heinz Krummanecher (eds), *Facing Global Environment Change: Environmental, Human, Energy, Food, Health and Water Security Concepts*. Berlin: Springer Verlag.

Ouédraogo, Jean-Bernard (2005). Africa: human nature as historical process. In Nadia Tazi (ed.) *Keywords: Nature: For a Different Kind of Globalization*. New York: Other Press, pp. 1–35.

Parenti, Christian (2011). *Tropic of Chaos: Climate Change and the New Geography of Violence*. New York: Nation Books.

Parker, Ian (2007). *Revolution in Psychology: Alienation to Emancipation*. London: Pluto Press.

Parker, Ian (editor) (2011a). *Critical Psychology: Critical Concepts in Psychology (4 Vols)*. London and New York: Routledge.

Parker, Ian (2011b). *Lacanian Psychoanalysis: Revolutions in Subjectivity*. London and New York: Routledge.

Piaget, Jean (1977). *The Moral Judgement of the Child* (originally published 1932). Harmondsworth, UK: Penguin.

Pini, Barbara and Leech, Belinda (eds) (2011). *Reshaping Gender and Class in Rural Spaces*. Farnham, UK: Ashgate.

Polanyi, K. (2002). *The Great Transformation: The Political and Economic Origins of Our Time* (originally published 1944). New York: Beacon.

Prügl, Elisabeth (2008). Gender and the making of global markets: an exploration of the agricultural sector. In Shirin Rai and Geogina Whelan (eds), *Global Governance: Feminist Perspectives*. London: Palgrave.

Puleo, Alicia (2005). Lo Personal es Político: El Surgimiento del Feminismo Radical. In Celia Amorós and Ana De Miguel (eds), *Teoría feminista: De la Ilustración a la Globalización. Del feminismo liberal a la posmodernidad*. Madrid: Minerva, pp. 121–152.

Rajchman, John (1991). *Truth and Eros: Foucault, Lacan, and the Question of Ethics*. New York: Routledge.

Robertson, Ann (1998). Critical reflections on the politics of need: implications for public health. *Social Science and Medicine*, **47**(10), 1419–1430.

Seshadri-Crooks, Kalpana (2000). *Desiring Whiteness: A Lacanian Analysis of Race*. London and New York: Routledge.

Schaffrin, André (2011). No measure without concept: a critical review on the conceptualization and measurement of environmental concern. *International Review of Social Research*, **1**(3), 11–31.

Shao-Ming, Chen (2005). Zi Ran (nature): A word that (re)structures thought and life. In Nadia Tazi (ed.), *Keywords: Nature: For a Different Kind of Globalization*. New York: Other Press, pp. 93–111.

Slovic, Paul (1999). Trust, emotion, sex, politics, and science: surveying the risk-assessment battlefield. *Risk Analysis*, **19**(4), 689–701.

Spash, Clive (1997). Ethics and environmental attitudes with implications for economic valuation. *Journal of Environmental Management*, **50**, 403–416.

Srivastava, Vinay (2005). On the concept of nature. In Nadia Tazi (ed.), *Keywords:*

Nature: For a Different Kind of Globalization. New York: Other Press, pp. 141–185.

Tanuro, Daniel (2010). Marxism, energy and ecology: the moment of truth. *Capitalism Nature Socialism*, **21**(4), 89–101.

Tazi, Nadia (ed.) (2004a). *Keywords: Gender: For a Different Kind of Globalization*. New York: Other Press.

Tazi, Nadia (ed.) (2004b). *Keywords: Truth: For a Different Kind of Globalization*. New York: Other Press.

Thomas, Jennifer and Joanne Evans (2010). There's more to life than GDP but how can we measure it? *Economic and Labour Market Review*, **4**(9), 29–36.

Tuhiwai Smith, Linda (1999). *Decolonizing Methodologies: Research and Indigenous Peoples*. London: Zed Books.

United Nations (1966). International covenant on economic, social and cultural rights. Available at: http://www2.ohchr.org/english/law/pdf/cescr.pdf (accessed 2 November 2011).

Went, Robert (2000). *Globalization: Neoliberal Challenge, Radical Responses*. London: Pluto.

Woolcock, Michael and Deepa Narayan (2000). Social capital: implications for development theory, research and policy. *World Bank Research Observer*, **15**, 225–249.

Yuval-Davis, Nira (1997). *Gender and Nation*. London: Sage.

Zaoui, Pierre (2005). Fables of nature. In Nadia Tazi (ed.), *Keywords: Nature: For a Different Kind of Globalization*. New York: Other Press, pp. 113–140.

13. Lessons learned and conclusions
Pushpam Kumar

From the examples, case studies and subsequent analysis provided in this volume, it can be concluded that payments for ecosystem services (PES) are one of the most cost-effective response measures for managing ecosystems. The analysis confirms that PES schemes offer financial incentives for local actors who provide ecosystem services outside of normal market transactions. Many PES schemes are now being implemented around the world, and cover four main ecosystem services: watershed services, carbon sequestration, landscape beauty and biodiversity conservation.

Arriagada and Perrings (Chapter 2) explain that for PES schemes to succeed, they must successfully complete several steps: potential service providers must enrol in the scheme; providers must comply with the terms of their contract; and compliance must result in a change in the provision of the ecosystem service compared with what would have happened without the scheme.

The effectiveness of PES schemes depends on whether they are able to deliver outcomes, measured in terms of the flow of services, which are better than the outcomes in the absence of such schemes. The design of PES schemes should be complemented by the measurement of the ecosystem services produced as a result of the scheme. The authors discuss how PES schemes should also avoid negative spillovers (leakage), or at least provide benefits sufficient to offset unavoidable spillovers. Spillovers, both positive and negative, will need to be taken into account in calculating the net benefits (additionality) of the scheme. Just as important is the necessity to ensure that PES schemes deliver additional benefits relative to the status quo. Relatively few PES schemes do well if evaluated against the status quo, but those that do are useful guides to the design of new schemes.

Increasingly, PES schemes will need to assure provision of services that provide large-scale benefits. The authors use the example of forests and carbon. Because of its focus on an ecosystem service with global benefits, Reducing Emissions from Deforestation and Forest Degradation (REDD) could gain access to a large pool of global stakeholders willing to pay to maintain carbon in forests, and because forests offer a number of benefits aside from carbon, the scheme could potentially support the supply of ecosystem services to local stakeholders who would otherwise be unable to afford them.

For international ecosystem goods and services, the analysis in these pages concludes that of all the Millennium Ecosystem Assessment ecosystem service categories, regulating services are most often supplied as international environmental public goods. Many international public goods are jointly produced with local public goods (co-benefits). An important feature of international environmental public goods is that their spatial extent depends partly on natural hydrological and atmospheric flows, and partly on the social linkages between countries, i.e. the flow of goods, people and information (Fisher et al., Chapter 3; Arriagada and Perrings, Chapter 9). International environmental public goods generate benefits that spill over national borders, so that the benefits (or costs) of those goods extend beyond the country of origin.

The failure of markets to materialize positive externalities and public goods has serious repercussions for welfare. In the case of international environmental public goods, the lack of payment for these public goods results in underinvestment in the protection, management and establishment of ecosystems. International PES schemes are appropriate where non-marketed ecosystem services are privately supplied in one country, but offer benefits that are public and accrue elsewhere. As pressure mounts on governments to curtail spending and cut budget deficits, their ability to invest directly in the provision of public goods and services is compromised. The costs and benefits associated with many human activities spill over jurisdictional boundaries, thereby generating externalities that are often reciprocal and quantitatively significant. The establishment of international payments for the local provision of public goods that offer wide public benefits can promote efficient provision at the global scale (Arriagada and Perrings, Chapter 9). It is important that the design of PES schemes fits the diagnosis of the public goods problem and the technology of public goods supply.

The analysis provided by the authors of several chapters makes it clear that successful implementation of PES requires clarity of the fundamental basis of provision of ecosystem services. For example, systematic approaches to measuring, modelling and mapping of ecosystem services, as well as governance analysis and valuation, are needed urgently. The endpoints that have a direct affect on human welfare are important in valuation exercises. There are spatial relationships between where an ecosystem service is produced and where the benefit is felt, and this 'flow' of services changes through space (see Fisher et al., Chapter 3). Ecosystem service provision changes its ecological processes, magnitude and beneficiaries, and costs across space, and an appreciation of this is critical for any valuation process. Ecological conditions or processes change over time, and society's preferences change over time, but any ecosystem service assessment occurs at a point in time. Focusing, too, on single provisioning ecosystem services in isolation from regulating services can result in policy failures, which has happened frequently.

The extensive use of the Earth's surface for agriculture severely affects the generation of many regulating ecosystem services that underlie human well-being. Elmqvist et al. (Chapter 4) discuss the need to understand how trade-offs among services can be addressed and to what extent new insights in ecology and innovations in institutions and governance may help to reduce some of the most undesirable trade-offs. Multiple failures can be avoided if appropriate knowledge, incentives and institutions are to hand. Even though many trade-off options are available, most do not currently provide an improved economy for the producer, and hence do not present an economic incentive. When a producer reduces yields to make room for biodiversity and promotion of regulating services, and is compensated by higher prices from, for example, certification schemes, the situation could be a zero-sum game for the producer, but a real improvement to the landscape as a whole.

Market price can, in some cases, be a poor approximation of value (Fisher et al., Chapter 3). The fundamental problem facing any economic analysis is one of measurement, i.e. how do we measure the value or utility provided by any given goods? This is especially true in the case of benefits from ecosystem services. The importance of providing policy-makers with timely and robust estimates of the value and benefits of well functioning ecosystems has never been more critical. Progress, conceptually and analytically, is being made in understanding ecosystem functioning and the provision of benefits to humans, such that estimates and recommendations can be delivered to decision-makers.

In developing countries, where PES schemes are implemented as part of institutional arrangements, it is usually expected that PES will also yield benefits of poverty reduction. In that context, the analysis of Rodriguez et al. (Chapter 6) suggests that while PES schemes are conservation instruments, they can indirectly affect people's livelihoods. However, they do not explicitly target poor landholders. Importantly, the amounts paid to beneficiaries are larger in government or publically funded schemes than in private schemes. The imperative of climate change means that carbon sequestration schemes generally provide larger payments to beneficiaries than biodiversity and watershed schemes, while payments for jointly supplied services might be substantially higher.

In areas chosen for implementing PES owing to their high conservation value, a household targeting exercise might increase the social benefits of the interventions by selecting poor households able to provide the desired level of environmental services. Targeting is an expensive activity, such that an appropriate method should be selected in order to avoid a situation where the costs of targeting exceed any gains in efficiency. Household-level findings highlight the linkages between the environmental and socio-economic systems and their contribution to people's well-being. Linkages between environmental and socio-economic systems are becoming more evident for policy and decision-makers. Environmental degradation is a major barrier to poverty reduction and, at the same time, reaching environmental conservation goals requires progress in the eradication of poverty. In order to be successful, poverty reduction and environmental initiatives could be linked and implemented together. Converting PES into poverty reduction programmes might be inefficient from a conservation perspective if there is no correlation between poverty incidence and provision of environmental services. However, in regions with such a correlation, the level of benefits provided to landholders by PES schemes seems to have the economic potential to contribute to poverty alleviation efforts, if properly targeted and designed.

Donors, research and development organizations and field practitioners are constantly looking to refine the selection, design and implementation of PES schemes to improve their cost-effectiveness and to jointly promote environmental conservation and poverty alleviation. Numerous Conditional Cash Transfers (CCT) and PES initiatives have provided increasing evidence of positive side effects of social interventions on the environment, and similarly, positive impacts from environmental interventions on social systems. Estimated transferred amounts per household tend to indicate that the level of benefits from CCT is comparable with that of the transfers from PES schemes, suggesting the high potential of PES to contribute to poverty alleviation initiatives if properly targeted to achieve that goal. Unified CCT–PES-type payment schemes are not expected to follow simple 'one size fits all' designs. To reach both environmental and social objectives, unified payment schemes, which have been called payments for environmental and poverty alleviation services (PEPAS), should define an appropriate transfer amount consistent with the need to reduce the poverty level of the beneficiaries and cover the required investments in conservation (see Rodriguez et al., Chapter 7). The transferred amount in a PEPAS scheme should cover the direct costs to landholders for participating. These will include, for example, the costs of sending children to school, the costs of implementing certain land management techniques, and opportunity costs such as income foregone by children who attend school, and thus do not work, and the revenues foregone by landholders adopting conservation activities.

Identifying overlapping areas where CCT and PES may operate jointly might help to select regions for designing unified payment schemes. A PEPAS scheme should be implemented if there are clear correlations between the provision of environmental services and poverty incidence in a region. If this condition does not exist, independent PES and CCT schemes should be preferred instead. As in individual PES and CCT schemes, PEPAS schemes could be linked to a set of environmental and social conditions designed to adjust individual behaviour towards the delivery of conservation and poverty alleviation objectives.

In the case of agricultural ecosystems, an adapted PES scheme can help to conserve endangered genetic traits, i.e. the unique genes that occur in rare varieties of plants, animals and fungi. Farmers tend to replace traditional crop and livestock varieties with financially more profitable 'improved' breeds. From an economics perspective, the conservation of agrobiodiversity requires that, where significant public values exist, these should be recognized and mechanisms put in place to permit the capture of those values by the farmers who incur the conservation costs. Farmers contributing to conservation activities provide these local and global public values as a positive externality that they are not rewarded for, while other members of society are able to benefit from these public conservation services. On-farm conservation would result in the sustained utilization of threatened crop varieties or livestock breeds within traditional farming systems, and thus the conservation focus is not on certain land uses, as in most existing PES schemes, but on a specific type of agricultural practice. This extension of the PES concept is discussed in Chapter 8 (Narloch et al.).

On-farm conservation does not only imply the cultivation of a certain land area and thus the generation and conservation of seeds and livestock breeds, but also the maintenance of local traditions and agricultural knowledge. High opportunity costs can make it expensive to reintroduce neglected varieties and breeds in highly intensive agricultural systems, Payments for Agrobiodiversity Conservation Services (PACS) schemes might therefore be recommended where de facto conservation of threatened plant and animal genetic resources (PAGR) is still carried out. Contrary to the genuine PES concept, such schemes would be government financed rather than user financed.

There are many factors and underlying dynamics that would affect the definition of a safe minimum standard (SMS) for PAGR (i.e. a minimum population size that ensures long-term survival), and PACS schemes will need to draw on sound interdisciplinary research into socio-ecological dynamics in order to determine scientifically justifiable conservation goals. PACS schemes should attain their conservation goal (effectiveness) at least-cost (efficiency), while ideally involving fair distributional outcomes (equity in outcome) (Pascual et al., Chapter 5). If the social dimensions of PACS schemes are ignored, it might prove to be destructive for poor farming communities or even create a 'PES curse' by undermining the success and legitimacy of PES as a concept. Nevertheless, PACS should be considered as a useful potential tool for policy-makers. There is much further research and development to be carried out before PACS can become established in the policy-makers' toolbox.

In terms of institutional arrangements, a consensus is emerging among environmental and ecological economists that, if human economic activity is to become sustainable, damages to the Earth's supporting ecological and physical systems must be reversed, a concept discussed by Gowdy and Salman (Chapter 10). The gross domestic product (GDP) of the poor is undervalued because so much of it depends on unpriced inputs from nature. Estimates from India suggest that ecosystem services add only 7 per cent to measured GDP but add 57 per cent to the GDP of the country's poorest (Sukhdev and Bishop 2008). It is generally the rich who decide whether to preserve biodiversity and ecosystem functions, based on market rates of interest and investment opportunities. The poor have a larger stake in preserving ecosystem functions and the resulting flows of ecosystem services, but a large percentage of the world's disadvantaged are in such a desperate position that they must sometimes sacrifice these services for immediate gain.

The values of ecosystems and biodiversity can be seen as levels in a hierarchy moving from market value (supporting, provisioning and regulating) to non-market value (cultural and psychological), underlain by the role of biodiversity and complexity in preserving ecosystem resilience. These levels of value point to the need for a pluralistic and flexible methodology to determine appropriate policies for ecosystem use and preservation of ecosystem functions. Although evidence from contingent valuation, hedonic pricing and other economic valuation tools underscores the importance of biodiversity and ecosystem functions, these methods give incomplete, lower-bound estimates of their values.

The well-being of the world's poorest communities depends, to a large extent, on the direct services of ecosystems and on maintaining ecosystem functions. For example, mangrove ecosystems are critical to the well-being of a large portion of the world's poorest communities because of their important ecosystem functions, and their direct economic uses in tropical areas around the world. Gowdy and Salman (Chapter 10) assert that mangrove forests are, in many ways, a major focal point in the intersection between poverty, declining ecosystems and global climate change. Ecosystem functions are under threat both from worldwide environmental changes and from inadequate institutional responses to these changes. Development policies are unlikely to be effective unless these institutional factors are understood and targeted in development initiatives. Resource conflicts among the world's poorest communities are happening throughout the coastal areas of South Asia and will probably get much worse as climate change disrupts local ecosystems. The implications are sobering, not only for the future of the environment, but also for social stability worldwide.

Ecosystem-based adaptation to reduce the physical impacts of climate change demonstrates a double dividend of both short-term and long-term benefits, but while the cost of interventions can be estimated with a relatively high degree of accuracy, the ensuing benefits are difficult to estimate or predict. In Chapter 11, Blignaut shows how a shrewd adaptation policy is one that renders benefits irrespective of the occurrence of a future extreme weather event. In so doing, the ongoing benefits could pay for the adaptation policy and, in the process, offer useful adaptation measures. Ecosystem-based adaptation is the use of ecosystem goods and services within a climate change adaptation programme. Ecosystems have a meaningful contribution to make towards climate change adaptation. Human well-being is an essential objective of achieving sustainable economies. Better protection and management of key habitats and natural resources can benefit poor, marginalized, and indigenous communities by protecting ecosystem services and maintaining access to resources even during difficult times, including drought and disaster.

Traditional engineered solutions often work against nature, particularly when they aim to constrain regular ecological cycles, such as annual river flooding and coastal erosion, and could further threaten ecosystem services if the construction of dams, seawalls, and flood canals leads to habitat loss. Restoration also contributes towards mitigation through long-term carbon storage and sequestration. It is unlikely that ecosystems undergoing restoration will sequester the same amount of carbon as intact ecosystems in the same area. Ecosystems with high restoration costs have the lowest internal rates of return, while ecosystems that have shown the best internal rates of return following restoration are those that produce a high level of ecosystem goods and services, or that have a low restoration cost. Investments that render both immediate and ongoing adaptation benefits and that contribute to socio-economic development have to be considered as extremely favourable.

Where there is a major science-policy gap, it can lead to further deterioration of environmental quality and quality of life, and can add to the increased vulnerability of people, because of a reduced ability to adapt and face the impacts of climate change. Not only does ecosystem degradation compromise the inherent integrity of ecosystems, thereby limiting their ability to function as an integral unit (hence, becoming more vulnerable and liable to collapse over time), it also compromises human welfare through the loss of services rendered by ecosystems. The only viable way to reverse the global trend of degradation is to protect and augment the currently declining supply of natural capital through investing in restoration.

On a different but critically important plank, Parker (Chapter 12) purports that the process of valuation of resources for individual, community and global well-being, and the rendering of such valuation into economics, rests upon what must be presumed to be an engaged evaluation by the individual living within the environment. There are good conceptual grounds for suspecting that such engaged and interested evaluation is made impossible by the already existing relations between individuals and the contemporary economic forms of life they inhabit. Attempts to conceptualize natural capital have attempted to overcome the gap between natural production and economic valuation by including quantitative and qualitative studies of the ecological, economic and socio-cultural dimensions of health and well-being. This is then supplemented in modelling of ecological resilience, which emphasises sustainability and aims to bring ecosystems back to a balanced state, but which does not adequately include the diverse meanings given to those ecosystems by people.

Studies of PES schemes have drawn attention to the problem of 'commodity fetishism' in conditions where ecosystem services are subject to trade. There is then necessarily the expectation that there is a standard unit of exchange, and that relationships are mediated by supply and demand. The enclosure of cultivable land, such that a difference opens up between the value given to it based on the worst product it may realize for the owner and the actual surplus produced, as 'differential rent', is a source not only of intensification of production at the expense of the environment, but also of a utilitarian conception of nature itself. Economic valuation is embedded in a specific ethical system, utilitarianism, that is but one of the dominant moral frameworks used by Western subjects to evaluate their own behaviour and the behaviour of others. There is a need to reflect on the underlying concepts that are used to grasp the nature of ecosystems and biodiversity if we are to understand how they function within a societal framework, and how we might wish them to function.

The measurement of well-being and sustainability by governments has attempted to go beyond standard economic models to bring in indicators based on subjectivity, but such indicators reflect the assumptions built into the measurement instruments, for example questionnaires, and the demand characteristics of those who complete them as study subjects. Critical interventions into debates over the valuation of ecosystem services that have sought to bring in a 'psycho-cultural' perspective have noted the inadequacy of market-based methods in estimating the aspects of ecosystems that have meaning for subjects. These have stressed that the 'ecological identity' of individuals can only be approached by also questioning dominant psychological models of the individual.

The problem of 'second nature' (i.e. received cultural assumptions) from within feminist and social theory provides a different perspective on how the elements of an ecosystem may then be abstracted as part of the particular culturally and ethically limited task of economic valuation. The attempt to recruit subjects as individuals who are invited to give an economic valuation of an ecosystem also recruits those subjects into a particular and biased representation of the ecosystem. Policy operates within ethical paradigms that also insist upon a certain view of the world; one that confirms and sustains the work of the policy-maker. Dominant available systems of ethics, whether they be Aristotelian, Kantian or Benthamite, limit the policy-maker to a particular sub-cultural paradigm, and a dominant system also invites them to treat this paradigm as the only one, or at least, as the best one.

Stepping back from the partial, limited and reductive notion of economic valuation would not mean that one single alternative paradigm should be implemented in its place. Rather, it means that policy itself would need to be elaborated so that it can be contested by social actors who are able to construct different foundations for a democratic and inclusive cultural diversity, albeit one that is itself without foundations.

CONCLUSIONS AND LESSONS LEARNED

Payments for Ecosystem Services

PES are not '*silver bullets*' for conservation or development (Ferraro and Kiss 2002; Engel et al. 2008). Economic incentives often may not be suitable to manage all environmental problems and may not be a sufficient guarantee for ecosystem service provision (Perrot-Maitre 2006: Engel et al. 2008). Nonetheless, where direct payments for marketed environmental services are appropriate, conservation can be made a more competitive land-use over conversion (Asquith et al. 2008).

Few, if any, current PES schemes strictly adhere to Wunder's (2005) definition (Tacconi et al. 2011; Bond 2007). Their success is dependent on a number of variables and, ultimately, on the overall objective of the project itself. Lessons can be learned from early PES schemes, such as the projects discussed above, which highlight the complexity of designing and implementing programmes. As emphasized by the Vittel PES scheme in France, the primary reason for success may not always be financial (Perrot-Maitre 2006).

Certain pre-requisites are needed for successful implementation, these include:

Prior to establishing a PES scheme, a scientific understanding of the ecosystem services at multiple scales is essential in order to effectively map, model and ultimately value ecosystem services (Farley and Costanza 2010). Outlining the full range of services from an ecosystem can also reduce the incidence of perverse incentives.
A clear view of the social, economic, political and institutional setting in which PES schemes are situated is also needed. This can enable targeted programmes and higher participation, and enables PES to work alongside other policy instruments that may be in place (Farley and Costanza 2010).
Defining the links between land-use and service provision will ensure that incentives can target the right land-use changes to support specific ecosystem services without payments becoming a flat subsidy (Wunder et al. 2008).
Trust-building during initial stakeholder negotiations and implementation of the PES scheme is important. Some schemes, such as the Vittel water project, were successful by using a trusted intermediary institution to implement the programme. This highlights the need for supporting institutions to match the spatial and temporal scale of ecosystem service provision (Farley and Costanza 2010).

Links between incentives and land tenure are one of the biggest hurdles for PES schemes. Establishing property rights, in particular in developing countries, is often unclear and can create conflicts over determining who should be paid (Pascual et al., Chapter 5). This can lead to tension when clearly defined boundaries of service provision, such as those in the CAMPFIRE programme (Child 1996; Frost and Bond 2008), may exclude other communities that are outside the targeted area.

Additionally, the implementation of PES can potentially lead to de facto privatisation of resources and elite benefit capture (Perrot-Maitre 2006). The design of PES schemes should ensure that those most reliant on common property resources such as ecosystem services are included, thus facilitating greater equity in the distribution of costs and benefits of PES (Farley and Costanza 2010).

Compliance

While attracting potential ecosystem service providers may be easy, ensuring their compliance to the conditions of their participation in the scheme requires regular monitoring, and sanctions for non-compliance (Wunder et al. 2008). This may reduce the potential for additionality, and limit the impact of PES on long-term service provision (Wunder 2005). Furthermore, monitoring is required to ensure leakage does not occur through the displacement of damaged land uses at a local level, or indirectly at a broader level (Wunder et al. 2008). Short-term contracts supported by monitoring, may encourage compliance, thus avoiding some of the obstacles illustrated in the PROFAFOR, Ecuador, case study (Wunder and Albán 2008).

Financing of PES Schemes

PES can be a direct way to target conservation, and can entail comparatively low implementation costs, however, initial start-up costs can be high. These include the project's design and the collection of baseline data on ecosystem services. Other contextual factors include negotiations with stakeholders, and monitoring and transaction costs that enable service providers to enter the market (Wunder et al. 2008). These costs have implications for the sustainability, efficiency, effectiveness and equity of schemes, which have impacts on the distribution of ecosystem services and payments to providers (Farley and Costanza 2010).

Payments themselves, either in-kind or cash, should also be appropriate for the scheme participants. As seen with the Los Negros programme in Bolivia, inflexible in-kind payments have the potential to exclude some participants who receive lower gains owing to a lack of skills. Payments need to be high enough to justify both investment and the lost opportunity costs of alternative land uses (Perrot-Maitre 2006). Negotiated incentives must therefore be sufficient to maintain levels of landowner income to counteract this opportunity cost. Differentiating payments according to opportunity cost per provider, such as the Vittel payments, may aid in achieving more environmentally effective compensation. However, it is clear from the CAMPFIRE scheme that benefit disparity can also lead to tensions between beneficiaries.

REFERENCES

Asquith, N.M., M.T. Vargas and S. Wunder (2008). Environmental services: in-kind payments for bird habitat and watershed protection in Los Negros, Bolivia. *Ecological Economics*, **65**, 646–685.
Bond I. (2007). Payments for watershed services: A review of literature. IIED Report G02522, London.
Child, B. (1996). The practice and principles of community-based wildlife management in Zimbabwe: the CAMPFIRE programme. *Biodiversity and Conservation*, **5**, 369–398.
Engel, S., S. Pagiola and S. Wunder (2008). Designing payments for environmental services in theory and practice: an overview of the issues. *Ecological Economics*, **65**, 663–675.
Farley, J. and R. Costanza (2010). Payments for Ecosystem Services: from local to global. *Ecological Economics*, **69**, 2060–2068.
Ferraro, P.J. and Kiss A. (2002). Direct payments to conserve biodiversity. *Science*, **298**, 1718–1719.
Frost, P.G.H. and I. Bond (2008). The CAMPFIRE programme in Zimbabwe: payments for wildlife services. *Ecological Economics*, **65**, 776–787.
Perrot-Maitre, D. (2006). The Vittel Payments for Ecosystem Services: a 'perfect' PES case? Project Paper No. 3. IIED, London.
Sukhdev, P. and J. Bishop (2008). The economics of ecosystems and biodiversity (TEEB): A step towards biodiversity markets? Workshop on 'Capitalizing on natural resources: new dynamics in financial markets'. Rüschlikon, Germany, September 10, 2008.
Tacconi, L., S. Mahanty and H. Suich (2011). Forests, payments for environmental services and livelihoods. In L.Tacconi, S. Mahanty and H. Suich (eds), *Payments for Environmental Services, Forest Conservation and Climate Change. Livelihoods in the REDD?* Cheltenham, UK and Northampton, MA, USA: Edward Elgar.
Wunder S. (2005). Payments for environmental services: some nuts and bolts. CIFOR Occasional Paper No. 42, Center for International Forestry Research, Bogor.
Wunder, S., S. Engel and S. Pagiola (2008). Taking stock: a comparative analysis of payments for environmental services programs in developed and developing countries. *Ecological Economics*, **65**(4), 834–852.
Wunder, S. and M. Albán (2008). Decentralized payments for environmental services: the cases of Pimampiro and PROFAFOR in Ecuador. *Ecological Economics*, **65**, 685–698.

Index

Adams, W. 1, 20
Africa
   carbon sequestration and restoration projects, survey of 228–33
   PES and CCT comparison 113, 130–31
   representations of nature 262
   see also individual countries
Agrawal, A. 200
agrobiodiversity conservation, in-situ, PES potential 150–71, 283–4
   biodiversity conservation on the farm 152–3
   conservation goals, defining 156–9
   direct payments, benefits of 150–51
   direct private use value to farmer 152
   eco-labelling and certification schemes 152
   ex-situ conservation strategies 153
   future research 156, 166, 167
   global responsibilities 154
   marketing and service purchaser role 154
   payment mechanisms 23–4, 150–51
   plant and animal genetic resources (PAGR) 150, 151, 152, 153, 156–61, 163, 165
   risk-averse reasons for using agrobiodiversity 152
   see also institution and ecosystem functions
agrobiodiversity conservation, PACS schemes (payments for agrobiodiversity conservation services) 151
   assessment of possible outcomes 159–65
   competitive tendering 163
   cost–benefit assessment 160–61, 162–4
   design and implementation constraints 154–6
   ecological effectiveness assessment 159–62
   equity and efficiency trade-offs 161, 164–5
   geographical targeting 153–4
   institutional requirements 155–6
   intermediaries, involvement of 155
   land tenure issues 155
   leakage considerations 161–2
   monitoring and enforcement 155–6, 164
   opportunity costs 163
   private sector funding considerations 162
   pro-poor outcomes and social equity 164–5
   scientific foundations, need for strong 156, 157
   supply and demand, matching 153–4
   sustainability 162
   threshold effects 161
   transaction costs 163–4
   wealth status of benefiting farmers, effects of 165
Ahlheim, M. 18
Ahmed, A. 134
Albán, M. 5, 41, 42, 46, 98, 112, 124, 134, 136, 289
Alongi, D. 197
Alvarez, C. 136
Alviar, C. 113, 130
Anand, M. 77
Anand, P. 172
Anderson, K. 198
Andrade, G. 20
Angelsen, A. 47
Anthoff, D. 217, 227
Archer, D. 198
Armbrecht, I. 77

Aronson, J. 12, 221, 222, 228, 233, 250–51
Arriagada, Rodrigo 16–57, 172–91
Asia
  Arab representations of nature 262
  PES and CCT comparison 112–13, 129–30
  see also individual countries
Asquith, N. 3, 5, 6, 134, 135, 144, 287
Asselin, L. 110
Australia 239, 242
  Bush Tender Program 27, 136
  National Landcare Program 22
Aylward, B. 19

Badiou, A. 267, 270, 273
Bali, A. 77
Balmford, A. 58
Bangladesh, Vulnerable Group Development project 134
Bann, C. 20, 101
Banuri, T. 200
Banzhaf, S. 59
Barbier, E. 195
Barnosky, A. 196
Barrera-Osorio, F. 133
Barrett, S. 172
Bateman, Ian 58–69
Batina, R. 177
Bawa, K. 20
Bellon, M. 152, 153, 156, 157, 158
Bengtsson, J. 82
Bennett, M. 3, 6, 26
Beria, L. 113, 130
Berkes, E. 20
Bernet, T. 152
Biermann, E. 47
biodiversity
  agrobiodiversity *see* agrobiodiversity conservation, *in-situ*, PES potential
  protection 3, 17, 194–6
Bishop, J. 23, 193, 284
Blignaut, James 215–58
Bloch, F. 186
Bloodhart, B. 267
Bohensky, E. 71
Bolivia 167
  Los Negros programme 5, 30, 134, 289

Bond, I. 6, 133, 288
Borkhataria, R. 77
Bows, A. 198
Boyd, J. 59
Braat, L. 82
Brand, F. 265
Brazil 112, 125–6
Brink, P. ten 82
Brock, R. 196
Bromley, D. 196
Brush, S. 153, 158
Brussaard, L. 75
Bulte, E. 19, 120, 121
Bush Tender Program, Australia 27, 136
Butler, J. 259

Cambodia 112, 129, 137
Cameroon 233
CAMPFIRE scheme, Zimbabwe 6, 133, 288, 289
Canada 239
carbon sequestration 3, 98, 245–6, 247
  and land management 234–43
  and restoration projects 226–33
Cardenas, J. 48
Carpenter, S. 71, 72
Carr, D. 75
Cartwright, A. 249
certification schemes with price premiums 80, 81
Champetier de Ribes, A. 51
Chapman, K. 120
Chavas, J. 152
Chen, S.-M. 263
Chichilnisky, G. 195
Chiesura, A. 265
Child, B. 6, 288
China 25, 242, 263
  bee extinction, Maoxian County 195
  Grain to Green programme (GTGP) 25, 26, 32, 40, 46, 167
  Natural Forest Conservation Program (NFCP) 25, 32
  Sloping Land Conversion Programme 3, 6, 25
Claassen, R. 26–7
Clements, T. 112, 129
climate change concerns 63–5, 74, 201–2, 202, 205–6, 210

adaptation and restoration of
  natural capital 216–20
and regulation 181–5, 186
South Asian ecosystems *see*
  institution and ecosystem
  functions, climate change and
  South Asian ecosystems
Coady, D. 115
coffee production 76–80
Colombia 112, 124–5, 134, 168
conditional cash transfers (CCT) 102,
  109, 110–11, 115, 116
  PES comparison *see* environmental
    and social protection, unifying,
    and PES and CCT comparison
conflict in social systems 93, 98,
  100–101, 263
Corbera, E. 25, 32, 44, 45, 47, 93, 112,
  121, 127, 155, 156, 164, 266
Cornes, R. 180, 182
Costa Rica 35, 168
  FONAFIFO forest conservation
    PES 98
  PSAH programme 6, 37–8, 45, 132,
    133
Costanza, R. 220, 288, 289
cost–benefit assessment
  environmental and social protection,
    PES and CCT comparison
    132–3
  equity and efficiency in payments for
    ecosystem services 103–4
  and internal rate of return (IRR) 231
  PACS schemes (payments for
    agrobiodiversity conservation
    services) 160–61, 162–4
  transaction costs 44, 142, 143, 163–4
  see also value
Cowling, R. 84
cropland management, carbon gain
  234–5
Cuesta, J. 121–2, 141
cultural diversity *see* ethical
  foundations of cultural diversity

Dahdouh-Guebas, F. 220
Daily, G. 2, 19, 58, 61, 70
Daly, H. 192
Das, A. 77
Das, J. 110, 121, 136

Dasgupta, C. 199
Dasgupta, P. 120, 192, 195, 217
Davis, B. 110, 121, 133, 135
De Groot, R. 221, 222, 223, 265
Deng, L. 25
Denich, M. 76
Denmark 21
Dennis, E. 155
Devooght, K. 96
Di Falco, S. 152, 154
Diaz, S. 75
Dobbs, T. 27
Dolia, J. 77
Dombrowsky, I. 175
Donnelly, A. 245–6
Downing, J. 196
Drucker, Adam, G. 150–71
Duraiappah, Anantha K. 51, 90–108

Echavarria, M. 144
eco-labelling and certification schemes
  152
ecological effectiveness assessment,
  PACS schemes (payments for
  agrobiodiversity conservation
  services) 159–62
ecosystem functions *see* institution and
  ecosystem functions
Ecosystem Services Districts,
  development suggestion 84
ecosystem services, payments for
  see environmental and social
  protection, unifying, and PES
  and CCT comparison; equity
  and efficiency in payments for
  ecosystem services, relationship
  between; payments for ecosystem
  services (PES)
ecosystem-based restoration benefits
  215–58, 285–6
  carbon sequestration *see* carbon
    sequestration
  climate change concerns *see* climate
    change concerns
  disproportionate impact dilemma
    218
  global damage cost of carbon,
    estimation of, and pure rate of
    time preference (PRTP) 217
  Hartwell Paper 215–16, 249

294   *Values, payments and institutions for ecosystem management*

intervention costs, assessment problems 220
mitigation options 218, 219, 226–8, 230–32
ecosystem-based restoration benefits, adaptation
costs 233–48
costs, and continuous benefits 248–51
costs, McKinsey adaptation options 248, 249
focus, need for 218–20, 224
ongoing benefits, assessing 220
planning problems 220
and precautionary principle 218–19
recommendations 252
ecosystem-based restoration benefits, natural capital restoration value 221–33
cost–benefit ratios and internal rate of return (IRR) 231
cultural services 224
degradation trend, reversing 221–3
ecosystem-based adaptation 223–6
and human well-being 222–3, 225–6
provisioning services 223
regulating services 223–4
restoration activities 222
restoration value 226–33
supporting services 224–5
will to change, importance of 225–6
Ecuador 112, 124
Pimampiro programme 30, 41, 46
PROFAFOR programme 5–6, 30, 42, 46, 98, 112, 124, 133, 134, 136, 137, 289
efficiency concerns 37–8, 39–49, 177–9
and equity in payments *see* equity and efficiency in payments for ecosystem services, relationship between
El-Bizri, N. 262
Elmqvist, Thomas 70–89, 161
endangered species, conservation of 177–9
Engel, S. 2, 20, 23, 34, 39, 90, 115, 120, 121, 132, 151, 154, 164, 180, 287
environmental additionality concept 93–4, 103–4

environmental public goods *see* international environmental public goods, paying for
environmental and social protection, PEPAS (Payments for Environmental and Poverty Alleviation Services) 137–43
conditions, setting 142–3
eligibility criteria 139–42
free-rider problem 138
payment vehicle and frequency of transfers 138–9
poverty alleviation 140–42
targeting mechanism 142, 143
environmental and social protection, unifying, and PES and CCT comparison 120–49, 282–3
community level impact 133
compensation types 132–5
compliance conditions 123–31, 289
conditional cash transfers (CCT) criticisms of 121
conditional cash transfers (CCT) as social protection mechanisms 121–2, 123
cost-effectiveness of programme 132–3
eligibility criteria 135–6
geographic targeting 123–31, 135
landholder compensation 133, 134
market-based interventions 122–32
participation conditions 136–7
payment frequency effects 134
payment vehicle 123, 133–4, 289
payments for ecosystem services (PES) criticisms of 121
payments for ecosystem services (PES) as economic transfers 120–22, 123
positive externalities, internalizing 122–32
poverty alleviation 132–3, 136, 137, 140–42
proxy means testing and environmental metrics 135–6, 140
public funding 123, 132
targeting mechanism 123
see also payments for ecosystem services (PES)

Environmentally Sensitive Areas (ESA) 21, 27
equity and efficiency in payments for ecosystem services, relationship between 90–108, 288–9
　actual provision fairness criterion 97, 98, 102
　compensation fairness criterion 97, 98, 100, 101, 102
　conditional cash transfers (CCT) see conditional cash transfers (CCT)
　conflicts between social groups, avoidance of 93, 98, 100–101, 263
　consensus fairness criterion 97, 98, 102
　cost–benefit evaluation 103–4
　disadvantaged groups, involvement of 101, 102
　distributional aspects 91
　economic fairness and welfare evaluative criteria 96–8, 99, 101–2
　efficiency criteria 94–5
　efficiency and equity, trade-offs between 92
　egalitarian fairness criterion 97, 98, 102
　environmental additionality concept 93–4, 103–4
　equity criteria 94–5, 96, 99
　evaluation method, effect of 103–4
　expected provision fairness criterion 97, 98, 102
　individual capabilities, evaluation of 99, 101
　individual wealth, effect of 103
　intermediary role in implementation of scheme 100–101
　internalizing externalities 91–2
　land-management decisions 98, 101, 102
　maxi-min fairness criterion 97, 98, 101
　planning and implementation process 100
　policy trade-off curve (PTOC) 94, 95, 96, 102, 104
　power imbalance among social groups 98
　pro-poor effects 91, 93, 100, 101–2
　property rights identification 92, 95, 105
　quality of life opportunities 100
　stakeholder effect 96, 99–100, 103
　status quo fairness criterion 97, 102
　uncertainty implications 105–6
　value of services, effects of 103
　welfare outcomes, evaluation of 98–103
　willingness to accept (WTA) and willingness to pay (WTP) 94–5, 104
equity and efficiency trade-offs, PACS schemes (payments for agrobiodiversity conservation services) 161, 164–5
ethical foundations of cultural diversity 259–78, 286–7
　commodification of natural resources, effects of 266–7
　conflict in social systems, role of 263
　consensual conceptions of societal systems 261
　ecofeminists and separation from nature 267
　economic valuations, foundations for alternatives 271–3
　ethics and economics 267–71
　ethics and economics, distribution of goods 269–71
　ethics and economics, 'good' as ideal standard 267–8
　human culture as ecosystem 260
　human nature characteristics 272–3
　moment of production of nature as economic activity 265
　morality and subject's duty to do good 269
　nature as such, separation from 266–7
　payments for ecosystem services (PES) and commodity fetishism 266
　policy implications 274
　'second nature' problem 271–2
　social dimension, lack of consensus of meaning 265

social functions of 'nature', and competing cultural conceptions 262–3
social realm, separation in and from 263–7
social-systemic problems and public health 266
societal diversity, contradictory conceptions of 261–2
'subject' concept in different cultures 270
unity and diversity within cultural and economic systems 261–3
see also welfare
Ethiopia 76, 80
 Cash for Relief Programme 141
 Global Environment Facility-funded project 151
Europe 19, 262
 payments for threatened livestock breeds 151
 see also individual countries
Evans, J. 268
externalities, effects of 20–21, 91–2, 122–32, 181–2
Eyzaguirre, P. 155

Farley, J. 116, 288, 289
Farrington, J. 120
Ferraro, P. 2, 16, 20, 21, 33, 34, 39, 91, 109, 121, 138–9, 150, 155, 163, 180, 287
Ferroni, M. 177
Field, C. 83
Fischer, J. 82
Fisher, Brendan 4, 58–69
Fizbein, A. 134
Folke, C. 81
FONAFIFO forest conservation PES, Costa Rica 98
forestry
 carbon sequestration 247
 Clean Development Mechanism (CDM) projects 25
 damage cost avoided approach 63–5
 deforestation 75–6, 181–2
 deforestation and REDD mechanism 16, 45–7, 181–5, 186
 FONAFIFO forest conservation PES, Costa Rica 98
 forestland management, carbon gain 235, 236, 237
 and international public goods 175
 mangrove *see* institution and ecosystem functions, Keti Bunder, Pakistan
 Natural Forest Conservation Program (NFCP), China 25, 32
 water-related services, Mexico 103–4
 see also land management practices
France 21, 245
 Vittel PES 3, 29
free-rider problem 34, 138
Frey, B. 200
Frost, P. 6, 288
future research 34, 71, 156, 166, 167
Fuwa, N. 151

Gadgil, M. 175, 178
Georgescu-Roegen, N. 192, 196
Geras, N. 272
Gilbert, N. 192
Glaser, M. 265
Glewwe, P. 135
Global Environment Facility (GEF) 151, 180, 187, 188
Godfray, H. 63
Gökalp, Y. 133
Goldman, R. 84
Goldstein, J. 180
Gordon, C. 74, 77, 79–80
Göschl, T. 153, 154
Goulder, L. 2
Gove, A. 76
Gowdy, John M. 192–214
Grain to Green programme (GTGP), China 25, 26, 32, 40, 46, 167
grassland practices and carbon gain 235, 242–3, 245–6
Grieg-Gran, M. 2, 20, 23, 33, 91, 101, 109, 121, 165
Grosh, M. 115, 138, 139
Gruère, G. 150, 152

Hajjar, R. 150
Halder, S. 140
Handa, S. 110, 121, 133, 135
Haq, M. 200
Harcourt, W. 262

Harding, S. 272
Hardner, J. 16
Hartshorn, G. 19, 37
Hartwell Paper 215–16, 249
Harvey, D. 262, 273
Haug, F. 260
Haywood, A. 198
Heal, G. 195
Heisey, P. 152
Hermann, M. 152
Hiittermann, A. 83
Hodge, I. 20
Holm-Müller, K. 28
Holzinger, K. 175
Honduras 98, 135
Hooper, D. 61
Houghton, R. 83
Huberman, D. 1
Hughes, G. 233
Hulme, D. 139
human benefits and ecosystem processes, relationships between 59–60
human culture as ecosystem *see* ethical foundations of cultural diversity
Human Development Index (HDI
human well-being *see* welfare
Hylander, Kristoffer 70–89

Ihori, T. 177
Imbach, P. 135
India 25, 77, 262
    climate change effects *see* institution and ecosystem functions, climate change and South Asian ecosystems
individual capabilities, evaluation of 99, 101
individual preferences, temporal context 67–8
individual wealth, effect of 103
Indonesia 112, 129
Inglehart, R. 193
institution and ecosystem functions 192–214, 284–5
    capability poverty measure (CPM) 200
    costs avoided by ecosystem preservation, estimation of 195
    ecological economics 193–4, 196

economic value of ecosystems and biodiversity 194–6
ecosystem integrity as economic development concern 193–4
environmental concern and subjective well-being, links between 201
Human Development Index (HDI) 200
poverty implications 193–4, 200
resilience and biodiversity, relationship between 196
'shadow price' and net national product (NNP) accounting framework 194–5
socio-cultural value of ecosystems and biodiversity 196
spiritual ties, significance of 196
values of ecosystem services 194–7
see also agrobiodiversity conservation, *in-situ*, PES potential
institution and ecosystem functions, climate change and South Asian ecosystems 197–200
    agricultural production loss 199
    climate-induced migration 199
    higher average temperatures 198
    human well-being *see* welfare
    rainfall pattern changes and extreme weather events 198–9
    sustainable well-being, economics of 200–201
institution and ecosystem functions, Keti Bunder, Pakistan 201–11
    climate change concerns 201–2, 202, 205–6, 210
    as designated wildlife sanctuary 202
    economic conditions 204–5
    fishing as livelihood, unsustainability of 204, 206–8, 210
    fresh water supply 203–4, 209
    livestock farming reduction 209
    mangrove forest degradation 197, 202–3, 204, 205–6, 210
    migration, limited 209
    resource conflicts 210–11
institutional requirements, PACS schemes (payments for

agrobiodiversity conservation
  services) 155–6
Intergovernmental Panel on Climate
  Change (IPCC) 63, 81
  see also UN Framework Convention
    on Climate Change (UNFCCC)
international environmental public
  goods, paying for 172–91, 280
  adequacy criterion 180
  'best shot' cases 175, 176–7, 186
  classification 174–5
  deforestation cases and REDD
    mechanism 16, 45–7, 181–5, 186
  ecosystem services 173–4
  efficient provision of conservation
    effort 177–9
  endangered species, conservation of
    177–9
  externalities, effects of 181–2
  Global Environment Facility (GEF)
    151, 180, 187, 188
  market failure problems 179
  PES schemes 180–81, 188
  public goods theory 185–6
  pure public good and private good,
    difference between 177–8, 185
  'simple sum' cases 175
  technology of supply 174–5, 176,
    186–7
  underprovision explanation 176–80
  'weakest link' cases 175
international environmental public
  goods, paying for, policy options
  180–85
  direct investment to specific
    countries 187–8
  global climate regulation 181–5,
    186
  implications 185–8
  international coordination or
    cooperation 187
  market forces and multi-site
    production 187
  public goods failure, diagnosis of
    186–7
international-level PES schemes 24–5,
  47–8
Italy 21
  Lazio Rural Development
    Programme 168

Jack, K. 20
Jackson, J. 192
Jameson, F. 273
Jianchu, X. 199
Jiang, D. 198
Jindal, R. 113, 131
Jones, M. 245–6

Kagan, C. 262
Kahneman, D. 200
Kallis, G. 193
Kamin, L. 270
Kanbur, R. 177
Kapos, V. 16, 45, 47, 185
Kareiva, P. 1, 19, 33
Karsenty, A. 121
Katz, C. 264
Kaul, I. 172, 173, 174, 175, 185
Kelt, D. 196
Kemkes, R. 2, 4, 5
Kennedy, D. 2
Kenya 113, 116, 130–31, 167
  Cash Transfer for Orphan and
    Vulnerable Children 113, 130,
    137
Khan, S. 200
Kinzig, A. 16, 17
Kiss, A. 16, 34, 180, 287
Kleijn, D. 32
Klein, A. 75, 76, 77
Konow, J. 96
Kontoleon, U. 154, 157
Kosoy, N. 98, 121, 141, 155, 156,
  266
Kothari, U. 264
Kovel, J. 264, 271
Krall, L. 193
Kreuter, U. 84
Krishna, V. 152
Krishnaswamy, Jagdish 70–89
Krutilla, J. 193
Kuik, O. 217
Kumar, B. 77
Kumar, M. 1, 4, 276
Kumar, Pushpam 1–15, 181, 261, 276,
  279–90

land management practices 22–3, 98,
  101, 102, 155, 234–43, 244, 245–6
  coffee production 76–80

forestry *see* forestry
landholder compensation 133, 134
soil quality maintenance and erosion prevention 75, 77, 78, 81
water protection schemes *see* water protection schemes
wetland management 66–7, 74, 237
Landell-Mills, N. 2, 20, 151, 155, 179, 181
Larsen, T. 75
Latacz-Lohmann, U. 163
Latin America 77, 112, 124–8, 242–3
see also individual countries
Lavelle, P. 75
Layton, D. 32, 33
Lazio Rural Development Programme, Italy 168
Leech, B. 260
Lévi-Strauss, C. 261
Lipper, L. 151
Liu, J. 25, 26–32, 38
Liverman, D. 32
Los Negros programme, Bolivia 5, 30, 134, 289
Losey, J. 75
Luck, G. 71

McClafferty, B. 136
McDaniel, C. 201
McKinsey adaptation options 248, 249
McLeod, E. 197
McShane, T. 1
Madsen, B. 154, 172
Malaysia 233
Maler, K.-G. 195, 200
mangrove forests *see* institution and ecosystem functions, Keti Bunder, Pakistan
market forces 2, 16–17, 19, 61, 62, 122–32, 154, 179, 187
Martinez-Alier, J. 193
Marx, L. 263
Massingarela, C. 113, 131
Mastrandrea, M. 58
Mather, A. 48
Matin, I. 139, 140, 141
Matson, P. 19
Maxted, N. 153
Mayrand, K. 19

means testing and environmental metrics 135–6, 140
Medeiros, M. 115, 121
Melillo, J. 195
Metzger, J. 83
Mexico 79–80, 98, 103–4, 112, 127
 Payments for Hydrological Environmental Services 29, 43, 46, 133
 PROGRESA (Oportunidades) programme 112, 127, 132, 134, 136
 PSA-CABSA programme 112, 127, 137
 Scolel Té Project 31, 42–3, 46
Meyerson, L. 71
Miles, L. 16, 45, 47, 185
Miranda, M. 116, 133
mitigation options, ecosystem-based restoration benefits 218, 219, 226–8, 230–32
Mody, A. 177
Moguel, P. 77
Moldovia 25
Morrissey, O. 174
Mozambique 113, 131
Munasinghe, M. 198
Munawir, S. 112, 129
Múñoz-Piña, C. 29, 43, 46, 103–4, 133
Muradian, Roldan 1, 4, 90–149, 151, 156
Myers, R. 58

Nadar, L. 261
Naidoo, R. 70, 71
Nair, P. 77
Narayan, D. 266
Narloch, Ulf 150–71
natural capital restoration value *see* ecosystem-based restoration benefits, natural capital restoration value
Nautiyal, S. 152
Neef, A. 18
Nelson, E. 1
Nemomissa, S. 76, 80
Netherlands 5–6, 245
Neumayer, E. 67
Nhate, V. 113, 131

Nicaragua, Silvopastoral Project 28, 41, 46, 112, 128, 168
Niesten, E. 16, 109, 121, 143
Nordhaus, W. 175, 217
Norton, B. 196
Norway 239
Nunes, P. 195, 259
Nussbaum, M. 200

O'Connor, D. 16, 45
OECD, agricultural policies 21–2
O'Hara, S. 271
Olchev, A. 77
Olinto, P. 135
Ollman, B. 264
Orr, D. 196
Ostrom, E. 47, 84, 174, 265
Oswald Spring, U. 260
Ouédraogo, J.-B. 262

PACS schemes *see* agrobiodiversity conservation, PACS schemes (payments for agrobiodiversity conservation services)
Pagiola, S. 5, 16, 19–20, 22–3, 28, 32–3, 39, 41, 46, 90–91, 98, 110, 112, 116–17, 120–21, 125, 128, 132, 134–5, 151, 154, 162, 164–5, 180–81
Pakistan, institution and ecosystem functions *see* institution and ecosystem functions, Keti Bunder, Pakistan
Paquin, M. 19
Paraguay, Tekopora project 134
Parenti, C. 269
Parker, Ian 259–78
Parry, M. 248
Pascual, Unai 51, 90–171
Pattanayak, S. 34, 39
payments for ecosystem services (PES) 1–57, 279–80
  agro-biodiversity schemes 23–4
  agrobiodiversity conservation *see* agrobiodiversity conservation, *in-situ*, PES potential
  biodiversity protection 3, 17
  bundling of services 34
  buyers' identity, significance of 23
  carbon sequestration PES 3
  characteristics of 25–39
  and commodity fetishism 266
  conceptual confusion 20
  conditional payments 2–3, 39
  counterfactual outcome, estimation of 35–6
  definition 18–25
  deforestation *see* forestry
  development benefits, potential to deliver 33
  economic evaluation 3, 4–6
  efficiency concerns 37–8, 39–49
  endogenous selection effect 35–6, 37
  environmental services, logic of payments for 22–3
  equity and efficiency 90–91
  free-rider problem 34, 138
  future research 34
  government-financed programmes 34, 39, 47
  human well-being *see* welfare
  impact estimation 35–6
  implementation conditions 34–9, 288
  indirect externalities, focus on internalizing 20–21
  integrated conservation–development projects 19–20
  international environmental public goods, paying for 180–81, 188
  international-level schemes 24–5, 47–8
  land management practices, incentives to follow 22–3
  landholders, private decisions of 32
  landscape beauty PES 3
  long-term service provision, lack of guarantee of 5
  marginal benefits, estimation of 44–5
  OECD agricultural policy origins 21–2
  and poverty *see* poverty
  price-based mechanisms, efficiency of 32–3
  'purchasers should be users' requirement 20
  redistributive mechanism, PES as 4
  spillover effects 45, 47–8
  success and dependent variables 287–8
  transaction costs, estimation of 44

user-financed programmes 34, 47
water protection schemes 3, 23, 24
see also environmental and social protection, and PES and CCT comparison; equity and efficiency in payments for ecosystem services, relationship between
Pazmino, Nathalie 120–49
Pearce, D. 51, 63, 64, 180
Pearson, R. 113, 130
PEPAS (Payments for Environmental and Poverty Alleviation Services) *see* environmental and social protection, PEPAS (Payments for Environmental and Poverty Alleviation Services)
Perales, H. 153
Perez-Ribas, R. 135, 144
Perfecto, I. 78, 79, 80
Perrings, Charles 16–57, 150, 151, 152, 172–91
Perrot-Maitre, D. 3, 116, 287, 288, 289
Peru 135
Pfaff, A. 37, 48
Phelps, J. 184
Philippines 112–13, 130, 151, 167
Philpott, S. 77, 80
Pimampiro programme, Ecuador 30, 41, 46
Pini, B. 260
Pirard, R. 19, 20
Platais, G. 5, 23, 154, 162
Polanyi, K. 267
Polasky, S. 83
policy implications
    ethical foundations of cultural diversity 274
    international environmental public goods *see* international environmental public goods, paying for, policy options
    policy trade-off curve (PTOC) 94, 95, 96, 102, 104
    regulation of ecosystem services *see* trade-offs in ecosystem services, management of, regulation of ecosystem services
Porras, I. 2, 20, 151, 155, 179, 181
Porter, G. 51

poverty
    capability poverty measure (CPM) 200
    pro-poor effects 91, 93, 100, 101–2, 164–5
    reduction 6, 19, 20, 25, 136, 137, 140–42, 193–4, 200
poverty alleviation efforts and payment levels 109–19, 281–2, 284
    conditional cash transfers (CCT), purpose of 115, 116
    conditional cash transfers (CCT) use 109, 110, 115
    conditional cash transfers (CCT) use, payment assessment 110–11
    environmental and socio-economic systems, links between 116–17
    geographic targeting 115
    incentives to beneficiaries 115
    payments for ecosystem services (PES) 109–10, 113–17
    payments for ecosystem services (PES), bundled ecosystem services, cost-effectiveness of 116
    payments for ecosystem services (PES), purpose of 115, 116
    PES and CCT amounts, comparison of 111–13, 115
    poverty reduction concept 110–11
Pretty, J. 27
Proctor, W. 91
PROFAFOR programme, Ecuador 5–6, 30, 42, 46, 98, 112, 124, 133, 134, 136, 137, 289
PROGRESA (Oportunidades) programme, Mexico 112, 127, 132, 134, 136
property rights identification 92, 95, 105
Prügl, E. 267
PSA-CABSA programme, Mexico 112, 127, 137
PSAH programme, Costa Rica 6, 37–8, 45, 132, 133
public goods, environmental *see* international environmental public goods, paying for
Puleo, A. 259

Rafey, W. 217
Rajchman, J. 267
Rangel, A. 201
Rawlings, L. 109, 121, 132
Rayden, T. 232
Redford, K. 1
regulation *see* policy implications
Reid, W. 33, 58, 109, 116
Reist-Marti, S. 157, 158
restoration benefits, ecosystem-based *see* ecosystem-based restoration benefits
Rice, R. 16, 109, 121, 143
Ricketts, T. 77
Righelato, R. 82
Rios, A. 112, 125, 128
Robalino, J. 48
Robertson, A. 266
Robertson, N. 39
Robinson, G. 196
Rockstrom, J. 74
Rockström, J. 81
Rodriguez, Luis C. 90–149
Rojas, M. 19
Romero, C. 20
Rosegrant, M. 58
Roy, M. 51
Ruane, J. 157
Rudel, T. 48
Ruijs, A. 120
Russman, E. 37

Sachs, J. 33, 58, 109, 116
Sajise, A. 151
Sala, O. 195
Salm, R. 197
Salman, Aneel 192–214
Salzman, J. 18
Samuelson, P. 180
Sanchez-Azofeifa, G. 16, 37
Sandler, T. 174, 175, 180
Schady, N. 134, 135
Schaffrin, A. 268
Scheer, S. 102
Scheffer, M. 74
Scherr, S. 16, 109
Schilizzi, S. 163
Schmitt, C. 76
Schneider, S. 58, 198
Schokkaert, E. 96

Schröter, D. 58, 71
Schubert, B. 120, 121, 136
scientific foundations, agrobiodiversity conservation 156, 157
Scolel Té Project, Mexico 31, 42–3, 46
Semu-Banda, P. 140
Sen, A. 99, 100, 200
Senbeta, F. 76
Settele, J. 4
'shadow price' and net national product (NNP) accounting framework 194–5
Shapouri, S. 102
Sharma, E. 1
Shennan, C. 75
Sidle, R. 76
Sierra, R. 37
Siikamaki, J. 32, 33
Silvopastoral Project, Nicaragua 28, 41, 46, 112, 128, 168
Simpson, R. 20, 21, 33, 34, 91, 109, 121, 139, 150, 155, 180
Singh, S. 196
Skoufias, E. 121, 136
Slater, R. 120, 121, 136
Slovic, P. 264
Smale, M. 152, 154, 163, 165
Smith, J. 109
social systems, conflict in 93, 98, 100–101, 263
society, and cultural diversity *see* ethical foundations of cultural diversity
Soto-Pinto, L. 78
South Africa, Working for Water (WfW) Program 29, 40, 46, 47
Sovacool, B. 217
Spain 21
Spain, A. 75
Spangenberg, J. 4
Spash, C. 267
Spracklen, D. 82
Sri Lanka 233
Srivastava, V. 262
Stafford, M. 249
Stage, R. 217
stakeholder effect 96, 99–100, 103
Standing, G. 121, 141, 143
Stavins, R. 181

Steffan-Dewenter, I. 77
Steinfel, H. 72
Stern, N. 197, 217, 227, 233
Stickler, C. 186
Stoneham, G. 136, 163
Stromberg, P. 158
Stutzer, A. 200
Sugden, R. 200
Sukhdev, P. 193, 284
Summer, D. 51
sustainability issues 162, 200–201, 204, 206–8, 210
Suyanto, S. 109, 121
Swallow, B. 16
Swanson, T. 153, 154
Sweden, Payments for Wildlife Conservation 28
Swim, J. 267
Switzerland 245

Tacconi, L. 288
Tallis, H. 1, 2, 19, 23, 25, 33, 70
Tanuro, D. 266
targeting mechanism
  environmental and social protection 123, 142, 143
  geographical 123–31, 135, 153–4
Tazi, N. 267
technology of supply, international environmental public goods 174–5, 176, 186–7
Tekopora project, Paraguay 134
Terakunpisut, J. 233
Thailand 233
Thiaw, Ibrahim 1–15
Thomas, J. 268
Thompson, M. 203
Tietenberg, T. 16
Tilman, D. 196
Tipper, R. 42–3, 46
Tirmizi, N. 203
Tol, R. 217, 227
Toledo, V. 77
Tomich, T. 39
Touza, J. 172, 175, 177, 186
trade-offs in ecosystem services, management of 70–89, 281
  Ecosystem Services Districts, development suggestion 84
  framework proposal 80–83

  framework proposal, conservation landscapes 82
  framework proposal, degraded landscapes 82–3
  framework proposal, intensively used agricultural landscapes 81–2
  future research 71
  multiple services, interlinking of 71
  provisioning ecosystem services in isolation, focusing on 71–2
  provisioning and regulating services 72–3
  provisioning services definition 70
  service-providing unit (SPU) 71
trade-offs in ecosystem services, management of, regulation of ecosystem services 73–80
  biological control 75, 77, 78
  certification schemes with price premiums 80, 81
  climate regulation and forecasting problems 74
  coffee production 76–80
  pollination 75, 77, 78
  soil quality maintenance 75–6, 77, 78, 81
  water regulation 74, 77
transaction costs 44, 142, 143, 163–4
  see also cost–benefit assessment
Tucker, C. 47
Tuhiwai Smith, L. 262
Turner, R. Kerry 58–69
Turpie, J. 29, 40, 46, 47, 101
Tuvendal, Magnus 70–89

UK 21, 27, 239, 245
Umali-Deininger, D. 102
UN Framework Convention on Climate Change (UNFCCC)
  Convention on Long Range Transboundary Air Pollution 187
  Kyoto Protocol, carbon sequestration markets 24–5
  Reduced Emissions from Deforestation and forest Degradation (REDD) scheme 16, 45–7, 181–5, 186
  see also Intergovernmental Panel on Climate Change (IPCC)

US 5, 239, 242, 246, 263
  Catskill watershed valuation 195
  Conservation Reserve Program
    (CRP) 21, 26–7

value
  economic valuations, foundations for
    alternatives, ethical foundations
    of cultural diversity 271–3
  natural capital restoration *see*
    ecosystem-based restoration
    benefits, natural capital
    restoration value
  see also cost–benefit assessment
valuing ecosystem services 4–6, 58–69,
    280–81
  adjusted market prices and
    willingness to pay (WTP) 62
  damage cost avoided approach 63–5
  discount function, decline of 68
  equity and efficiency in payments
    103
  'here and now' versus 'there and
    then' 65–8
  human benefits and ecosystem
    processes, relationships between
    59–60
  individual preferences, temporal
    context 67–8
  institution and ecosystem functions
    194–7
  intergenerational opportunities 67
  intermediate and final services–
    benefits scheme 59–61
  market price as poor approximation
    of value 61
  preference changes 67
  prices versus values 61–5
  productivity methods 62
  replacement costs as proxy for loss
    of existing services 65
  revealed preference methods 62
  services versus benefits 59–61
  stated preference methods and
    willingness to pay (WTP) 63, 66
  supply and demand and consumer
    surplus 61
  temporal and spatial context 65–8
  value measurement 61–2
  water regulation, spatial context
    65–6
  wetland restoration and temporal
    changes 66–7
Van de Wouw, M. 156
Van den Bergh, J. 195, 259
Van der Hamsvoort, C. 163
Vaughan, M. 75
Venter, O. 116
Veras Soares, F. 121
Vermeulen, S. 112, 129
Viana, V. 112, 125
Vittel PES, France 3, 29

Wagner, M. 84
Wale, E. 151
Walker, B. 196
Wallace, K. 59
Wang, H. 198
water protection schemes 3, 23, 24,
    65–6, 74, 77, 98, 103–4
  Working for Water (WfW) Program,
    South Africa 29, 40, 46, 47
Weitzman, M. 33, 157, 159–61
welfare
  human well-being and health 17,
    44, 199–200, 201, 203–4, 209,
    222–3, 225–6
  outcomes, evaluation of 98–103
  social protection mechanisms 121–2,
    123
  see also ethical foundations of
    cultural diversity

DATE DUE

PRINTED IN U.S.A.